# Understanding Organisational Context

Pearson
Education

# Understanding Organisational Context

Claire Capon

School of Business and Finance
Sheffield Hallam University

with contributions from Andrew Disbury

*An imprint of* Pearson Education

Harlow, England · London · New York · Reading, Massachusetts · San Francisco · Toronto · Don Mills, Ontario · Sydney
Tokyo · Singapore · Hong Kong · Seoul · Taipei · Cape Town · Madrid · Mexico City · Amsterdam · Munich · Paris · Milan

**Pearson Education Limited**
Edinburgh Gate
Harlow
Essex CM20 2JE
England

and Associated Companies throughout the world

*Visit us on the World Wide Web at*
*www.pearsoneduc.com*

---

First published in Great Britain in 2000



ISBN 0 273 63162 4

*British Library Cataloguing-in-Publication Data*
A CIP catalogue record for this book can be obtained from the British Library.

*Library of Congress Cataloging-in-Publication Data*
A catalog record for this book can be obtained from the Library of Congress.

10 9 8 7 6 5 4 3 2
04 03 02 01 00

Typeset by 35 in 9.5/13pt Stone Serif
Printed and bound in Great Britain by T J International, Padstow, Cornwall

# Contents

# List of case studies

# Preface

The idea for *Understanding Organisational Context* arose out of the desire to provide first-year undergraduate students studying broad-based business units with a book covering both the external and internal environments, unlike the many business environment textbooks currently on the market.

It is envisaged that students on business studies courses (e.g. BA Business Studies and HND Business Studies) and those studying business as part of a degree course (such as BSc Business and Technology) will find this book appealing and useful.

Andrew Disbury was involved in the early stages of developing this book.

A *Lecturer's Guide* is available from the publishers free of charge to lecturers adopting this book.

*Claire Capon*

# Teaching with this book

This textbook is designed to be used on level 1 units on courses in business studies or on courses with a significant element of business in them. Examples of such courses would be BA Business Studies, BA Business Administration, HND/HNC Business Studies or courses such as BSc/HND Business and Technology.

This book is suitable for units that examine the context in which organisations operate. The organisational context model, shown at the start of each chapter, summarises all that is covered in this book and provides a useful diagrammatic overview of all that organisations have to consider.

The general view taken in *Understanding Organisational Context* is that of outside inwards – namely, organisations have to understand their external environment (covered in Chapters 1–3) before they can begin to understand what it is the organisation should be concentrating on internally and in response to the external environment. Chapter 1 examines why organisations analyse their external environment and how such analysis may be undertaken. Chapter 2 builds on Chapter 1 and explores in great detail the elements that could go to make up the external environment of an organisation. Chapter 3 continues with the theme of the external environment, but considers an extremely important constituent of any organisation's external environment, namely competition.

Key internal issues for all organisations are the management of resources, structure and culture, and these are examined in Chapters 4 and 5. The idea of an organisation as a resource convertor and how organisations may arrange themselves are both considered in Chapter 4. The idea of organisational culture linked to structure is also explored briefly in that chapter. Correspondingly, Chapter 5 examines the influence of personal and national culture on doing business.

The four key functions or areas of activity for organisations are covered in Chapters 6 to 9 (marketing, operations management, finance and human resource management). Chapter 6 looks at the development of marketing, the discipline as it is today and some marketing tools. Chapter 7 presents an overview of the operations management activities that both manufacturing and service providers will undertake. Chapter 8 covers finance by looking at the key areas of financial management and management accounting, along with financial reporting and financial stakeholders (for more on stakeholders, *see* Chapter 11). Human resource (HR) management is examined in Chapter 9, covering the impact of the external environment on the HR function and the recruitment process.

The later part of the book examines tools and techniques for organisations to analyse and manage their position with respect to both their internal and external environments. Chapter 10 is devoted to TOWS analysis, reverse SWOT analysis; this reflects the underpinning concept of this book, that organisations must first examine and understand their external environment before deciding on internal issues. Stakeholder analysis is the theme of Chapter 11 and examines how an organisation can identify and manage those other organisations and people, both external and internal, that may have a role to play in the future of the organisation. In common with Chapters 10 and 11, Chapter 12 considers tools and techniques for analysing organisations with the aim of gaining an improved understanding of the organisation's situation when managing a changing external or internal environment.

Finally, some future trends for the proactive organisational member to consider and be aware of are briefly discussed.

### Designing a schedule of study

In units/modules designed to cover both the internal and external environments of organisations, there are several ways in which this book can be used. The most common semester length is 12 weeks and the most obvious programme of study using this book would be to cover one chapter per week. However, there are many possible combinations in which the chapters of this book could be studied during a 12-week teaching semester.

A possible schedule of study for an organisational context-type unit is shown below. This schedule of study takes two weeks to cover Chapter 2, due to its length.

In considering the four key functional areas, tutors may choose to omit one as it is covered elsewhere on the students' course. For example, many business studies students will cover finance and accounting in a separate module or unit. This is illustrated in the design of the schedule of study shown below.

Alternatively, if an area is not covered anywhere else on the first year of a business studies programme, it may be that more than one week will be spent on it in the organisational context-type unit. The example used on the schedule of study shown below is operations management, which takes two weeks.

The remaining weeks of such a unit or module may be used to cover the chapters presenting a variety of different tools and techniques (Chapters 10, 11 and 12). Alternatively, the last week could be used to run discussion and debate-style sessions in which students examine future trends and their impact on organisations and their stakeholders.

**Alternative schedule of study for an organisational context-type unit**

| | | |
|---|---|---|
| Week 1 | Chapter 1 | The external environment |
| Week 2 | Chapter 2 | The composition of the external environment |
| Week 3 | Chapter 2 | The composition of the external environment |
| Week 4 | Chapter 3 | The competitive environment |

| | | |
|---|---|---|
| Week 5 | Chapter 4 | Inside organisations |
| Week 6 | Chapter 5 | Culture and organisations |
| Week 7 | Chapter 6 | Marketing |
| Week 8 | Chapter 7 | Operations management |
| Week 9 | Chapter 7 | Operations management |
| Week 10 | Chapter 9 | Human resource management |
| Week 11 | Chapter 10 | TOWS analysis |
| Week 12 | Chapter 12 | Managing a changing environment |

**Unit/module delivery**

To aid the tutor in delivering organisational context-type units, a number of features appear in each chapter.

**Clear chapter structure**

Each chapter has the same layout: the organisational context model; learning outcomes; entry case study; introduction; main text; conclusion; ethical issues case study and questions; exit case study and questions; short-answer questions; assignment questions; references; and further reading.

**Organisational context model**

The organisational context model at the start of each chapter shows by shading the areas examined in that particular chapter. This allows readers to see at a glance what is covered in a chapter.

**Learning outcomes**

Detailed learning outcomes allow tutors to check that they have covered everything they intended with a class and they also allow students to check that they have achieved the knowledge and skills covered by a particular chapter.

**Case studies**

There are three case studies in each chapter. Most of the case studies are copyright extracts from the *Financial Times* and are reproduced with its kind permission.

The entry case studies provide an example of the topic(s) covered in a chapter and are often referred to in the main text of the chapter by way of a real-life example. Therefore students should be encouraged to read the entry case study before starting to read the chapter in detail. Students should also be encouraged to refer back to the entry case study, if necessary, when it is discussed or referred to in the main text.

The ethical issues case studies appear at the end of chapters and have been included to provide a vehicle for discussion of some of the ethical or less clear-cut issues surrounding a topic. In contrast, the exit case studies have been chosen to allow students to apply the knowledge and skills gained to a real-life situation.

**Questions at end of chapter**

The short-answer questions found at the end of each chapter have several uses. They could constitute a quick testing mechanism with students to see if they have learnt basic facts about a topic. This could be done in the form of an in-class quiz. Alternatively, if the formal examination for the organisational context-type unit contains a section of short-answer questions, then those provided in the book allow students an opportunity to practise answering short-answer questions.

Finally, the assignment questions are intended to be used for formal assessed coursework. The normal length of report or essay that a student should be able to produce in response to such questions is around 2000 words.

***Lecturer's Guide***

A *Lecturer's Guide* is available to lecturers adopting this book.

**About the author**

Claire Capon is a senior lecturer in Strategic Management in the School of Business and Finance at Sheffield Hallam University. She previously worked at Huddersfield Polytechnic and UMIST as a researcher in the areas of strategic management and the use of design by SMEs.

# Acknowledgements

I would like to thank the following family and friends for their support and encouragement during the long time this manuscript has been in preparation. They are Brian Capon, Julia Capon, Ruth Capon, Joanne Duberley, Chris Duggan, Anthea Gregory, Janet Kirkham, Susan Richardson, Jane Thomas, John Varney and Graham Worsdale.

Thanks go to Linda Purdy of the Learning Centre at Sheffield Hallam University for her work on permissions.

Thanks also to Penelope Woolf of Pearson Education who managed the project in its earlier stages and got it off to a good start. Finally, a deep debt of gratitude is owed to Sadie McClelland of Pearson Education, who saw the project through the rough times and to its conclusion.

# 1 The external environment

## Organisational context model

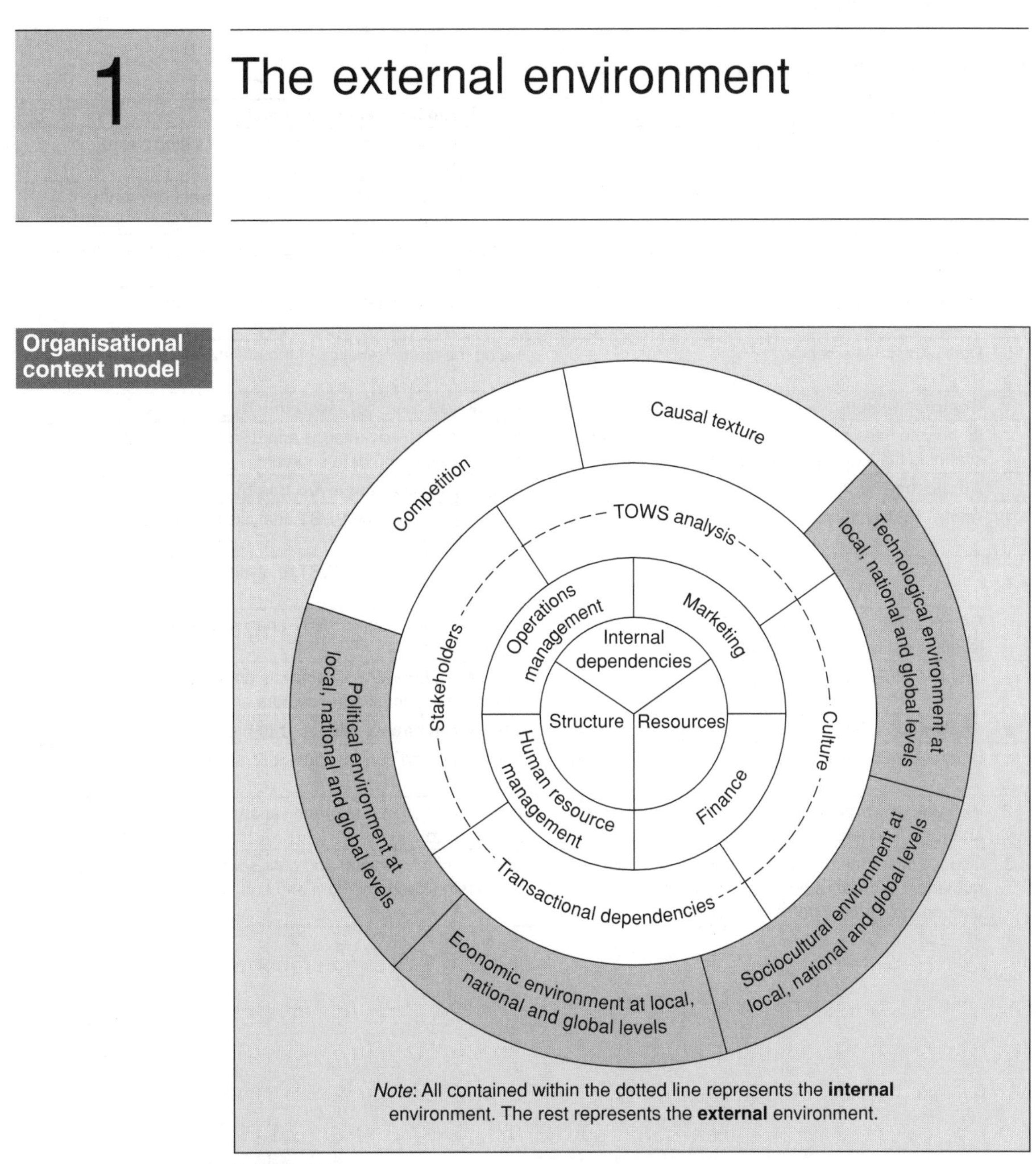


*Note*: All contained within the dotted line represents the **internal** environment. The rest represents the **external** environment.

**Exhibit 1.1**

## Learning outcomes

While reading this chapter and engaging in the activities, bear in mind that there are certain specific outcomes you should be aiming to achieve as set out in Exhibit 1.2 (overleaf).

**Exhibit 1.2 Learning outcomes**

| | Knowledge | Check you have achieved this by |
|---|---|---|
| 1 | Define the external environment | stating in one sentence what an organisation's external environment could be said to be |
| 2 | List elements of the external environment | reviewing generic elements of the external environment from memory |
| 3 | Name organisations and individuals comprising the external environment | identifying the elements of specific organisations' external environments |
| 4 | Define the public sector | listing the characteristics of the public sector |
| 5 | Define the private sector | listing the characteristics of the private sector |
| | **Comprehension** | **Check you have achieved this by** |
| 1 | Identify sources of information and data that make scanning the external environment possible | discussing the advantages and disadvantages of a variety of information and data sources |
| | **Application** | **Check you have achieved this by** |
| 1 | Apply PEST analysis to general situations | performing a generic PEST analysis for a hypothetical organisation |
| 2 | Apply PEST analysis to specific situations | performing a specific PEST analysis for a real organisation |
| 3 | Demonstrate ability to apply PEST analysis to LoNG levels of external environment | modifying the specific PEST analysis above to incorporate LoNG levels |
| 4 | Practise PEST analysis | performing LoNGPEST analysis on a variety of organisations in differing sectors and contexts |
| | **Analysis** | **Check you have achieved this by** |
| 1 | Differentiate between the public and private sectors | comparing and contrasting public-sector and private-sector activity |
| 2 | Examine why organisations need to understand and analyse the external environment | testing the notion that organisations could exist in splendid isolation |
| 3 | Choose information and data sources that best assist managers in organisations with scanning and understanding the external environment | comparing information and data sources and evaluating the most suitable medium and source |

ENTRY CASE STUDY 1.1

# ScotRail bid backed by Strathclyde officials

**By James Buxton**

Officials of Strathclyde passenger transport authority, which provides about half the annual £250m subsidy of ScotRail, are recommending that it agrees to National Express being granted the train operating franchise.

Members are being told the authority has secured protection of current services from the Office of Passenger Rail Franchising (Opraf) and managed to obtain substantial improvements at no extra cost.

The PTA will be able to dictate what services National Express operates in the greater Glasgow area, the timetable and fares, but will be relieved of having to contribute to almost all their cost.

Although National Express was named preferred bidder for ScotRail last week, Strathclyde PTA still has to agree to granting the seven-year franchise. It will consider the matter on Friday.

The PTA, which is composed of 12 Labour local authorities which have made no secret of their dislike for rail privatisation, has succeeded in delaying the franchising of ScotRail for more than two years by holding out for concessions from Opraf.

It could still refuse to accept National Express but Sir George Young, the transport secretary, has the power to overrule it.

Under the deal negotiated by the PTA and Opraf, the government has agreed to continue a grant of about £105m a year to ScotRail for operating services. It will also take over paying £35m a year which the PTA levied from council tax payers for rail services.

The PTA will receive £1.25m a year to cover the direct cost of managing the franchise, and will be reimbursed for the cost of consultants.

National Express will replace 101 diesel multiple units and 303 electric multiple units by March 2000 at no cost to the authority.

Members of the authority have been given a report by KPMG, the accountants, which recommends National Express as the franchisee in preference to the rival bidders, Stagecoach, Prism, Go Via and the ScotRail management.

• ScotRail passengers face four more days of strike action by ticket collectors and conductors. Although a dispute over productivity payments and working conditions which began last August was settled in November, the union is protesting against the disciplining of 17 of its members who ScotRail says put heavy pressure on railway staff to take part in the strike.

Three 24-hour strikes have already been held since the original dispute was settled, and strikes are planned for February 22 and 24, and March 8 and 10. However the strikes do not affect all trains and ScotRail claims to have been able to operate about 70 per cent of services.

*Source*: *Financial Times*, 18 February 1997. Reproduced with permission.

## Introduction

We all live in an increasingly complex and dynamic world. Therefore, the purpose of this chapter is to examine how organisations can make sense of this volatile world and also take appropriate management and business decisions. This introductory chapter looks at the public and private sectors and the different types of organisation faced with making management and business decisions. The changing world or external environment in which both public- and private-sector organisations must operate is also examined. In contemplating the complex and dynamic world in which they operate, organisations have to consider many influences and issues. The entry case study of National Express and ScotRail is a good example of how organisations can influence and create complexity in the external environment of another organisation. The outside world and the matter of the ScotRail franchise create complexity in the external environment faced by the coach company National Express. This complexity is determined by the fact that many organisations, all holding differing views on the matter, are seeking to influence the outcome of the ScotRail franchise and whether or not it is issued. The influencing organisations creating the complexity include the Strathclyde Passenger Transport Authority, comprised of 12 Labour local authorities; Office of Passenger Rail Franchising (OPRAF); a major private-sector accounting firm, KPMG; rival bidders for the franchise, including Stagecoach and ScotRail management; ScotRail; and National Express itself. All the organisations seeking to influence the external environment faced by National Express do so with the intention of gaining differing benefits depending on whether or not the granting of the ScotRail franchise goes ahead.

In order to understand the complexity in the outside world, organisations need to analyse their external environment. This can be done at a broad general level by use of PEST and LoNGPEST analyses, which are covered in this chapter and Chapter 2. Further examination of other components of the external environment needs to occur for a full understanding of the external environment to be possible. These other components include competition, competitors, customers and the marketplace. If National Express is successful in gaining the ScotRail franchise, as referred to in the entry case study, then the geographic marketplace it faces will include the Glasgow area, northwards to Loch Lomond and southwards down the Clyde coast to Ayr and Largs, popular seaside towns. Hence National Express customers would include weekday commuters into and out of Glasgow; schoolchildren and shoppers using trains for short local journeys; weekend shoppers; theatre and cinema goers and nightclubbers travelling into Glasgow in the evening; day trippers and holiday makers going down the Clyde coast, particularly during the school summer holidays.

To be able to compete effectively, organisations need to understand who their competitors are and the best way to compete with them. The competition faced by National Express for the customers described above will arise from people using their own car and bus services, some of which will be provided by Strathclyde Passenger Transport Authority, the body granting the ScotRail franchise.

Organisations may choose to compete by offering low-priced, good-value products and services, or by offering a luxury product at a higher price. The rail customers that National Express would seek to retain and attract are outlined above, and National Express will have to address *how* their retention and attraction may be achieved. For many customers a train service that runs at the right time and on time, along with value for money and comfort, are likely to be key criteria that they will consider when deciding whether to use the National Express rail service. It is on these key criteria that National Express will have to endeavour to perform more successfully than its competitors, the bus companies, if it is to succeed in operating train services in Strathclyde.

The issues of competition are looked at in Chapter 3. Equally, the understanding of competitors and their behaviour ties in directly with the need to understand the marketplace; who your customers are; where they are located;

and how they can be persuaded to buy the products and services offered by your company rather than those of one of your competitors. This is explored in Chapter 6.

The tool of PEST analysis is examined and developed for the purpose of analysing the external environment or outside world. Analysis of the external environment is an ongoing process for organisations that take the dynamic and changing nature of the external environment seriously. For ease of understanding, this chapter will assume the position of an organisation in the UK, and thus will apply analyses of the external environment from this perspective. Reference will be made to public-sector organisations but, inevitably in the context of business studies models, there will be a focus on private-sector companies.

In summary, this chapter examines:

- the public and private sector;
- the elements and levels of the external environment faced by organisations;
- the benefits of external environmental analysis;
- the guidelines for undertaking LoNGPEST analysis;
- in brief, the part that competition plays in the organisation's external environment.

In order to be able to analyse the external environment in which both public- and private-sector organisations operate, it is essential to define briefly the public and private sectors.

## The public and private sectors

**Characteristics of the public sector**

The public sector consists of enterprises in which the whole or a majority stake is owned by either local or national government, or another publicly owned body established by government. Exhibit 1.3 gives some examples of the more important organisations in UK that remain in the public sector in the late twentieth century.

State schools deliver compulsory education free to children in the local area. Traditionally, local schools have been funded from and run by local government. However, in an effort to promote higher standards and to circumvent local authorities where the party in power was opposed to that in central government, the Conservatives introduced a system to allow for schools to opt out of local control and be funded directly by central government.

**Exhibit 1.3 Public-sector organisations**

- National government
  - ministries and departments
  - agencies
  - civil service
- DVLA
- NTVLRO
- Local government
- National Health Service
- Armed services
- Police forces
- Fire and rescue services
- Universities and colleges
- State schools
- Royal Mail
- Post Office

Some organisations that have been part of the public sector have had a mixed history. Some companies were private, then were nationalised, i.e. taken into public ownership, by previous Labour governments, and are now privately owned again. The largest wave of nationalisations occurred under the first post-war administration when Labour was swept into power in the 1945 general election. The railway system in Britain was nationalised after the Second World War and British Rail was formed. This was followed about 40 years later by the privatisation of the railway system in the early 1990s under a Conservative government. This privatisation saw the formation of Railtrack and many train operating companies. Railtrack is responsible for running and maintaining track, signals and stations. The train operating companies, of which there are many (e.g. Midland Mainline and Virgin), run train services. The right to operate a train service on a particular route is determined by the train operating company purchasing the franchise to do so. For example, Midland Mainline run services between London and Sheffield, while services from London to Glasgow are operated by Virgin on the west coast and GNER on the east coast.

As the public sector is owned and regulated by government itself, it operates in markedly different ways from the private sector. Public-sector organisations tend to provide a service over which they have a monopoly, or certainly most of the responsibility for provision. Public-sector organisations tend to be large and bureaucratic entities, governed by strict rules and procedures that are often prescribed in law.

**Market forces**

The Conservative governments between 1979 and 1997 believed strongly that the private sector, and competition in particular, had certain advantages over the monopolistic public sector. They believed that market forces and competition could bring better value for money in terms of public expenditure and for private individuals. Many organisations that had been in the public sector since nationalisation were privatised, i.e. sold off to the mass public in what Margaret Thatcher referred to as the creation of a 'shareholder democracy'. It was felt that a deregulated private sector operating in a free market would be able to provide all the services and products needed at the market price. Examples of privatised organisations are shown in Exhibit 1.4. This also led to considerable one-off cash revenues for the government, which were used in other parts of the government's fiscal policy.

**Exhibit 1.4 Privatised organisations**

- Deregulated regional and local bus companies
- Regional Electricity Boards
- Regional Water Boards
- British Telecom
- British Gas
- British Airways
- British Aerospace
- British Rail
- British Steel
- British Coal

An internal market was introduced to the NHS by the Conservative government in the late 1980s in an attempt to introduce some of the perceived benefits of private-sector operation, such as flexibility of decision making and market forces effects on cost and pricing strategy. This allowed for general practitioners to become fundholders, deploying funds along with their referral of patients to hospitals, so that the number of referrals to a certain hospital became a key source of revenue. Hospitals, previously run by locally appointed boards on which locally elected councillors held a majority of seats, were able to opt out of this system and establish themselves as independent NHS trusts, directly funded by central government. This internal healthcare market is the subject of review by the current Labour government.

Public-sector organisations are sometimes run at a loss to government or are sometimes required to recoup their operating costs through income generation. However, public-sector enterprises rarely operate in order to make a profit.

### Characteristics of the private sector

The private sector consists of privately owned companies and businesses. These may be owned by individuals, families or groups, or they may be large organisations that are quoted publicly on the stock exchange, with ownership residing with the shareholders. The differences between limited companies and public limited companies are discussed in Chapter 4, but for now it is enough to understand that a 'public limited company' has nothing to do with the public sector. Private-sector companies are regulated by laws and regulations introduced by the local, national and global levels of the external political environment.

Some services provided by the public sector also have private-sector providers, including hospitals and schools. Just as 'plcs' are not in the public sector, the term 'public school' refers to privately owned and run schools for which parents pay tuition fees, and not the free state-run counterparts.

As mentioned above, public-sector organisations are sometimes funded partly through their own income-generation activities, but their main source of income is budgets awarded by government from its tax revenues. For example, as outlined in the entry case study, Strathclyde Passenger Transport Authority received £35 million per annum from council tax, collected by local government, to subsidise rail services. Therefore, private-sector organisations making money and paying taxes in fact fund activities in the public sector. Private-sector organisations look to recoup their costs and make profits from their business activities.

### Public and private partnership

The election of the Labour government in 1997 did not see the reversal of privatisation policies or a commitment to renationalise already privatised industries. Further privatisations, such as attempts to privatise the Post Office or the Royal Mail – which had been split into two for this purpose by the Conservatives – were not to be pursued. Instead, the message became the formation of direct partnerships between the public and private sectors. The

Conservatives began this idea with 'Private Finance Initiatives', enabling public-sector organisations to raise money for projects in private money markets, and it has been continued by Labour. For example, private monies continue to be available for public road-building schemes.

## The elements and levels of the external environment

The external environment is literally the big wide world in which both public- and private-sector organisations operate. Whatever the nature of their business, organisations do not and cannot exist in splendid isolation from the other organisations or individuals around them, be they customers, employees or suppliers. It is therefore clear that the external environment of any organisation is a large and complex place.

The term 'environment' in this case refers to much more than the ecological, 'green' issues that the word commonly evokes. 'Environment' here is more appropriately interpreted as the external context in which organisations find themselves undertaking their activities. Each organisation has a unique external environment that has unique impacts on the organisation, due to the fact that organisations are located in different places and are involved in different business activities, with different products, services, customers and so on. In addition to this unique context, individual organisations will all have their own distinctive view of the world surrounding them, leading them to interpret what is happening in the external environment potentially correctly or incorrectly, depending on their ability to understand the external forces affecting them. This begins to suggest how crucial it is for organisations to undertake external environmental analysis and to aim to get it right.

Careful and accurate analysis of the external environment benefits organisations by providing overall greater understanding and an appreciation of the context in which the organisation operates. The key benefits of external environmental analysis are best realised if it is undertaken on a long-term, ongoing basis. The key benefits of analysing the external environment can be summarised as follows:

- managers in the organisation achieve a greater understanding and appreciation of the external environment leading to improvement in long-term and strategic planning;
- highlighting of the principal external environmental influences generating change;
- anticipation of threats and opportunities within a timescale of long enough duration to allow responses to be considered.[1]

External environments can be defined and analysed using PEST analysis, examining Political, Economic, Sociocultural and Technological categories into which external influences on the organisation can be placed.

- *Political* influences on organisations encompass both those with a big and small letter 'p', i.e. Politics in the conventional sense, with the rules and regulations imposed by government, as well as the political influences on organisations of various trade associations, trade unions and chambers of commerce.
- *Economic* influences on organisations include the impact of banks, stock markets, the world money markets, and trading blocs such as the European Union.
- *Sociocultural* influences on organisations include changes in the age and structure of populations, the manner in which populations behave and the way in which the culture of a population or country changes and develops.
- *Technological* developments influence the magnitude and rate of change that organisations face, and how this affects their capacity to meet their customers' demands. Technological developments include information and communications technology and the application of technology by organisations. The preparation that organisations made to manage the 'millennium bug' was also a crucial issue in the late twentieth century.

Basic analysis of an organisation's external environment can be done by categorising the external influences on the organisation into the PEST categories and assessing the impact of the individual elements identified in each category.

However, there exists a second dimension to the external environment of organisations. This is the level at which the influences occur. There are three levels that will be considered alongside the PEST categories. The levels are Local, National and Global (LoNG).

- The *local* level can be said to be the immediate town, city or region in which the organisation operates.
- The *national* level is then the home country in which an organisation identifies its headquarters.
- The *global* level then becomes anything outside the local and national levels.

A company operating in and being influenced by the global level of the external environment will be trading in a foreign country, be it right next door or on the other side of the world. A company in this situation is not only subject to the laws and culture of its local and national environments, but also to the laws and culture of the foreign country in which it is trading. In addition to national rules and regulations, there are also the laws and procedures of both home and host countries specifically governing importing/exporting and foreign direct investment activity to consider.

The literature of international business clearly differentiates the terms 'international' and 'global'. Some of the issues considered here are 'international', i.e. they are issues that occur between nations. However, the third level will be described as 'global', i.e. affecting all parts of the world in similar and simultaneous ways. This is because many of the issues of globalisation at the

**Exhibit 1.5 Generic LoNGPEST grid**

| | Political | Economic | Sociocultural | Technological |
|---|---|---|---|---|
| **Local** | • Local government<br>• Local offices of national government<br>• Local associations<br>– Chambers of Commerce<br>– Business Link | • Local bank branches<br>• Local economy | • Local community | • Communications technology<br>– Mobile phones and faxes<br>– Video conferencing<br>– Internet and world wide web |
| **National** | • National government<br>• Devolution for Scotland and Wales<br>• National bodies<br>– Employers' bodies<br>– Employees' bodies | • Central bank<br>– Bank of England<br>• Stock market<br>– London | • Demographic change<br>• Social change | • Organisations and the application of technology<br>– The personal computer<br>– The banking and financial services industry |
| **Global** | • Alliances and agreements<br>– UK and USA<br>– UK and China<br>– EU<br>– Cold War<br>– CIS<br>– CBSS<br>– Eastern Europe<br>• International bodies<br>– The Commonwealth<br>– NATO | • Trading blocs and bodies<br>– EU<br>– EFTA<br>– OECD<br>– NAFTA<br>– ASEAN<br>• World money markets<br>• WTO | • Cross-cultural issues<br>– Language<br>– Behaviour<br>– Culture shock | • Millennium bug |

end of the twentieth century increasingly affect the local and national levels of organisations' external environments.

The traditional PEST analysis, then, is a short, one-dimensional view of the external environment. This two-dimensional analysis will be referred to in this text as a LoNG (Local, National and Global) PEST analysis. A generic LoNGPEST analysis could look like the grid shown in Exhibit 1.5.

The grid represents the view that these external influential elements, whether political, economic, sociocultural or technological, all exist at local, national and global levels. The political, economic and sociocultural influences are easily identified at the three different levels. However, it could be argued that all types of technology affect and influence organisations at all levels of the environment.

It should be noted that not every organisation will identify strongly influential elements in all four PEST categories at all three levels all the time, but the possibility should be considered for any organisation when carrying out a LoNGPEST analysis, because elements have to be identified before they can be evaluated and discounted.

The next section of this chapter offers some guidelines for use when carrying out external environmental analysis. The grid in Exhibit 1.5 shows the possible different external environmental influences on an organisation. These external environmental influences are discussed in more detail in Chapter 2.

## Performing external environmental analysis

As has previously been stated, the external environment is an immensely complex and dynamic place. Therefore, performing an analysis of the external environment of an organisation requires access to a wide range of information. Information concerning the external environment may already exist within the organisation or it may have to be sought, collected and collated from other sources. Sources of information within an organisation will encompass information from the four key functional areas of marketing, production, finance and human resource management. This will include sales reports; customer/client survey results; reports on staff skills and availability; and budgets and cashflow statements detailing the amount and availability of cash. In addition, information should be available on the systems in place in the organisation, including that on their capability and capacity, and efficiency and effectiveness. This type of information provided by internal sources, if it is up to date, will provide details of the resources available to deal with current influences in the external environment and indicate the level of resources needed to respond to possible future influences from the external environment.

External information sources are compiled by organisations other than that undertaking analysis of its external environment. The external information sources most widely available and accessible to everyone are the press, television and radio. In the UK the most familiar press includes the daily and Sunday broadsheet newspapers, which all contain business pages or sections in addition to reporting the general political and economic news. These are the *Independent*; the *Daily Telegraph*; the *Times*; the *Guardian*; the *Financial Times*; the *Sunday Times*; the *Sunday Telegraph*; the *Observer* and the *Independent on Sunday*. Publications such as *The Economist* and *Management Today* supply more extensive coverage of the economy and the latest developments in the world of trade and commerce than that provided by the daily and Sunday press. The annual report and accounts of a company also provide a summary of recent activities and may offer clues or an indication of future activities. The annual report and accounts of publicly quoted companies are readily available from the companies themselves, so are easily obtainable by competitor companies. Specific information concerning an industry will be available from industry- or trade-specific publications.

Current affairs programmes on radio and television cover economic and political news as well as reporting company and industry news and events. For example, Radio 4's *Today* programme comprises items of economic, political and business news and is broadcast at breakfast time on weekday mornings. In addition, economic, political and business news is reported on television news programmes, for example on ITV and BBC1 at teatime and in the late evening. Daily news broadcasts, such as *Newsnight* on BBC2 and Channel 4 News are longer news programmes than the broadcasts on BBC1 and ITV and consequently devote more time to economic, political and business news stories.

There are other television programmes devoted to business, economic and political stories and issues. For example, *Working Lunch* on BBC2 is broadcast every weekday at 12.30 pm and other weekly programmes include *Question Time*, which focuses on political and economic issues. Programmes examining political, economic and business issues are broadcast in abundance on Sundays in the UK. These include BBC1's *Breakfast with Frost* programme, which has a political focus, followed by *On the Record* on BBC1 or *Jonathan Dimbleby* on ITV at lunchtime, followed by *The Money Programme* on Sunday evenings. Other television programmes that examine issues of relevance to business or organisations include documentary programmes such as *Panorama*.

The Internet and electronic information are other sources of information which are widely available in organisations. An immense and extensive amount of information is available on the Internet, although one must be aware of who or which organisation originated the information, as this will affect its reliability and accuracy.

The other method of accessing information electronically is via the use of CD-ROMs, available in libraries that have purchased the CDs and have the necessary computers to read them. Publications, including back issues, can be stored efficiently on CD-ROMs and easily accessed. Publications such as the *Financial Times* and MINTEL reports are available in this format.

The use of the LoNGPEST framework and information will allow analysis of the external environment of an organisation to be completed. The following guidelines will help in applying the LoNGPEST framework. It is suggested that a blank grid like the one in Exhibit 1.6 be used when carrying out external environmental analysis.

### Guidelines for external environmental analysis

1 Identify the influences affecting the organisation.
2 Categorise the influences by using the LoNGPEST grid.
   (a) First, decide whether the influence is Political, Economic, Sociocultural or Technological.
   (b) Second, decide whether the influence is Local, National or Global.
3 Make sure you can explain how and why a particular influence is affecting an organisation. Remember, elements in the external environment do not exist in isolation at any level and can impinge on and influence one another. See the entry case study at the start of this chapter for a good example of elements of the external environment seeking to influence each other.
4 Select and judge which categories are most important to the company, for example technological influences at the global level or economic influences at the national level.
5 Select key individual influences from the important categories. For example, the important category economic influences at the national level may contain the crucial influence of falling interest rates.

**Exhibit 1.6 Blank grid for LoNGPEST analysis**

| | Political | Economic | Sociocultural | Technological |
|---|---|---|---|---|
| Local | | | | |
| National | | | | |
| Global | | | | |

6 Consider the important categories and influences you have identified. Do any of these pose threats or opportunities to which the company must react immediately or in the longer term, when anticipating and planning the future?

7 How should the organisation react to and deal with the opportunities and threats? Do short-term opportunities take priority over long-term threats or vice versa?

## Conclusion

This introductory chapter has explored the different types of organisations that exist, the need for external environmental analysis and how it can be carried out. Chapter 2 discusses in greater detail the complexity and composition of the external environment. Competition and the marketplace are key elements of the external environment and these are discussed in Chapters 3 and 6 respectively.

Later chapters look at means by which the organisation determines how its unique external environment is affecting it, and how it should respond to the external environment. We will also see that the internal functions of an organisation can be very shortsighted about their role in relation to other functions and to the external environment, and that lack of objectivity and functional convergence can lead to problems for the organisation as a whole.

ETHICAL ISSUES CASE STUDY 1.2

FT

# Leading supermarket chains to ban 'alcopops'

**By David Wighton, Liam Halligan and Robert Wright**

Leading supermarket chains yesterday moved to ban 'alcopops' from their stores as Mr Tony Blair, the prime minister, underlined the government's determination to crack down on under-age drinking.

The Co-op and Iceland said they would stop selling the alcoholic soft drinks which critics claim are deliberately marketed to appeal to under-18s. Other large chains said they would end all promotions while reviewing their policy.

The moves follow this week's fierce attack on alcopop manufacturers by Mr Frank Dobson, the health secretary, who said the government would have to contemplate a ban.

Mr Blair yesterday used his first 'meet the people' question and answer session to back the tough stand. 'It is important that we enforce responsible behaviour in relation to this and you will find we will tackle this particularly clearly.'

The Co-op Retail Trading Group said its own ban on alcopops was prompted by concern over the popularity of the drinks among under-18s. 'We believe these drinks are designed specifically to appeal to young people and are, in fact, largely consumed by under-18s who cannot legally buy them,' it said.

Mr Malcolm Walker, chairman of Iceland, said the company was reacting to customer concerns. 'While commercially this decision will hurt, as a family company we must act responsibly and reflect the views of our customers.'

There was clear evidence that alcopops were encouraging under-age drinking. But Bass, whose Hooper's Hooch is leader in the £250 m-a-year market, denied the claim and insisted it took its social responsibilities seriously. 'There is no consistent, objective evidence that under-age drinking has increased since alcopops came on the market.'

Mr George Howarth, the Home office minister who is heading a government inquiry into alcopops, welcomed the Co-op's move. 'I am pleased to see that the Co-op shares ministers' concerns about alcopops and I welcome their readiness to address this serious issue.'

The Home Office said the review would consider all options, including an outright ban. However, the industry argues that this would be difficult without outlawing other drinks such as ready mixed gin and tonics.

The Portman Group, a voluntary watchdog funded by the drinks industry, has acknowledged the need for further restrictions, blaming problems on a few companies that have ignored its guidelines.

Tesco announced it would stop selling alcoholic milk drinks and the controversial sachets of alcohol, but stopped short of an outright ban.

Mr John Gildersleeve, a Tesco director, said: 'After listening to our customers, it is clear the majority do not favour a total ban. However, they do want further action.' Tesco joined other retailers in pressing for further packaging and labelling changes.

*Source: Financial Times*, 14 June 1997. Reprinted with permission.

**Questions for ethical issues case study 1.2**

1 Do you think that other large supermarket chains should follow the example of the Co-op and Iceland and stop selling alcoholic soft drinks or follow a policy of ending promotion of alcopops and press for additional packaging and labelling changes. Explain your answer.

2 Draw up a set of guidelines for manufacturers and retail outlets selling alcopops. The guidelines, if followed, should result in the retail outlets not selling alcopops to under-age drinkers.

## EXIT CASE STUDY 1.3

FT

# Council gives away shopping area

**By Alan Pike**

Enfield council in north London is to give away what it believes is the UK's largest municipally-owned shopping centre to a development consortium under a £100m regeneration deal.

The Labour-controlled council has agreed on arrangements for the consortium, which includes Wimpey Homes, Metropolitan Housing Trust and London & Quadrant Housing Association, to take over the Edmonton Green shopping centre and surrounding housing estates.

The scheme – which the council says is intended to 'set a benchmark for other regeneration projects to meet' – is a striking illustration of a growing interest by local authorities in releasing assets to finance development.

A report to Enfield's policy committee, which was considering final details of the plan last night, acknowledges that regeneration of the Edmonton area required investment that the authority was unable to fund. In a similar move another London council, Hackney, is to transfer assets to Peabody Trust housing association in order to fund a £10m regeneration initiative.

Under the Enfield proposals, the council will transfer the 1960s shopping centre and housing estates to the consortium at nil value. The new owners will refurbish the shopping centre and part of the estates, and provide new social housing.

Plans involve nearly 1300 new or refurbished homes, £10m direct investment in the shopping centre, and the creation of 450 permanent jobs. The £100m regeneration costs will include about £80m of private sector funds.

It is estimated that the council's loss of income from giving up ownership of the shopping centre will be matched by no longer having to maintain it and the surrounding estates, at an estimated £30m cost over the next 30 years.

The transfer of the property has been preceded by a far-reaching exercise in public consultation. A community jury, including council tenants, retailers and street traders, was formed to evaluate the regeneration plans of four competing teams of developers with the council providing training and specialist advisers for community organisations.

Mr David Mason, who chairs Enfield's housing committee, said the winning consortium was the first choice with both council tenants and business representatives. Council tenants will have to vote on the transfer of their homes to housing association ownership.

*Source*: *Financial Times*, 18 February 1997. Reprinted with permission.

**Question for exit case study 1.3**

1 (a) Perform a PEST analysis of the external environment faced by the consortium which is to take over the management and regeneration of Edmonton Green shopping centre and surrounding housing estates.

(b) Clearly identify and evaluate the key factors and influences and comment on the necessity to deal with them in the short term or longer term.

(c) Indicate how the consortium should manage both the longer- and shorter-term key factors and influences that you have identified.

## Short-answer questions

1 Outline the characteristics of a public-sector organisation.

2 Outline the characteristics of a private-sector organisation.

3 Name three public-sector organisations.

4 Name three private-sector organisations.

5 Define the term 'external environment'.

6 Explain the terms 'PEST' and 'LoNGPEST'.

**7** Name three sources of information for external environmental analysis.

**8** Write down the three key benefits to organisations of undertaking external environmental analysis.

## Assignment questions

**1** (a) Refer to the section 'Performing external environmental analysis' in this chapter. Identify and collect appropriate information to perform a LoNGPEST analysis of your own college, university or organisation. Your LoNGPEST analysis should contain a completed grid and relevant discussion and explanation.

or

(b) Refer to the section 'Performing external environmental analysis' in this chapter. Identify and collect appropriate information on a public- or private-sector organisation of your choice. Perform a LoNGPEST analysis on your chosen organisation. Your LoNGPEST analysis should contain a completed grid and relevant discussion and explanation.

**2** External environmental analysis is an ongoing process for any organisation. Advise an organisation wishing to undertake external environmental analysis for the first time and seeking to set up a system to allow external environmental analysis to take place on an ongoing basis in the future. Present your advice in the form of 2000-word report.

**3** Identify and collect appropriate information from a variety of media sources that show how a particular organisation is evolving. Evaluate and analyse the general usefulness and reliability of the information sources you identify as currently available. Comment on the usefulness of the information to managers within the organisation that is currently being affected by changes in the external environment. Your answer should take the form of a 2000-word essay.

## References

1 Duncan, Peter M, and Ginter, W Jack (1990) 'Macro-environmental analysis for strategic management', *Long Range Planning*, 23 (6), December.

## Further reading

The following books and articles all look at analysis of the external environment of organisations:

Cole, G A (1997) *Strategic Management*, Chapter 3, Letts.

Johnson, G and Scholes, K (1999) *Exploring Corporate Strategy*, 5th edn, Chapter 3, Harlow: Prentice Hall.

Lynch, R (2000) *Corporate Strategy*, 2nd edn, Chapter 3, Harlow: Financial Times Prentice Hall.

Thompson, J L (1997) *Strategic Management: Awareness and Change*, 3rd edn, Chapter 8, London: International Thomson Business Press.

Worthington, I and Britton, C (2000) *The Business Environment*, 3rd edn, Chapter 1, Harlow: Financial Times Prentice Hall.

# 2 The composition of the external environment

## Organisational context model

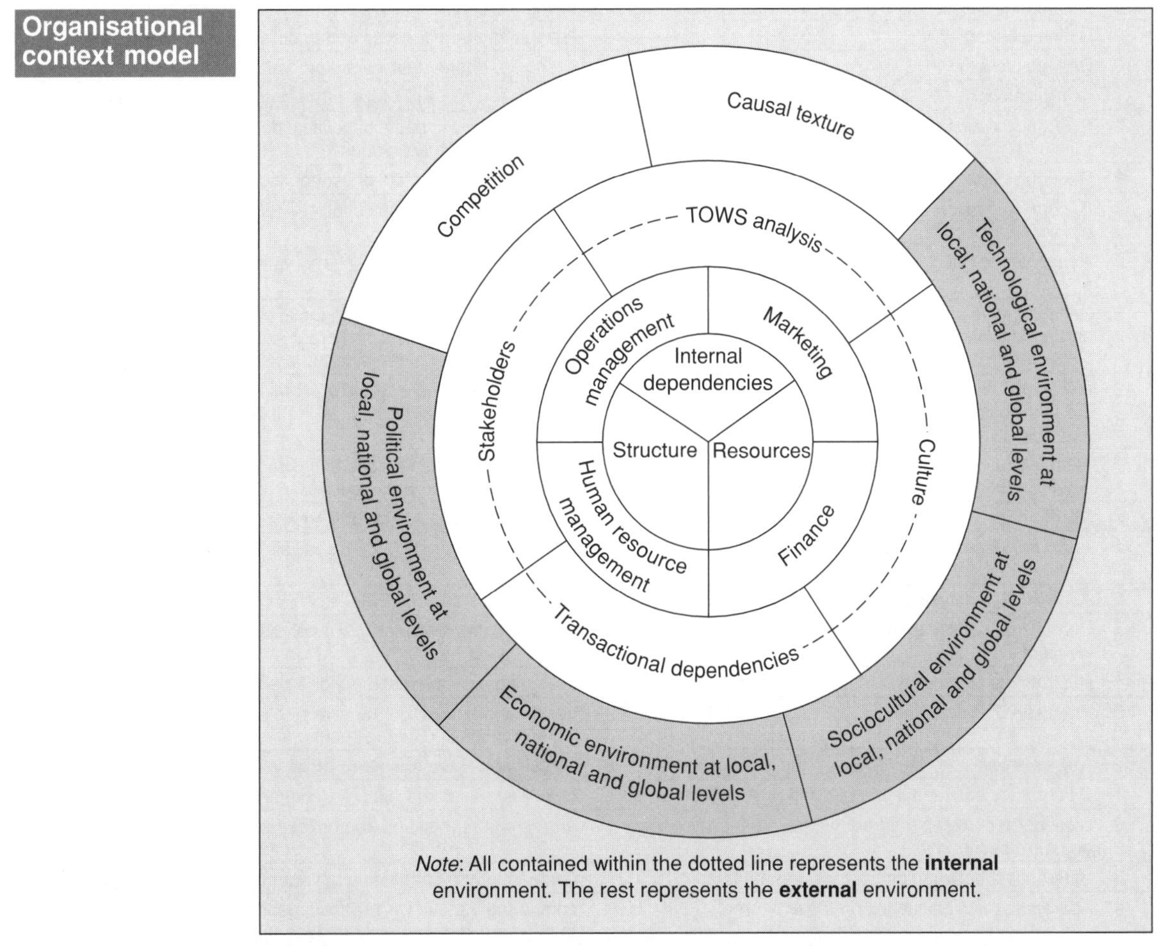


**Exhibit 2.1**

## Learning outcomes

While reading this chapter and engaging in the activities, bear in mind that there are certain specific outcomes you should be aiming to achieve as set out in Exhibit 2.2 (overleaf).

**Exhibit 2.2 Learning outcomes**

| | Knowledge | Check you have achieved this by |
|---|---|---|
| 1 | Define the external environment | stating in one sentence what an organisation's external environment could be said to be |
| 2 | List elements of the external environment | reviewing generic elements of the external environment from memory |
| 3 | Name organisations and individuals comprising the external environment | identifying the elements of specific organisations' external environments |
| 4 | Define the public sector | listing the characteristics of the public sector |
| 5 | Define the private sector | listing the characteristics of the private sector |
| 6 | Define the free market | listing the characteristics of the free market |
| 7 | Define the planned economy | listing the characteristics of the planned economy |
| 8 | Define the mixed economy | listing the characteristics of the mixed economy |
| 9 | Define democracy | listing the characteristics of democratic government |
| 10 | Define autocracy | listing the characteristics of autocratic government |
| | **Comprehension** | **Check you have achieved this by** |
| 1 | Recognise specific elements of the political environment | discussing the impacts of political decision making and legislation on organisations |
| 2 | Recognise specific elements of the economic environment | discussing the impacts of economic activity on organisations |
| 3 | Recognise specific elements of the sociocultural environment | discussing the impacts of societal attitudes on organisations |
| 4 | Recognise specific elements of the technological environment | discussing the impacts of changing technology on organisations |
| 5 | Recognise the specific context of political, economic and sociocultural issues at the local, national and global levels of the external environments | discussing how political, economic or sociocultural issues may have different effects on an organisation at the local, national and global levels of the external environment |
| 6 | Recognise that technological issues might affect organisations in a similar way at all levels of the external environment | explaining why technological issues might affect organisations in a similar way at all levels of the external environment |
| 7 | Identify sources of information and data that make scanning the external environment possible | discussing the advantages and disadvantages of a variety of information and data sources |
| | **Application** | **Check you have achieved this by** |
| 1 | Apply PEST analysis to general situations | performing a generic PEST analysis for a hypothetical organisation |
| 2 | Apply PEST analysis to specific situations | performing a specific PEST analysis for a real organisation |
| 3 | Demonstrate ability to apply PEST analysis to LoNG levels of external environment | modifying the specific PEST analysis above to incorporate LoNG levels |
| 4 | Illustrate the organisation in its environment | sketching a contextual map for a particular organisation |
| 5 | Practise PEST analysis | performing LoNGPEST analysis on a variety of organisations in differing sectors and contexts |
| | **Analysis** | **Check you have achieved this by** |
| 1 | Differentiate between the public and private sectors | comparing and contrasting public- and private-sector activity |
| 2 | Distinguish between the free market, planned economy and mixed economy | analysing deregulation and privatisation in the UK since 1979 |
| 3 | Differentiate between democracy and autocracy | comparing and contrasting governments in different countries |
| 4 | Appraise political influence on the economy | comparing various UK political agendas and how they intend to affect economic activity through legislation |
| 5 | Examine why organisations need to understand and analyse the external environment | testing the notion that organisations could exist in splendid isolation |
| | **Integration** | **Check you have achieved this by** |
| 1 | Create your own model of organisations in their external environments | proposing arguments in support of your model |
| | **Evaluation** | **Check you have achieved this by** |
| 1 | Choose information and data sources that best assist managers in organisations with scanning and understanding the external environment | comparing information and data sources and evaluating the most suitable medium and source |
| 2 | Evaluate PEST analysis as a management tool for understanding the external environment | assessing the benefits and limitations of the model when applied to real organisations |

ENTRY CASE STUDY 2.1

FT

# Into the east at full throttle

## Low costs and fast growth are persuading some leading carmakers to open plants in central and eastern Europe, says **Haig Simonian**

Mr Peter Bognar, a 28-year-old maintenance engineer, earns the equivalent of DM700 (£250) a month at Audi's engine plant in Györ, western Hungary. In the Czech Republic to the north, Mrs Vaclava Buriánková, a 48-year-old body-shop worker, takes home the equivalent of about DM650 at Skoda's car plant in Mlada Boleslav.

Both could make around eight times as much in Germany. Such huge disparities are one of the reasons why some of the world's leading car companies have invested in central and eastern Europe since the collapse of communism in the early 1990s.

'For west European carmakers, faced with near permanent overcapacity and low trend growth, the prospects in the east are too tempting to pass up,' says Mr Simon Miller, automotive analyst at UBS Securities in London.

More than $8.9bn (£5.3bn) has been committed to new plants in the region in the past six years by the four biggest investors: Germany's Volkswagen; Adam Opel, the General Motors German subsidiary; Fiat of Italy; and South Korea's Daewoo. Spending by the less ambitious Suzuki of Japan and Ford of the US takes the total to more than $9bn.

Low pay is not the main reason why they have headed for central and eastern Europe. While investment grants and tax incentives have played a part, soaring demand on the back of generally buoyant economic growth has been the main draw when sales in traditional markets such as the US, western Europe and Japan have been sluggish.

The explosion in sales in the former East Germany shows the potential. Car registrations exceeded 777 000 in 1992 – the peak year of a post-unification boom – compared with a fraction of that in the former German Democratic Republic. Although sales have fallen since, they have stabilised at about 600 000 vehicles.

And although demand in Poland, Hungary, the Czech Republic and Slovakia has also been volatile, the trend has been decisively upwards. Poland – the biggest and most populous country in the region – was Europe's fastest growing car market last year. Sales jumped 41 per cent to 373 542, making it Europe's eighth biggest new car market, according to Samar, a local consultancy. Registrations should hit 450 000 by 2000, it reckons.

Sales in the Czech Republic reached 128 701 in 1995 from 104 142 the previous year. In Hungary, sales have yet to exceed their peak of 90 000 in 1994. But after collapsing to 68 800 following a government austerity package in 1995, registrations climbed last year and should reach 76 000 in 1997.

Volkswagen, Europe's biggest carmaker, has led the investors into this new market. This has been partly the result of political pressure after unification – the group is 20 per cent owned by the west German state government of Lower Saxony. It will have spent DM3.2bn on new car and engine plants in the former east Germany by 2000.

VW has also invested elsewhere in the region. It has allocated DM3.7bn to acquire and modernise Skoda, the Czech Republic's leading carmaker, by 2000. In Hungary, VW's Audi subsidiary expects to have spent DM1bn over the same period. More modestly, the group has also invested in Slovakia and Poland.

Opel has been almost as ambitious. It has ploughed DM500m into a showcase new 'lean production' car factory at Eisenach in the former East Germany. The lessons learnt at Eisenach will be incorporated at the DM500m plant under construction in Poland, to supersede a DM30m assembly unit in Warsaw. A further DM700m has been invested to assemble cars and engines in Hungary.

Fiat and Daewoo have concentrated on Poland. The Italian company has spent $1bn on acquiring 80 per cent of FSM – now Poland's biggest carmaker. Another $800m is going into two new models by 2002.

Daewoo's ambitions are so big it has bought two Polish vehicle-makers since 1995. The largest, FSO, is being groomed to build a new generation of Daewoos now taking to the roads in Korea.

The boldness of these four carmakers contrasts with the caution of the other car companies. Although others concede that demand will eventually rise significantly, they are much more pessimistic about the immediate potential for sales growth.

Such arguments often disguise self-interest: many of the financially strapped European carmakers face overcapacity in their home countries. Many would prefer to export vehicles from their domestic plants rather than to build new factories in central and eastern Europe. They argue that it will be easier to export to the region once countries such as Poland and Hungary join the European Union and have to cut high tariffs on imports.

They also point out that the advantages of investing in low-wage economies are limited: pay accounts for no more than 10 per cent of the costs in a capital-intensive car plant; and wage levels are rising fast in these countries.

The big investors acknowledge salaries are climbing. But they say it will be years before they reach Spanish or UK levels, let alone those in western Germany. Even after unification, pay in the former communist east is still 15–25

**Entry case study 2.1 *continued***

per cent below that in the west – and that ignores the more generous fringe benefits in the latter.

Companies investing heavily emphasise the importance of gaining an early foothold in these growing markets by establishing local plants. 'A local presence is essential to building goodwill,' says Mr Albert Lidauer, managing director of Opel Hungary.

However, Opel and VW expect their investments to serve a wider region than just the countries in which they are located. General Motors last month said its Hungarian operation would lead its push into the Balkans. Both companies also have their eyes on a bigger prize – the vast but still untapped markets of the former Soviet Union.

Meanwhile, Daewoo sees a double benefit in Poland: local production will provide a back-door to sales in western Europe if exports from Korea were ever threatened, as well as a bridgehead to the east.

And the evidence suggests the spoils will indeed go to local manufacturers. Opel has led Hungary's car market for the past five years, with more than 20 per cent of sales in 1996. Suzuki, another local manufacturer, came second with 19 per cent.

In Poland, Fiat took 42 per cent of sales last year. Together with Daewoo (26 per cent) and GM, the three biggest brands accounted for more than 75 per cent of sales in 1996.

Such data may explain why Toyota, Japan's biggest carmaker, which has so far steered clear of building cars in the region, is having second thoughts. Mr Akira Yokoi, its head of international operations, says it is now looking closely at local production.

But there is a special reason why German carmakers have been in the vanguard. Their greenfield factories in eastern and central Europe are convenient test-beds for more flexible production methods, which would be resisted by their unionised workers in Germany.

'When you start from scratch, you can be very efficient,' says Mr Lorenz Köstner, head of VW's east German engine plant.

Moreover, the threat of shifting investment to the east has given managements a lever to negotiate productivity concessions from their traditionally pampered domestic workforce.

The Chemnitz plant, which has been VW's most productive engine unit for the past two years, contracts out dozens of tasks which would be done by VW workers in the west, to cheaper third parties. Similarly, Audi chose to assemble its new sports cars in Hungary because its highly flexible contracts suit a model for which demand will be very seasonal, says Mr Karl Huebser, managing director of Audi Hungaria Motor.

VW and Opel know they must tread carefully to avoid being accused of 'exporting' jobs to the east or of 'blackmailing' their west German workers into giving up hard-won privileges on pay and conditions.

Such concerns can provoke some extraordinary corporate doublespeak. Last year, Mr Adreas Schleef, Audi's head of personnel, justified the decision to shift work to Györ because it corresponded to the company's strategy of 'enhancing our competitiveness, corporate growth and safeguarding employment, each of which influences the others in a kind of magic triangle'.

But there is no alchemy behind the companies' ultimate intentions. They believe much of what is being learnt at the new plants in the east will eventually be applied at home. 'Whether for VW or Opel, the aim is to transfer the experience back to western Germany,' says one executive who has worked for both companies.

Mr David Herman. Opel's chairman and a tough-talking former lawyer, is characteristically succinct. 'We have expressed, through our decisions, the element that time is pressing. Having a new plant on your doorstep is different from having it in Indonesia.'

High-cost manufacturers like VW and Opel believe they have no other way to fight the threat of rising imports from lower-cost Asian rivals in western Europe. For them, the factories in the east are vital lifelines to developing new markets for tomorrow.

*Source*: *Financial Times*, 13 February 1997. Reprinted with permission.

## Introduction

This chapter follows on directly from Chapter 1 and seeks to examine in much greater detail the generic LoNGPEST model, *see* Exhibit 2.3. The constituent elements of the external environment and their influence on organisations are discussed here. The chapter examines the nature and influence on organisation of politics, economics, society and culture and technology, all at the local, national and global levels.

**Exhibit 2.3 Generic LoNGPEST grid**

| | Political | Economic | Sociocultural | Technological |
|---|---|---|---|---|
| **Local** | • Local government<br>• Local offices of national government<br>• Local associations<br>– Chambers of Commerce<br>– TECs<br>– Business Link | • Local bank branches<br>• Local economy | • Local community | • Communications technology<br>– Mobile phones and faxes<br>– Video conferencing<br>– Internet and world wide web |
| **National** | • National government<br>• Devolution for Scotland and Wales<br>• National bodies<br>– Employers' bodies<br>– Employees' bodies | • Central bank<br>– Bank of England<br>• Stock market<br>– London | • Demographic change<br>• Social change | • Organisations and the application of technology<br>– The personal computer<br>– The banking and financial services industry |
| **Global** | • Alliances and agreements<br>– UK and USA<br>– UK and China<br>– EU<br>– Cold War<br>– CIS<br>– CBSS<br>– Eastern Europe<br>• International bodies<br>– The Commonwealth<br>– NATO | • Trading blocs and bodies<br>– EU<br>– EFTA<br>– OECD<br>– NAFTA<br>– ASEAN<br>• World money markets<br>• WTO | • Cross-cultural issues<br>– Language<br>– Behaviour<br>– Culture shock | • Millennium bug |

## The political elements of the external environment

The political external environment comprises local and national government, local associations, trade unions and employers' bodies. All of these are likely to hold a political viewpoint. Countries and organisations that are the result of political agreement and co-operation are also considered in this section. Because national and global economies are largely organised and run by governments, there are many areas of overlap between the political and economic environments. In this chapter the use of separate sections attempts to clarify and analyse political and economic environments individually. There are many political elements in the external environment, so analysing political elements of the external environment is sometimes confusing. This section principally examines the effects that external environmental elements have on organisations, but inevitably it will become clear that organisations also affect the external environment around them. This is perhaps especially so for employers' organisations, as will be seen later.

## The local political environment

At the local level, the political elements of the external environment are local government, local offices of national government and local associations; *see* Exhibit 2.4.

**Local government**

Local government has a significant impact on the businesses within its boundaries. It has responsibilities for regulating private-sector business activity and for directly providing a wide range of services. Some of these responsibilities have a very direct effect on businesses and others a more minimal influence. Local authorities have a statutory responsibility, i.e. they are bound by law, to uphold trading standards and environmental health. As such, they hold great powers over all organisations within their area, and have the authority to close them down if severe breaches of law are discovered.

The services provided by the local authority range from general social services such as care of the elderly or education, to services used by private householders and organisations alike, such as refuse collection, street cleaning, transport and planning permission for new buildings or extensions to existing premises. Clearly, a local authority's policies on trade refuse collection, street cleaning and transport to and from the city centre are all going to affect local businesses and shops in the city centre. A local transport policy aimed at reducing the number of cars in a particular street or shopping area is likely to have a detrimental effect on the amount of passing trade that shops in

**Exhibit 2.4 The local political environment**

| | Political | Economic | Sociocultural | Technological |
|---|---|---|---|---|
| Local | • Local government<br>• Local offices of national government<br>• Local associations<br>– Chambers of Commerce<br>– Business Link | | | |
| National | | | | |
| Global | | | | |

the street will obtain. Provision and promotion of effective and cheap public transport to those areas, however, ought to counterbalance these detrimental effects.

Local government is also responsible for setting and collecting council tax from residents and business rates from commercial organisations. The money raised via council tax and business rates will partially fund the services that local authorities provide, although the main source of funding for local government is money provided by national government. For the purpose of environmental analysis, this illustrates neatly that the levels of the external environment are as interdependent as organisations and their external environments.

It is also important to remember that a local authority is a large public-sector organisation operating in its own external environment. In many UK regions, cities, towns and villages, the local authority is the biggest local employer and the largest organisation. It too has customers and suppliers, and with the introduction of Compulsory Competitive Tendering (CCT) by the Conservative governments of the 1980s, local authorities now also have competitors for the provision of services in the locality where previously they had the monopoly. Here we begin to observe one of the complexities of environmental analysis. Clearly demonstrated here are the inter-dependencies between different elements of the environment: one organisation has a complex external environment consisting of other organisations and individuals; it in turn is an element in each of these organisations' individual external environments.

**Local offices of national government**

As national government has responsibility for the direct provision of many welfare services, the local offices of national government departments play a key role in the lives of local organisations and the local people. There is a local interface between the Department of Social Security's Benefits Agency, the local offices of the Inland Revenue, and the Department for Employment and Education's local Employment Services. The current Labour government introduced a review of welfare provision geared towards removing the disincentives to finding work that the welfare system can engender, and towards promoting the opportunities that unemployed people can offer organisations. The 1998 green paper 'A New Contract for Welfare' lays out proposals to introduce the first wholesale reform of the welfare state since its introduction in 1945. This will affect organisations in the local environment by providing funding for training and employment of those who are currently workless through various initiatives under what is called the 'New Deal'.

**Local associations**

For private-sector organisations in the model, local associations are defined as trade and business associations operating in the geographic vicinity.

**Chambers of Commerce**

Chambers of Commerce are a membership scheme operating as a private company and funded through subscription from local members. The Chamber of Commerce can provide information to local managers on opportunities at home and abroad. Chambers also offer short courses for local businesses, which are a cheap and effective source of high-quality training for their employees. Chambers also act as a forum for members of local businesses to meet and discuss local issues that directly affect their activities. Local Chambers of Commerce are affiliated nationally and collectively and offer a powerful voice in terms of affecting member companies' behaviour, and also influence the external environment through political activity. Chambers of Commerce act collectively to put forward their views and opinions on matters affecting their member businesses with a view to influencing both local and national government.

**Business Link**

Business Links are privately led partnerships between Training and Enterprise Councils, Chambers of Commerce, Enterprise Agencies, local authorities and government. Business Link partnerships are involved in providing advice via a one-stop shop for small and medium-sized businesses. Business Link partnerships operate very much on a regional and local level; for example in South Yorkshire, Business Link services are available for Barnsley, Doncaster, Rotherham and Sheffield.

Business Link partnerships generally provide advice and expertise in the areas of raising finance, entering export markets, employee training and managing change. A Business Link partnership aims to provide advice and expertise at a price that small and medium-sized businesses can afford. The partnerships are funded by the Department of Trade and Industry (DTI), with the condition that by the fifth year of existence the partnership must be at least 25 per cent self-funding.[1]

## The national political environment

The national government and national bodies constitute the national political environment; *see* Exhibit 2.5.

**National government**

The ways in which the national government manages the economic environment are dealt with under the relevant section below. As a result of the UK's membership of the European Union, the distinction of a national political environment is becoming blurred. Acts that appear on the surface to have been the British Parliament making laws governing British businesses turn out to have originated in Brussels and required all member countries to implement them. Examples of European decisions enacted by the national British government are the fitting of seatbelts to coaches, vans and lorries, adding costs to transport companies; the export ban on British beef in the wake of the BSE crisis and its effects on British beef farmers; and the abolition of duty-free goods within the European Union, meaning that tax-free shopping has come to an end except for inter-continental flights.

**Exhibit 2.5 The national political environment**

| | Political | Economic | Sociocultural | Technological |
|---|---|---|---|---|
| Local | | | | |
| National | • National government<br>• Devolution for Scotland and Wales<br>• National bodies<br>– Employers' bodies<br>– Employees' bodies | | | |
| Global | | | | |

Examples of the purely national political environment affecting organisational activity are therefore increasingly hard to find, but largely focus on the legislation passed regarding permissible commercial activity, production and service functions and human resource strategies. This latter is dealt with in Chapter 8, but to give one example, the 1995 Disability Discrimination Act (DDA) made it illegal for organisations not to provide access for disabled workers or customers on their premises. The direct effect of this legislation, which was stronger than anything preceding it, caused many organisations to commission alterations to their existing facilities in order to comply with the new law. The DDA covers equal access to goods, facilities and services, and employment and education.

Although its effects could be said to be largely economic, the national government's political decision making affects most citizens directly through its annual budget. Government has to raise money to provide for its services. It does this via taxation, revenue that it uses to pay for publicly funded services such as education, the National Health Service, public transport subsidies and social and welfare services such as the state pension for those too old to be economically active. Some of the money raised by taxation is distributed via local government and some is distributed and spent directly by national government. This is covered in more detail under the national economic environment, later in this chapter.

Decisions with an economic effect on business, then, may be classified as political when taken by government. One such decision was that in

November 1992 by Norman Lamont, the then Chancellor of the Exchequer (Finance Minister) in the Conservative government. The pound sterling had been entered into the Exchange Rate Mechanism (ERM), the European Union's mechanism for preparing its member nations' currencies for European Monetary Union (EMU) and eventual merger into a single European Currency Unit (ECU). The ERM was designed to bring all its currencies to a level value, by setting a range of exchange rate limits beyond which a currency should not be allowed to rise or fall. Norman Lamont took the decision to withdraw the pound from ERM and hence effectively took Britain out of the first group of countries to enter EMU. The pound was removed from the ERM because its value fell outside the exchange rate range within which it was supposed to remain. However, the decision was also taken for political reasons, as the Conservative party was showing early signs of the internal divisions over European policy that partially led to its defeat at the 1997 elections.

The withdrawal of the pound from the ERM resulted in what the City referred to subsequently as 'Black Wednesday', when there was an immediate 10–15 per cent devaluation of the pound in relation to other currencies. Consequently, businesses in Britain trading with overseas companies were affected. Companies importing goods, components or raw materials into the UK from overseas had to pay more for them. In 1992 the UK economy was depressed and importing companies faced the dilemma of whether to pass the increased costs on to consumers or to absorb the increased cost in whole or in part, via reduced profit margins. On the day the pound was removed from the ERM, Yorkshire Television news reported on a company based in Hull that was facing exactly this dilemma. The company imported high-quality, up-market German kitchens and kitchen appliances, manufactured by companies such as AEG, Siemens and Bosch. In contrast, companies exporting UK manufactured goods found it easier as the lower cost of production made UK goods more price competitive abroad. This partly contributed to the growth of the UK economy in 1993.

From this it can be seen that the link between the national political and economic environments is clear. Politicians are elected by the people and run the economy. In doing so, they decide the economic policy that affects both individuals and organisations operating at the national level of the environment.

### Devolution in Scotland and Wales

The Labour government elected on 1 May 1997 had promised in its election manifesto to hold referendums on devolution in both Scotland and Wales. This was duly implemented. In Wales the referendum was held on 18 September 1997 and 50.1 per cent of the electorate turned out to vote, with 50.3 per cent voting for a Welsh Assembly and 49.7 per cent voting against. In Scotland the referendum on 11 September had a turnout of 60.4 per cent, with 74.3 per cent of those who voted supporting the creation of a Scottish Parliament and 63.5 per cent voting to give the Scottish Parliament limited tax-varying powers.[2]

The Scottish Parliament consists of a total of 129 seats, of which 73 are directly elected and 56 seats allocated to additional members. Directly elected members were elected by a constituency to represent them. The additional members were elected by the electorate voting for a party and the additional seats given to members of the party voted for. The allocation of additional seats is complex, but is representative of the voting pattern in the electorate as a whole.[3]

The Scottish Parliament appointed a First Minister, Donald Dewar, who in turn appointed members to and heads the Scottish Executive. The First Minister and Scottish Executive are drawn from the party or group commanding the majority in the Scottish Parliament. The Scottish Parliament is able to pass laws in a number of areas for Scotland, including health; education, local government, housing, economic development, trade, transport, criminal and civil law, courts, prison, police and fire services, animals, the environment, agriculture, food standards, forestry, fisheries, sport and the arts. The Scottish Parliament also has powers to repeal legislation passed at Westminster as far as Scotland is affected.[4]

The Welsh Assembly contains 60 directly elected seats and 40 additional seats. In contrast to the Scottish Parliament, the Welsh Assembly does not have the same powers governing the establishment of legislation and the repeal of Westminster legislation. The areas of responsibility devolved to the Welsh Assembly include economic development, agriculture and food, industry and training, education, local government, health and personal social services, housing, environment, planning, transport and roads, arts and culture, the built heritage and sport and recreation. The Secretary of State for Wales has the power to make secondary legislation in these areas. For example, the Secretary of State for Wales is able to decide the school curriculum in Wales.[5]

## National bodies

National bodies represent businesses and employees and present views on employment and trade issues to one another, and to the government and opposition parties. Both employer and employee organisations try to shape and influence events in the external environment for the benefit of their members. Employer and employee national bodies are often closely aligned themselves with the government or a particular political party. Employers' organisations traditionally support the Conservative Party, while the Labour Party is linked financially and in terms of membership to the trade unions. For the Labour Party to achieve victory at the 1997 general election, it was thus crucial for it to have gained support from both sides of the negotiation table.

### Employers' bodies

Examples of national employers' bodies include employers' organisations such as the Institute of Directors (IOD) and the Confederation of British Industry

(CBI). These organisations provide a forum for employers or owners of businesses to put forward their views and be represented on issues affecting businesses, such as the national minimum wage and a single European currency. They frequently use the printed and broadcast media to do this. At the time of the election of the Labour government, the CBI was contributing to the minimum wage debate via the media. Labour had made the introduction of a minimum wage in the UK a manifesto pledge, and the CBI was suggesting a minimum wage of £3.50 per hour as an acceptable amount. It pitched this figure between the £4.29 per hour favoured by the Trades Union Congress (TUC) and the £3.00 per hour suggested by 'certain employers'.[6] The rate at which the minimum wage is set (currently £3.60 for workers aged over 21) is crucial to employers as there are implications for costs and profitability; arguments which the previous Conservative government had always used to abolish pay review bodies, which had existed for particular industries, and to reject calls for a minimum wage in the UK.

#### Employees' bodies

In contrast, trade unions are national bodies representing employees. Examples of Trade Unions are NATFHE, representing teachers in further and higher education, UDSAW, representing shop workers, and UNISON, representing public-sector employees. Trade unions represent their members in negotiations with employers on issues concerning pay and conditions, whether in the private or public sector. They are linked to the Labour Party, although reforms made to party membership after Labour's 1992 election defeat attempted to reduce the unions' impact and promote the concept of 'one member, one vote'. Most trade unions are affiliated to the Trades Union Congress, which is the largest voluntary organisation and the largest pressure group in UK. The TUC operates to represent its member unions collectively at a national level.

Trade union membership and powers were drastically curbed under the Conservative administrations led by Margaret Thatcher in the 1980s, which viewed Britain's industrial unrest of the 1970s as the cause of its industrial decline and evidence of the unions' rise to too much political power. In contrast, the Employment Relations Act 1999 established a new statutory procedure for recognition of independent trade unions in the workplace, although it did not include the adoption of European-style workers' councils. Union influence, however, is unlikely to rise to previous post-war levels.

**Exhibit 2.6 Trade union membership**

| Year | Membership | Percentage of workforce who are union members |
|---|---|---|
| 1995 | 7.3 million | 32% |
| 1979 | 13.3 million | 55% |

*Source*: http://bizednet.bris.ac.uk/compfact/tuc/tuc25.htm.

## The global political environment

### Alliances and agreements

The global political environment encompasses alliances and agreements between countries that have an effect on the international activities that each country's citizens may undertake. Two or more countries may come together to establish independent or semi-independent global bodies to oversee or regulate the conduct of international trade and commerce, as well as to work towards the improvement of social issues, such as health, poverty or human rights.

Alliances and agreements occur between two or more countries for mutual benefit. The countries involved concur to support each other in global politics or in bi-lateral or multi-lateral economic activity. In extreme cases, consent may be given to the merger of countries into a single entity, or one country may consent to divide into separate countries to fulfil ideals of cultural identity, national integrity or economic benefit.

#### The UK and the USA

A good example of an alliance in which two countries concur to support each other is the close and special relationship between the UK and the USA. Although there are few formal bi-lateral treaties, a special relationship exists as a result

**Exhibit 2.7**
**The global political environment**

| | Political | Economic | Sociocultural | Technological |
|---|---|---|---|---|
| **Local** | | | | |
| **National** | | | | |
| **Global** | • Alliances and agreements<br>– UK and USA<br>– UK and China<br>– EU<br>– Cold War<br>– CIS<br>– CBSS<br>– Eastern Europe<br>• International bodies<br>– The Commonwealth<br>– NATO | | | |

of the two countries' historical and linguistic ties. In the 1980s the relationship between the UK and the USA was personified by the close friendship and mutual support of Conservative Prime Minister Margaret Thatcher and US Republican President Ronald Reagan. Both headed conservative governments in their respective countries and combined radical free-market agendas with strong global politics. This caused some difficulty for President Reagan during the UK–Argentine Falklands War in 1982, as his special relationship with the UK required support of Britain's efforts, while US links with the rest of the American continent precluded actual aggression against Argentina. On a separate occasion in 1986, Margaret Thatcher approved the use of US airbases in the UK to launch bomb attacks against Libya.

The relationship and support continues, embodied in Labour Prime Minister Tony Blair and US Democratic President Bill Clinton. In February 1998 this relationship was illustrated by the agreement between the UK and the USA over the handling of the Iraq crisis, as the two countries stood firm against the opinion of other countries. There had been greater global consensus over the 1991 Gulf War, where international forces, UK and US among them, liberated Kuwait from Iraqi invasion.

### The UK and China

China is extremely important on the global political stage because of its sheer size, geographically, politically and economically. The Chinese market is an important part of Britain's overseas trade. UK exports to China totalled £117m while imports topped £452m in January 1998.[7] The return of Hong Kong is an example of an agreement at the global level of the external environment, which was designed to promote business stability and minimise political and economic risk for British business in China. On 1 July 1997, the British Colony of Hong Kong was handed back to China after 99 years of British rule under the terms of a lease forced from the Chinese at the height of Britain's imperial activity. This was unlike the fate of any of the UK's other colonial possessions, which all left the Empire to become independent countries in the Commonwealth; *see* Exhibit 2.12 on p. 35. Hong Kong, however, had not been independent before the colony was established, but had always been part of China, so had to be returned to it at the end of the lease.

The negotiations governing the return of Hong Kong to China were started in the early 1980s by Margaret Thatcher and Deng Xiaoping, China's then paramount leader. In the late 1970s Deng Xiaoping engineered economic reforms in China and allowed the slow development of a more market-based economy, while retaining strict political control over personal liberty. He died on 19 February 1997 prior to the deadline for Hong Kong's return to China, which duly went ahead. China described itself as 'one country – two systems', referring to the promise to continue unchanged for a minimum of 50 years Hong Kong's capitalist free-market economy under Chinese communist rule. This free-market economy has long been the gateway to the Chinese market for foreign businesses and for Chinese exports to the world. Over half the value of China's foreign trade has been directed via Hong Kong, so it is greatly in China's interest to maintain the status quo.

China's lack of a democratic political system should not have been a cause for concern, as the British colony had been ruled without recourse to democracy. Nevertheless, following Britain's long-held policy of establishing democratic systems before withdrawing from colonial possessions, the last Governor of Hong Kong, former Conservative politician Chris Patten, attempted to introduce some last-minute democracy to the colony. In the 1990s Hong Kong people were allowed to vote for the first time for a minority of members of Hong Kong's governing body, the Legislative Council (Legco). However, the return of Hong Kong to China in 1997 saw the dismantling of the Legco and the appointment by Beijing of a replacement administration and a new Chief Executive of the Hong Kong Special Administrative Region, Tung Chee-hwa, a wealthy local businessman who had previously been ennobled by the Queen.

### The European Union

In 1957 the European Economic Community was established by the Treaty of Rome. There were six founding countries (*see* Exhibit 2.8), with a further nine countries joining the European Union (EU) by 1995. In 1998 the EU's three main objectives were[8]:

- implementation of the Treaty of Amsterdam;
- expansion of the EU;
- the launching of the single currency.

The euro became legal currency for trade and financial markets on 1 January 1999 in all participating EU countries. Coins and notes will be introduced on 1 January 2002. (See the section on the European Union later in this chapter.)

The 1997 Treaty of Amsterdam covered new rights for EU subjects concerning issues such as freedom of movement and employment. The Agenda 2000 blueprint presented to the European Parliament in July 1997 allowed the EU to expand eastwards and for former Communist countries of Eastern Europe (*see* Exhibit 2.9) to become members. The political and economic conditions of applicant countries have to be judged satisfactory by the EU before it allows their admittance. In 1997 ten central and eastern European countries wanted to join the EU, with five of the ten countries expected to be admitted in the near future. The negotiations to admit new members are complex and examine

**Exhibit 2.8 European Union member countries**

| Year of entry | European Union member countries | | |
|---|---|---|---|
| 1957 | • Belgium<br>• Italy | • France<br>• Luxembourg | • Germany<br>• Netherlands |
| 1972 | • UK | • Denmark | • Ireland |
| 1981 | • Greece | | |
| 1986 | • Spain | • Portugal | |
| 1995 | • Austria | • Finland | • Sweden |

*Source*: *Financial Times*, 25 March 1997.

**Exhibit 2.9 European Union membership sought**

| *EU deems political and economic conditions satisfactory, EU membership expected in imminent future* | | |
|---|---|---|
| • Poland<br>• Slovenia | • Hungary<br>• Czech Republic | • Estonia |
| *EU deems political and economic conditions inadequate, membership in distant future* | | |
| • Latvia<br>• Romania | • Lithuania<br>• Slovakia | • Bulgaria |

*Source*: *Financial Times*, 17 July 1997.

how the significant differences in economic and social development between member and potential members states can be tackled.[9]

### The Cold War

World political events result in alliances being created and demolished, hence altering the political map of the world. In order to understand events in the late 1990s, it is essential to consider some political background. After the Second World War in 1939–45, the world was plunged into the Cold War, an ideological battle between democratic and communist political systems backed up by the technology of nuclear warfare, bringing for the first time in history the constant threat of a Third World War that would annihilate millions of civilians and destroy entire countries. The two sides of the Cold War, put simplistically, consisted of the US and Western Europe – with its institutions such as NATO, EFTA and the EU – and the then Soviet Union (USSR or Union of Soviet Socialist Republics) and Eastern Europe, with its equivalent institutions – the Warsaw Pact was the mutual defence treaty signed by Communist countries in the USSR sphere of influence, and Comecon was the alliance of planned economies.

The division of East and West was never more starkly evident than in the division of Germany on its defeat by the victorious Allies into two states, the western, democratic, free-market Federal Republic of Germany (FRG) and the eastern, communist, planned economy of the German Democratic Republic (GDR). The former capital of Berlin, geographically inside the GDR, was itself split into two halves and a wall built down the centre between East and West. The Berlin Wall came to symbolise the Cold War.

The Cold War split was mirrored in the Far East during the Korean War (1950–53), where US forces supported the democratic South Korea against the Chinese-backed communist north. The country remains divided into two administrations today and a state of civil war still exists officially. Similarly, the People's Republic of China (PRC) and the Republic of China on Taiwan both claim to be the legitimate government of China, after a similar civil war whose hostilities ended in 1949 but peace has never been negotiated. The US originally backed the Republic of China, and did not switch allegiances until president Nixon visited the PRC in 1973. In Vietnam, the American-backed south eventually fell to the communist north after many years of bloody war, uniting the country under a communist government in the early 1970s.

**Exhibit 2.10 Members of the Commonwealth of Independent States (CIS)**

| | | |
|---|---|---|
| • Armenia | • Azerbaijan | • Belarus |
| • Georgia | • Kazakhstan | • Kyrgyzstan |
| • Moldova | • Ukraine | • Uzbekistan |

In Europe in the late 1980s and early 1990s, the geopolitical map altered considerably and unexpectedly, as a result of the collapse of strict controls over personal liberty in many Communist-controlled countries. This had begun with the last ever President of the USSR, Mikhail Gorbachev, who introduced policies of *glasnost* (openness) and *perestroika* (economic restructuring) during his term in office. These alterations to the political map provided unprecedented opportunities for trade and commerce and began in the GDR, when the Communist regime lost control over the population and the Berlin Wall was suddenly dismantled in a popular uprising that was not opposed by the police or military forces. By 3 October 1990 East and West Germany had officially reunified under the former FRG's government. As a result of reunification, the German government has moved its parliament from Bonn, capital of FRG, back to Berlin. The opportunities for business in eastern Germany, now that the planned economy is over, are evident and illustrated in the entry case study for this chapter, 'Into the East at full throttle'.

### Commonwealth of Independent States

The ending of the Cold War resulted in other changes to the geopolitical map of Eastern Europe. During the years 1989–91 the USSR collapsed as the Soviet government in Moscow went the way of the GDR government, and the Soviet Union dissolved the federation into its component independent states, which then came together under the auspices of a looser association called the Commonwealth of Independent States (CIS); *see* Exhibit 2.10.

### Council of Baltic Sea States

The newly independent Baltic countries, Estonia, Latvia and Lithuania, wanted nothing more to do with their former Soviet masters and formed instead the Council of the Baltic Sea States (CBSS), established in March 1992; *see* Exhibit 2.11. The CBSS aims to be 'a forum for linking more closely together the emerging co-operation between the countries around the Baltic Sea following the revolutionary changes in Eastern Europe'.[10]

### Eastern Europe

Countries such as Poland, Hungary, Czechoslovakia, Romania and Yugoslavia all experienced cataclysmic change after the dissolution of the USSR. In

**Exhibit 2.11 Members of the Council of Baltic Sea States (CBSS)**

| | | |
|---|---|---|
| • Denmark | • Estonia | • Finland |
| • Germany | • Iceland | • Latvia |
| • Lithuania | • Norway | • Poland |
| • Russia | • Sweden | • European Commission |

*Source*: http:/www.baltinfo.org/CBSS.htm, 28 April 2000.

geographic terms Poland, Hungary and Romania remained unchanged, in contrast to Czechoslovakia and Yugoslavia, which altered significantly. Czechoslovakia, for reasons of national and economic identity, split into two countries, the Czech Republic and Slovakia. In the 1990s civil war split Yugoslavia into several independent countries: Croatia, Bosnia and Herzegovina, the Former Yugoslav Republic of Macedonia and Slovenia, with Montenegro and Serbia remaining as a rump Yugoslavian Federation. A multinational peace-keeping force continues to play a role in maintaining a fragile peace.

All these former eastern bloc countries have seen major changes to their political and economic systems, with elections taking place in many countries and greater opportunities for international trade and commerce becoming available. Some of the opportunities for trade and commerce allow western companies to invest and manufacture in countries such as Poland, Hungary, Slovakia and the Czech Republic, as illustrated in the entry case study for this chapter.

## International bodies

The international bodies discussed here are bodies that have a global political influence on the world political order. This is important, as peace and political stability are key elements in allowing a country to have an active economy and stable trade with other countries.

### The Commonwealth

The Commonwealth derives from Britain's imperial past and comprises countries from all regions of the world, including Europe and the Pacific, Asia, Africa and the Caribbean. The countries belonging to the Commonwealth are shown in Exhibit 2.12 and range from India, with a population of over 900m people, to Naura in the Pacific, with 8000 people. The Commonwealth consists of 53 countries, 52 of which are former colonies or protectorates of the UK, the exception being Mozambique, a former Portuguese colony. Mozambique was admitted as a member in November 1995 due to its close association with the Commonwealth in opposing apartheid in South Africa and because it wanted to reap the benefits of membership. Apartheid was the official system of keeping the white minority population in South Africa in a position of power and wealth, while the black and coloured populations had no access to money or political decision making. South Africa withdrew from the Commonwealth in 1961 following pressure from other member countries over apartheid. It was only readmitted in 1994 after the promulgation of a new, democratic and multi-racial constitution, with Nelson Mandela, a leading dissident of the old regime, released from imprisonment and elected president in the country's first ever democratic elections.

In contrast, Nigeria was suspended from the Commonwealth in November 1995 after its most recent military coup. The deteriorating relationship between Nigeria and the Commonwealth was caused by the Nigerian military cancelling the presidential elections in 1993 and executing in November 1995 nine minority rights leaders, including the author Ken Saro-Wiwa. The Commonwealth

**Exhibit 2.12 Members of the Commonwealth**

| Year of entry | Countries | | | |
|---|---|---|---|---|
| 1931 | • Australia | • Canada | • New Zealand | |
| 1947 | • India | | | |
| 1948 | • Sri Lanka | | | |
| 1957 | • Malaysia | • Ghana | | |
| 1960 | • Nigeria[1] | | | |
| 1961 | • Sierra Leone | • Cyprus | • Tanzania | |
| 1962 | • Uganda | • Jamaica | • Trinidad and Tobago | |
| 1963 | • Kenya | | | |
| 1964 | • Malawi | • Zambia | • Malta | |
| 1965 | • Singapore | • Gambia | | |
| 1966 | • Lesotho | • Botswana | • Barbados | • Guyana |
| 1968 | • Nauru | • Mauritius | • Swaziland | |
| 1970 | • Tonga | • Western Samoa | | |
| 1972 | • Bangladesh | | | |
| 1973 | • The Bahamas | | | |
| 1974 | • Grenada | | | |
| 1975 | • Papua New Guinea | | | |
| 1976 | • Seychelles | | | |
| 1978 | • Solomon Islands | • Tuvalu | • Dominica | |
| 1979 | • St Vincent and the Grenadines | • St Lucia | • Kiribati | |
| 1980 | • Zimbabwe | • Vanuatu | | |
| 1981 | • Antigua and Barbuda | • Belize | | |
| 1982 | • Maldives | | | |
| 1983 | • St Kitts and Nevis | | | |
| 1984 | • Brunei | | | |
| 1989 | • Pakistan[2] | | | |
| 1990 | • Namibia | | | |
| 1994 | • South Africa[3] | | | |
| 1995 | • Cameroon | • Mozambique | | |

*Notes*:
1 Suspended 1995.
2 Rejoined, had withdrawn in 1972.
3 Rejoined, had withdrawn in 1961.
*Source*: http://www.thecommonwealth.org/about/general/general1.htm1.

held a summit meeting in November 1995 in New Zealand and acted quickly and forcibly by suspending Nigeria from the Commonwealth. The EU and USA also imposed sanctions on Nigeria over the same democracy and human rights matters. However, the USA remains the largest buyer of Nigerian oil.[11, 12]

The two examples discussed above illustrate that the Commonwealth is an international body that can, along with other international bodies, affect and influence the status of a country with regard to participating in trade and commerce on a global level. Accordingly, in the late twentieth century the role of the Commonwealth is to further economic and social development, democracy and human rights in its member countries.[13]

**Exhibit 2.13**
**Members of NATO**

| Year of entry | Countries | | |
|---|---|---|---|
| 1949 | • Belgium<br>• France<br>• Luxembourg<br>• Portugal | • Canada<br>• Iceland<br>• Netherlands<br>• UK | • Denmark<br>• Italy<br>• Norway<br>• USA |
| 1952 | • Turkey | • Greece | |
| 1955 | • Germany | | |
| 1982 | • Spain | | |

*Source*: http://www.nato.int/welcome/home.htm.

### The North Atlantic Treaty Organization

The North Atlantic Treaty Organization (NATO) was formed in April 1949 and consisted of 12 members, with a further four countries joining between 1952 and 1982; *see* Exhibit 2.13. NATO's primary role is to ensure the security of its member countries and this largely involved countering the threat created by the Cold War until that came to an end in 1990/1. Since the ending of the Cold War NATO has reorganised and restructured to develop security arrangements for the whole of Europe and to allow 'peacekeeping and crisis management tasks undertaken in co-operation with countries which are not members of the Alliance and with other international organisations'.[14] The ending of the Cold War created an opportunity for the expansion of NATO and in July 1997 agreement was reached between member countries to allow the Czech Republic, Poland and Hungary to become members in 1999. Future likely members of NATO are Slovenia and Romania.[15]

## Political and economic links

The alliances and agreements between countries described in this section altered the political map and political systems for the countries involved. Alterations to their economic systems also occurred alongside the political changes. In terms of international business, the assessment of political and economic risks is a major factor when deciding which target market and mode of entry would be appropriate in various countries. Comparative analysis of various countries' political and economic stability, potential and future trends enables companies to take a judgement about whether or not to invest in or export to that destination.

**Democracy**

Democracy is the system of government most recognisable to those born and brought up in the West. Democracy comes from the Greek *demos*, 'the people', and *kratis*, power, and describes those systems where the people are able to choose, through a system of voting, those who represent them in the corridors of power and decision making. As it is impossible for every individual in society to be completely free and unfettered by the rules of the state, democratic systems are the mutual agreements under which people are able to live together collectively and yet express a collective opinion on periodic change via fair and proper elections. It would be understandable to think of all western governments as being equally democratic, yet not each country has the same system. Democratic systems divide broadly into those that have evolved and those that are the product of a sudden and cataclysmic change, e.g. popular revolution. The latter are those that have perhaps the greatest spread of democracy in their institutions.

For example, the UK is called a constitutional monarchy and has a parliamentary system. Despite a 40-year interruption under Oliver Cromwell, the UK's monarchy can be traced back a thousand years. The current system has evolved through a system of concessionary changes to the point where its system of hereditary figurehead monarch, popularly elected House of Commons and mix of hereditary and appointed second chamber, the House of Lords, is used as a blueprint of democratic government (there are now only 92 hereditary peers in the House of Lords, elected by their colleagues, pending further reform). Countries such as the USA, France and Germany, in contrast, have all had occasion to invent their political systems from scratch after revolution or war. These systems are based on written constitutions and codes of conduct, and feature elected government at all levels, including both chambers and head of state (president).

**Autocracy**

The exact opposite of democracy, autocratic systems are those where one person retains all power in his or her hands. The historical embodiment of autocracy was the absolute monarchy, where the king held all political and economic power. During the Cold War, this lack of democratic process was most keenly observed under Communism where the people were unable to choose the political party to represent them and had little or no personal liberty (these systems are more appropriately called oligarchies, since they were dominated by a powerful elite).

**The free market**

Providing in-depth knowledge of economics is not the purpose of this text, but to be able to discuss a variety of different political and economic contexts, it is necessary to define different economic systems. The free market is that situation in which there is little or no regulation of commercial activity on the part of political entities. In a free market, the market forces of demand and supply will lead to perfect competition, providing all that the people need or want at a price they can afford. If there is no demand for a product or service in a free market, then that product will not sell, no matter how cheaply it is priced.

**The planned economy**

In Communist countries, attempts were made to eradicate the free market and its differences between rich and poor, by planning and orchestrating all economic activity from central government. Thus people were not free to seek whatever job they wished, and organisations could not recruit whomever they wanted. All jobs, housing, production, services and food were organised by the state via its work units, or state-owned organisations. Market forces were denied, as consumers were only able to purchase products or services from the state-owned factories and companies, whether they were good or bad. No decision was left to the individual, no matter how small. Thus individuals' every need was catered for in a basic way, but individualism, innovation and creativity were stifled.

### The mixed economy

Most western economies are a mixture of free and planned economic activity. Where there is no regulation on business, employees and consumers alike are at risk. Regulation in the UK protects employees at work through health and safety and employment protection legislation, while the consumer is protected by the Sale of Goods and Trades Descriptions Acts.

### Links between countries' political and economic systems

When considering the global political environment (*see* Exhibit 2.7) or the global economic environment (*see* Exhibit 2.19), it is essential first to know something about the political and economic systems of various countries and regions in the world. Exhibit 2.15 is a matrix that plots various countries (*see* Exhibit 2.14) in relation to one another based on their level of democracy and the level of political interference in the economy. For the sake of this text, this matrix has been completed in a highly subjective and unsubstantiated manner, and indeed countries' positions will alter over time subject

**Exhibit 2.14 Country summaries**

| | |
|---|---|
| **Cuba** | After a Communist revolution over 30 years ago, dictator Fidel Castro has presided over a decaying planned economy. Following the dissolution of the USSR, Cuba's main source of political and economic support, Cuba has found it increasingly difficult to survive under the US trade embargo. |
| **Germany** | A federal republic, united finally in 1991, its mixed economy is one of the post-war European success stories, despite the costs of integrating the former Communist East. |
| **Japan** | Since the Second World War Japan has been a constitutional monarchy. However, the Liberal Democrats have been elected for most of those 50 years. The original successful Asian tiger economy, its success is also often attributed to the high levels of government intervention in the economy. |
| **India** | The largest democracy in the world, with a highly protectionist stance. |
| **Hong Kong** | Directly ruled from London for 99 years, Hong Kong reverted to the sovereignty of the People's Republic of China in 1997. A radical free market, it remains to be seen how this will continue under the 'one country, two systems' plan. |
| **New Zealand** | A Westminster system in the Commonwealth, it has seen a Labour government heavily interested in privatisation. |
| **Russia** | Since the collapse of the Soviet Union, the Russians have elected their leader for the first time in their history. The emergence of a mafia indicates a lack of regulation in the economy. |
| **People's Republic of China** | Controlled by the Chinese Communist Party following 'liberation' in 1949, China has liberalised its economy radically since the late 1980s without relinquishing any political power. |
| **United Kingdom of Great Britain and Northern Ireland** | Often called the 'Mother of Parliaments', the UK is a constitutional monarchy without a constitution, and only directly electing one of its two Houses of Parliament. There has been widescale privatisation since the early 1980s. |
| **United States of America** | With a written constitution, the Americans directly elect both Houses and their President. The federal system essentially allows for a variety in levels of free-market approach. |

**Exhibit 2.15 Government type and political involvement in the economy**

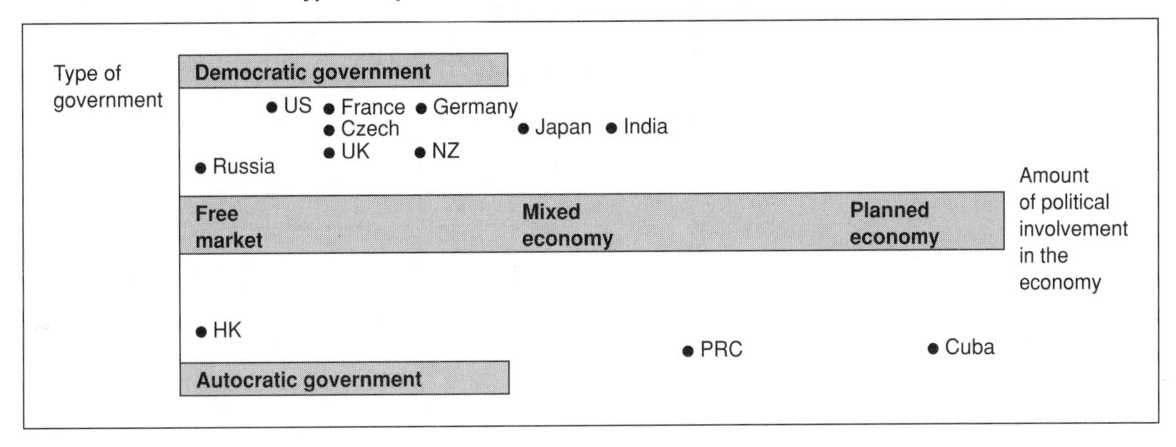


to political and economic change. In terms of assessing political and economic risk before entering new markets, organisations should invest in more objective forms of analysis, either done in-company or by external experts. The countries are plotted according to the background information given in Exhibit 2.14.

## The economic elements of the external environment

As has been stated, the economic environment faced by organisations is shaped and influenced by the political environment as well as by the economic bodies that are constituents of the external environment. The previous section examined the political elements of the external environment. In this section the role of economic bodies such as banks, stock markets, world money markets, trading blocs and bodies is considered.

## The local economic environment

**Local bank branches**

Banks are the key local economic environmental player for companies, as the relationship between a company, big or small, and its bank is a local relationship between the two organisations. The relationship between a company and its bank is not normally one of national or global importance or influence. Hence a company's relationship with the bank with which it holds a business account will tend to be governed by issues concerning the amount of money in the account, overdraft and loan facilities.

**The local economy**

The general state of the local economy also exerts a direct influence on organisations. If the local economy is buoyant, then the customer base for products and service will be wide and profitable. However, wages will be high and

**Exhibit 2.16**
**The local economic environment**

| | Political | Economic | Sociocultural | Technological |
|---|---|---|---|---|
| **Local** | | • Local bank branches<br>• Local economy | | |
| **National** | | | | |
| **Global** | | | | |

employees in short supply from among the local population. In areas with a depressed local economy, there will be little opportunity for selling products or services to people with little disposable income, but the workforce should be cheap and plentiful.

## The national economic environment

The main influences on businesses at a national level come from the actions of the country's national bank and stock market. These two organisations together influence the value of investments held by businesses and individuals and hence affect the amount of cash available in the economy.

Government actions in relation to the economy also obviously have a great influence. The most common forms of taxation are income tax, corporation tax, value added tax (VAT) and duty on imports/exports. VAT is payable on many goods and services and currently stands at 17.5 per cent for most goods in the UK. The Conservative government increased the rate of VAT across the board from 8 to 17.5 per cent and also increased the number of items on which VAT is charged. The most opposed introduction of VAT was VAT on domestic fuel, which had been zero rated and was to be charged at 8 per cent. This was not abolished but was reduced to 5 per cent by the Labour government in 1997. Basic food items and children's clothes are currently zero rated, i.e.

**Exhibit 2.17 The national economic environment**

| | Political | Economic | Sociocultural | Technological |
|---|---|---|---|---|
| Local | | | | |
| National | | • Central bank<br>– Bank of England<br>• Stock market<br>– London | | |
| Global | | | | |

VAT is chargeable but it currently stands at 0 per cent. Food items considered to be luxury items are subject to VAT at the standard rate.

The rate of taxation affects businesses as higher income tax lessens the amount of disposable income that people have available to spend on goods and services, and VAT makes goods and services more expensive to the consumer than the price set by the manufacturer or service deliverer. Therefore it is crucial to the standard of living of the individual citizen, to the profitability of private-sector organisations and to the standard of public services that a balance is achieved between taxation and public spending.

Businesses are affected by corporation tax as the higher the rate, the greater the amount of tax due on any profits made and hence the lower the amount of money available to pay dividends to shareholders and spend on activities such as marketing, product development, pay increases and updating equipment and facilities used to run the business.

Although much of the above is concerned with affecting the economic environment of the country, the decision making behind the management of the economy is essentially political, as it can be political parties' economic strategies that voters find key at the ballot box. Additionally, budgets from opposing political parties often show different approaches towards solving the same problems. In 1998 the Chancellor of the Exchequer Gordon Brown made the decisions shown in Exhibit 2.18 in relation to business organisations in the UK; this represents only a few of many decisions made in the Budget.

**Exhibit 2.18 Budget decisions 1998**

| Budget decision | Likely effect on businesses or individuals |
| --- | --- |
| Corporation tax was reduced and some smaller organisations were exempted | Increased profitability for businesses<br>Increased job opportunities for the unemployed |
| A venture capital fund for universities was established | Greater investment in higher education teaching and learning<br>Greater access to better facilities for students |
| Capital gains tax was reduced for long-term investments | Individuals will invest more for longer periods, providing investment, in terms of shares, for business |
| New Independent Savings Accounts were introduced with tax relief on long-term interest | As above |
| National Insurance would be reviewed in an effort to make it cheaper for organisations to employ low- and middle-wage earners | Increased profitability for businesses<br>Increased job opportunities for the unemployed |
| Working Family Tax Credit would replace Family Credit | Greater incentive to work, as tax relief is on earnings, instead of current incentive to remain in low-paid employment or additional welfare payments will be discontinued |
| VAT on energy-saving materials would be cut to 5 per cent | Individuals and organisations will invest in conservation and both will benefit from using less energy |
| Road fund licences would be made cheaper for the smallest, cheapest cars | Customers will have an incentive to buy the cheaper, cleaner cars, or to replace older, dirty cars with new ones |

### The central bank – Bank of England

In the UK the central bank is the Bank of England. In May 1997 the new Labour Chancellor of the Exchequer, Gordon Brown, gave independence in the setting of the base rate of interest to the Bank of England and its Governor Eddie George. This separated out the economic and political decision-making processes concerning interest rates for the first time in the UK's history. Under the previous system, the Chancellor of the Exchequer and the Governor of the Bank of England would meet regularly and jointly agree changes to the base rate of interest. The base rate directly influences businesses and individuals as repayment of overdrafts, loans or mortgages will increase with higher interest rates, hence reducing the amount of activity in the economy. For example, when interest rates are high businesses spend less on new equipment, advertising and product development and individuals spend less on the goods and services produced by businesses. The effects of the government handing over decision making on base interest rates to the central bank should be that the influence of short-term political need on economic management is removed. Interest rates, a key economic management tool, should be decided from a purely economic perspective. When a government minister is deciding interest rates, they can be manipulated for political advantage before an election, as Nigel Lawson did when Chancellor of the Exchequer before the 1987 election. The so-called 'Lawson Boom' that he engendered led directly to the recession of the early 1990s. The Bank of England, in contrast, should only take rational, objective decisions about long-term economic strategy. This is in line with other European countries' practice, and further prepares the UK for the possibility of entry to EMU at some date in the future.

**The London Stock Market**

The stock market is where shares of publicly quoted companies are traded. London is a centre of international, as well as financial, opportunity.

Company names and share prices are published in all the daily broadsheet newspapers. The shares of a company will trade on the stock market at a market price and the price will go up or down depending on the performance of the company and how it is viewed by the City. For example, a company that announces good profits is likely to see a rise in its share price and vice versa. However, good performance can be measured by factors other than profit. Manchester United, a profitable company, had a share price of £1.64 in February 1998; after being beaten 3–2 by Barnsley and knocked out of the FA cup on 25 February 1998, Manchester United shares fell 25p to £1.39. Continued good profits and better performance by the team on the pitch saw the share price climb steadily to stand at £1.52 in June 1998, and to £2.30 in January 2000.

The value of shares will also be altered by overall movements in the stock market. If the economy is doing well, and unemployment and inflation are low, then trading conditions will be viewed as generally favourable. In this type of situation the stock market will be buoyant and overall share prices will rise. However, if share prices rise too much or the economy performs less well, then share prices generally may fall.

## The global economic environment

The global economic environment is shaped by how countries decide to behave economically in relation to each other, for example countries may decide to group together on a regional basis and confer trade benefits and support on each other. The other economic influence at a global level is the effect of the world money markets.

**Trading blocs and bodies**

These are geographic blocs made up of a number of countries that agree to act together in some way with regard to trade and commerce. The countries may, for example, agree to allow special concessions on the taxing and movement of goods within the trading bloc or merely support each other with regard to economic issues.

### The European Union

Because of the close association between politics and the economy, it is necessary to mention the European Union (EU) again under trading blocs and bodies. The EU's Single European Market came into operation in 1993 and allows free movement of goods between member states. However, the issue of differing rates of taxation in different member states has yet to be dealt with fully. For example, the tax on wine, beer and cigarettes is much lower

**Exhibit 2.19 The global economic environment**

| | Political | Economic | Sociocultural | Technological |
|---|---|---|---|---|
| **Local** | | | | |
| **National** | | | | |
| **Global** | | • Trading blocs and bodies<br>– EU<br>– EFTA<br>– OECD<br>– NAFTA<br>– ASEAN<br>• World money markets<br>• WTO | | |

in France than in the UK and this is clearly shown by the popularity of day trips to France to stock up on cheap drink and cigarettes. The sale of drinks from pubs and retail outlets in the UK has been badly hit, particularly in the South East of England, with the closest and best access to France. There have been moves by the government to tackle the grievances of the UK drinks industry and this has included the abolition of duty-free sales within the EU. This led the government to face another fierce campaign, this time from the ferry companies, airports, travel companies, airlines, trade unions and the general public, all of whom were in favour of retaining duty-free sales. However, abolition will only partly tackle the difficulties faced by the UK drinks industry, and day trips to France to stock up on cheaper goods will remain popular until duty rates throughout the EU are harmonised.

The 1991 Maastricht Treaty on European union covers the issues of currency, immigration controls and defence policy. It set the timetable for the first round of European monetary union (EMU) that occurred in 1999. The deadline for meeting the necessary economic targets to join the first wave of European monetary union was the end of 1997. The key economic target that countries had to meet was the deficit:gdp (gross domestic product) ratio, which had to be 3 per cent or less. All member states except Greece met this target. However, Denmark, the UK and Sweden along with Greece did not join the single European currency in the first wave.[16] It is anticipated that EMU will generate an area of economic stability around France and Germany with low inflation and low interest rates.

### European Free Trade Area

The European Free Trade Area (EFTA) was established in 1960 by the Stockholm Convention. Initially EFTA had seven member states: Austria, Denmark, Norway, Portugal, Sweden, Switzerland and the UK. Eventually some states left EFTA to join the European Union and new states joined EFTA. EFTA currently has four member states: Iceland, Liechtenstein, Norway and Switzerland.[17]

The undertakings of EFTA are split into three essential areas:

1 The creation of the European Economic Area (EEA). On 1 January 1994 the three EFTA states, Iceland, Liechtenstein and Norway, and all EU states combined to create a large free-trade area known as the EEA. The EEA agreement allows free movement of people, goods and capital between participating nations.
2 The development of relationships with other countries. Since the end of the Cold War, EFTA has sought to develop relationships with Central and Eastern European countries. This has led to the establishment of free-trade agreements with countries such as the Czech Republic, Poland, Latvia, Lithuania, Estonia, Bulgaria, Romania and Slovakia. EFTA is also currently seeking to sign free-trade agreements with countries such as Malta, Tunisia, Albania and Egypt.
3 The monitoring and regulation of relationships between EFTA states and administering of the EFTA Development Fund for Portugal.[18]

### Organisation for Economic Co-operation and Development

The Organisation for Economic Co-operation and Development (OECD) was established in 1961 and is based in Paris (*see* Exhibit 2.20). On 5 June 1947 a speech by George C Marshall, US Secretary of State, gave rise to the post-war European Aid Program, commonly known as the Marshall plan. The view of the US government was that there had to be agreement among some or preferably all European countries as to the aid required and its use to reconstruct Western Europe. Therefore, European and North American countries came together to establish the Organisation for European Economic Co-operation (OEEC). Its chief role was to facilitate the planning, allocation and implementation of post-war aid to Western Europe. However, in 1961 the OEEC decided that the reconstruction of post-war Western Europe was complete and sought to convert the OEEC into the OECD. The OECD did not focus on the giving and use of aid but on making the economies of the member countries involved prosper. To this end the policies of the OECD seek to:

- achieve the highest sustainable growth and employment and a rising standard of living in member countries, while maintaining financial stability, and thus contribute to the development of the world economy;
- contribute to sound economic expansion in member as well as non-member countries in the process of economic development; and
- contribute to the expansion of world trade on a multilateral, non-discriminatory basis in accordance with international obligations.[19]

**Exhibit 2.20 Members of the OECD**

| Year of joining | Countries | | |
|---|---|---|---|
| 1961 | • Austria<br>• Denmark<br>• Greece<br>• Italy<br>• Norway<br>• Sweden<br>• UK | • Belgium<br>• France<br>• Iceland<br>• Luxembourg<br>• Portugal<br>• Switzerland<br>• USA | • Canada<br>• Germany<br>• Ireland<br>• Netherlands<br>• Spain<br>• Turkey |
| 1964 | • Japan | | |
| 1969 | • Finland | | |
| 1971 | • Australia | | |
| 1973 | • New Zealand | | |
| 1994 | • Mexico | | |
| 1995 | • Czech Republic | | |
| 1996 | • Hungary, Korea, Poland | | |

*Source*: http://www.oecd.org/about/whats.htm, 28 April 2000.

### North American Free Trade Agreement

In 1994 the North American Free Trade Agreement (NAFTA) between the governments of Canada, USA and Mexico was established. The principal strands of NAFTA relate to:

- encouraging goodwill and collaboration between NAFTA countries;
- establishment of a larger and more secure market for goods and services produced in NAFTA countries;
- strengthening and sharpening the competitiveness of companies from NAFTA countries in global markets;
- generation of additional employment opportunities in NAFTA countries;
- improvement of working conditions and living standards in NAFTA countries;
- protection and implementation of workers' basic rights.[20]

Trade and the movement of jobs and goods between NAFTA countries are key issues for the countries involved. The setting up of NAFTA was rapidly followed by a devaluation of the Mexican peso in December 1994. This devaluation resulted in manufacturing wage rates in Mexico falling to 10 per cent of comparable US wage rates. Therefore cheap labour in Mexico, combined with weak environmental laws and rare health and safety inspections, usually with advance warning, meant that US companies found Mexico an attractive choice for relocation.[21] For example, in Mexico the hourly wage rate for workers in the motor industry has been estimated to be as little as 70 cents per hour. Union leaders in the USA contend that low levels of pay in Mexico push down wage rates on both sides of the US/Mexico border.[22]

The establishment of NAFTA was expected to provide benefits for industries such as the motor industry. Giant US car manufacturer General Motors (GM) has gained enormously from the establishment of NAFTA, although not always in the manner envisaged. It was expected that NAFTA would permit a company such as GM to manufacture in Mexico and export one-third of vehicles back to the USA. The Mexican market for small cheap cars, like the Volkswagen Beetle, was not able to be sustained as the lower middle classes were severely affected by the peso devaluation. Nevertheless the GM Chevey, which is viewed as a mid-range car in Mexico, has been successful, which means that in Mexico GM has overtaken Volkswagen and Nissan in terms of sales figures for the first time. GM has also benefited from the availability of cheap labour in Mexico and has been able to manufacture cars, trucks and parts for export, usually back to the USA. In Mexico in 1995 domestic sales of vehicles were 185 000, with exports standing at 781 000 vehicles. In comparison in 1996, domestic sales of vehicles stood at 334 000 and exports at 975 000 vehicles. GM mirrored this trend in 1996 by exporting three-quarters of its Mexican production of 263 000 cars and trucks back to the USA.[23]

At the other geographic end of NAFTA, similar arguments about Canada taking jobs from the USA are also put forward, although they are less convincing. In Erie in Pennsylvania, local chief executives view the impact of NAFTA on local businesses and commerce as negligible, instead seeing NAFTA as the way forward. However, the trade unions dislike NAFTA and the lack of progress it has delivered in the South as regards improvements to the lives of Mexican workers. In Erie and surrounding areas the loss of fewer than 1000 out of 127 300 jobs has been directly attributed to NAFTA. Manufacturing jobs in Erie and surrounding areas make up 28 per cent of employment, compared with 18 per cent of jobs nationally in the USA, and this is due in part to the region making comprehensive attempts to re-equip people with the skills and abilities that companies require, particularly with the establishment in 1991 of a Technical Institute. The Regional Skills Centre has also helped retrain people by working with companies to train people for specific manufacturing jobs, such as maintenance technicians. Help is provided in the form of retraining assistance and prolonged unemployment benefits for people who are deemed, by the US Labor Department, to have lost their job due to trade with Mexico or Canada.

NAFTA's biggest effect is estimated to have been to compel small and medium-sized businesses to understand that persistent reinvestment and modernisation are required if competitiveness is to be maintained. People who choose not to exploit the opportunities to retrain often find lower-paid employment in shops and cinemas. A shopping centre on Eire's outskirts, for example, has 200 shops and a cinema with 17 screens and the attraction of tax-free shopping lures customers from a wide area, including New York, Ohio and Canada.[24]

### The Association of South East Asian Nations

The Association of South East Asian Nations (ASEAN) was formed in 1967 in Bangkok with five founding members and a further four members joining between 1984 and 1997 (*see* Exhibit 2.21). In July 1997 Burma and Laos were admitted, but Cambodia's entry was postponed due to violence that threatened civil war.[25]

**Exhibit 2.21**
**Members of ASEAN**

| Year of entry | Countries | | |
|---|---|---|---|
| 1967 | • Indonesia<br>• Singapore | • Malaysia<br>• Thailand | • Philippines |
| 1984 | • Brunei | | |
| 1995 | • Vietnam | | |
| 1997 | • Laos | • Myanmar (Burma) | |

*Source*: http://www.aseansec.org/history/asn_his2.htm.

The three main objectives of ASEAN are:

- to promote the economic, social and cultural development of the region through co-operative programmes;
- to safeguard the political and economic stability of the region against big power rivalry;
- to serve as a forum for the resolution of intra-regional differences.[26]

## The world money market

The world money market encompasses the world's stock markets and the exchange value of currencies against each other. The main stock markets that make up or have the biggest influence on the world money market are New York, London, Tokyo and Hong Kong. A collapse in one is quite likely to result in a drop in another, with the value of stocks falling and wiping value from the shares of companies.

This was clearly evident in the Asian economic crises of late 1997/early 1998, which saw the stock markets of Japan, South Korea, Indonesia and Hong Kong all plummet. This also meant the key exchange rate to the American dollar for the Japanese yen, South Korean won and Indonesian rupiah all fell. The Hong Kong dollar is pegged to the US dollar at a rate of HK$7.8 to US$1, and this peg remained in place.

The falling exchange rate against the US dollar for Far East currencies was mirrored in the latter currencies falling in value against other currencies. Hence, for example, the Malaysian dollar devalued very quickly against western currencies and halved in value, from around M$4 to £1 to M$8 to £1, hence making imported western goods in Malaysia twice as expensive. In contrast, the cost to overseas visitors from the West of holidaying in Malaysia became very cheap as hotels, restaurants, taxis and souvenirs, which had been reasonably priced before the currency crisis, halved in price. One other effect of the crisis was to slow economic activity in the Asian countries concerned. This meant that goods were more expensive, people felt poorer, and the prices of basic foodstuffs and fuel rose.

A further key effect was a rise in unemployment, in an area of the world that had traditionally been viewed as having very low unemployment. As a result, immigrant workers were sent back to the neighbouring countries from which they had come. This happened in Malaysia, where many legal and illegal migrant workers were sent back to Indonesia by the authorities.

The chasm into which the economies in the Far East were plunging became deeper when student demonstrations and rioting in Indonesia eventually led to the downfall of President Suharto after many years in power as an authoritarian head of state. This meant that there was widespread rioting and looting in the capital city, Jakarta, as people fed up with the authoritarian rule of President Suharto and his family vented their years of frustration of living in an authoritarian state with a poor economy. Foreign reserves in Indonesia were critically low and the rupiah virtually ceased trading, with little value to anyone. Interest rates rose to over 60 per cent and foreign debt was equivalent to £86 billion.

The full effects of the Asian economic crisis were felt in the West as Far East countries competed to export goods with falling prices to western countries. The other effect was that companies in western countries such as the UK which sold much produce to lucrative eastern markets faced a severe downturn in their export business. In May 1998 the CBI confirmed that British export orders were at their lowest levels since 1983.

The Asian economic crisis clearly demonstrates the effect that an economic crisis can have on its home markets and region as well as the knock-on effects on other parts of the world.

## The World Trade Organisation

The World Trade Organisation (WTO) is located in Geneva, Switzerland and officially came into existence on 1 January 1995, replacing its predecessor, the General Agreement on Tariffs and Trade (GATT). The WTO has 132 member countries and 34 observer governments, of which 33 have applied for membership.

The role of the WTO is summarised as:

- administering WTO trade agreements;
- forum for trade negotiations;
- handling trade disputes;
- monitoring national trade policies;
- technical assistance and training for developing countries;
- co-operation with other international organisations.[27]

Recent examples of WTO activities include handling a trade complaint against India, a WTO member, and consideration of China's application to join the WTO. The formal complaint against India was made by the USA, Canada, Australia and the European Union. The complaint objected to India's far-reaching import controls and payments on consumer goods and the slow speed at which these controls and payments were being phased out. India had offered to remove the import controls and payments over seven years, whereas the complainants wanted them phased out over three years. The complainants were supported by an International Monetary Fund (IMF) report concluding that India no longer had an acute balance of payments problem. Therefore the country's foreign exchange reserves were healthy and such import controls and payments could not be justified under WTO rules.[28]

China's drive to become a member of the WTO has been long and slow, since it first applied to join GATT in 1987. Consequently in 1997 the thawing of previously frosty political relations between China and the USA provided an opportunity for progress on China's entry to the WTO and negotiations took place in Geneva. Final agreement on the terms of China's entry to the WTO were reached on 15 November 1999. Admittance to the WTO is a significant step towards allowing China's fast-growing economy to face more international competition and allowing foreign exporters and investors access to the country. However, membership requires China to adopt a thorough approach to many complicated issues that establish barriers to trade and commerce. These include the dismantling of investment restrictions, quotas, subsidies and tariffs. China has agreed to accept WTO rules, open its economy to the rest of the world and to continue to engage in market-based reforms. However, there was significant opposition in China itself to joining the WTO, emanating from the state trading companies and the agricultural industry. The state trading companies will lose profitable monopolies and market share in the face of increased competition. There also exists the worry that more competition will lead to greater urban unemployment in China.

Similarly, the agricultural industry in China receives substantial state subsidies, and membership of the WTO, requiring China to open up agricultural markets to greater competition, will result in reduced state subsidies and more unemployment among agricultural workers. Consequently, the large agricultural exporters to China, including the USA, Canada and Australia, were all able to influence China's application for WTO membership. As China takes up its membership of the WTO, it will gain most favoured nation (MFN) trading status on an immediate and permanent basis.[29] MFN status ensures China's exporters access to US markets. Prior to membership of the WTO, China's MFN status was reviewed annually by the American Congress.

## The sociocultural elements of the external environment

The sociocultural elements of the external environment include the age and structure of a national population, the way a population or society behaves and elements determining how the culture of a population develops. This section will look at some of these and how businesses and organisations are affected. Greater detail on culture and organisational behaviour can be found in Chapter 5.

## The local sociocultural environment

**Local community**

The local community is part of an organisation's external environment. The local community may be passive as regards the organisation or the local community and organisations can have an active influence on each other. For example, a local transport company regularly burns old and broken wooden

**Exhibit 2.22 The local sociocultural environment**

| | Political | Economic | Sociocultural | Technological |
|---|---|---|---|---|
| Local | | | • Local community | |
| National | | | | |
| Global | | | | |

pallets in its yard and produces a dense black smoke, making the atmosphere unpleasant for people who live in nearby houses with gardens overlooking the firm's yard. This local company has a very direct influence on the quality of life in the local community. The local people may act either individually or collectively and complain to the local council department dealing with pollution. If appropriate action is taken to stop the pollution, this is a very good example of a local firm, its local community and local government affecting and being affected by each other's actions.

There may be other local community influences to which the organisation is subject. A school is a public-sector organisation that interacts significantly with its local community. The community is greatly dependent on how well the school performs due to the personal nature of the service that the school provides. This kind of inter-dependence is considered in Chapter 3.

## The national sociocultural external environment

**Demographic change**

Demographic changes are changes in the age and structure of a population. This section considers demographic changes in the population of the UK. There are three fundamental demographic changes.

First, the number of people aged 35–50 in the population is increasing and these are those born in the post-war baby boom years of 1950–64. If people in this group have held good jobs and experienced the consumer boom of

**Exhibit 2.23 The national sociocultural environment**

| | Political | Economic | Sociocultural | Technological |
|---|---|---|---|---|
| Local | | | | |
| National | | | • Demographic change<br>• Social change | |
| Global | | | | |

the 1980s, then they are likely to continue to be generous consumers. These people will become the 50+ population of tomorrow. Secondly, there is a shrinking youth population, defined as those aged 15–24 years. The youth population is viewed, in general, as not being affluent as many young people are unemployed, in training or in tertiary education, none of which provides a large disposable income.

Finally, there is an increasing older population, those aged 50+, in the United Kingdom. In 1961 there were 16 million people aged over 50 in the UK; by 2021 there will be 22.5m people, an increase of 40 per cent against an overall population increase of 15 per cent.[30] In 1995 people aged 50+ formed one-third of the UK's population; the number of people aged 75+ will have doubled by 2045.[31] A significant number of these older people, 3–4m, are white-collar workers of social class ABC1. This type of person will tend to have a good retirement income from an occupational pension scheme and savings and will own his or her home, with the mortgage paid off. Retired people in this situation are inclined to spend their relatively large disposable incomes on reading material (books, magazines and newspapers), gardening, visiting family and friends, eating out and drinking, insurance policies, home improvements and holidays. Consumers aged 50+ are the largest buyers of winter sun holidays and cruises.[32]

However, there is also a large group of less affluent older people, aged 60+, for whom the main source of income in old age is the state pension. These people, like the rest of the population, are likely to live longer and require medical treatment and a state pension during their longer life. The effect on the working population is an increase in their tax contributions. It is estimated that in the UK National Insurance contributions will have to rise from 12.5 per cent of the wage bill in 1990 to 18 per cent by 2030.[33] The increase in taxes paid by the employed and industry will prompt governments to alter the welfare system fundamentally to encourage more people to save for their own pension. Tony Blair's Labour government has targeted those currently in work but not contributing to an occupational pension scheme. This type of encouragement is crucial, as provision of retirement income needs to be boosted for the majority. This need is evidenced by the fact that 75 per cent of the 10m recipients of the state pension were not paying income tax in 1995, indicating a total income of less than £4800 (single person) or £11 876 for a couple.[34]

This increasing difference in income between different groups of older people is a reflection of what is happening more broadly in society, leading to an increasing income gap between rich and poor people. Later in this section there is further discussion of this income gap.

#### The effects of demographic changes on businesses

The effects of these last two changes in the population can be seen in employment policies and products/services provided by organisations. In the mid-1980s the DIY company B&Q recognised that it would be affected by changing demographics and the declining number of young people in the population. Therefore it decided to utilise the opportunity presented by the growing number of older people. In 1988 it decided to open a B&Q store staffed only by people aged 50+ in Macclesfield. Several years later the Macclesfield project was evaluated by an independent study, carried out by the American & Commonwealth Fund and Warwick University. The benefits identified were that, next to comparable stores, the Macclesfield store was 18 per cent more profitable, staff absenteeism was 39 per cent lower, staff theft was 59 per cent lower and employee turnover was six times lower. This led to B&Q opening a similar store in Exmouth and seeking to have at least 15 per cent of the workforce aged over 50.[35] This is also an indication of organisations appreciating different attitudes towards work among different generations. B&Q has benefited from older people's stricter work ethic and high skill levels when engaging in work.

The big brewers have sought to find a profitable alternative to the city centre pub, which traditionally made most of its money from young drinkers on Friday and Saturday nights. The concept that is evident in city centres around the UK is that of the café/bar, which is designed to be part wine bar, part café and continental in style. A café/bar will serve alcoholic and non-alcoholic drinks, teas and coffees and have a menu of plentiful choice. This type of outlet is designed to appeal to a wide range of male and female customers. Typical clients will include lunching businesspeople, students seeking a mid-morning coffee, shoppers wanting a mid-afternoon cup of tea, the business brigade going for drinks immediately after work and the pre-nightclub market. Attracting these consumers means that men and women of a wide range of ages will frequent the premises and this should help ensure that there will be customers throughout the day.

Another product designed and publicised to appeal to both older and younger consumers is the Renault Clio car, promoted by use of the 'Nicole and Papa' adverts. By advertising the car being driven by two generations of the same family, and by both sexes, Renault was able to show its potential appeal to different driver groups without the cost of separate advertising campaigns. This was important in establishing the Clio's market position in the first couple of years:

> In the recession older people had more money and the grey market helped to support the car during the first 18 months.[36]

## Social change

Social change covers the development of and alteration in the way society behaves. The changes in society examined in this section are the developing inequality of income and the evolving family and household groups present in UK society. The role and influence of government are also considered.

### Inequality of income in UK society

UK society is one of widening inequality: the well-off have got richer and the less well-off have become poorer. Income inequality has grown rapidly since 1977, fuelled to a large extent by the Conservative governments in power from 1979 to 1997. These governments introduced policies that led to large pay packets and rewards for successful people, in contrast to fewer and increasingly meagre social security benefits for the less well-off. Since 1981, state benefits have increased in line with prices and not average incomes. Hence the income distribution gap in the UK is one of the greatest among the industrialised countries of the world. This is confirmed by the statistics that 6 per cent of the population had an income less half the national average wage in 1970, while by 1990 over 20 per cent of the population had an income at this level.[37]

There are many ways in which government is reacting, as any organisation should, to these changes in the external environment. In terms of the ageing population, governments of both main parties take the view that more of the burden of social security, healthcare and pensions needs to be provided for privately as the population ages, since the workforce cannot keep up the level of provision that has been the case in the 50 years after the welfare state was introduced. The Blair government also expressed concern about the widening gap in UK society, and established a Social Exclusion Unit to look at ways of bringing into the mainstream of opportunity and care those who had been excluded for social or systematic reasons, with a view to including more disadvantaged groups in mainstream provision.

This widening inequality is apparent when looking at different areas of the country. In England the problems of poverty, ill health and unemployment are mainly concentrated in cities in the South and the depressed industrial North, with pockets of rural poverty in Cornwall, Kent, Cumbria and Northumbria.

In the South, outside of poor inner-city areas, there is generally a higher standard of living, with the problems of unemployment, poverty and ill health occurring less frequently than in impoverished areas. However, the stress of modern living is felt both in the city and the countryside, with fewer people working longer hours. This is illustrated by city workers attending working breakfasts and evening meetings and people in the countryside working longer hours on the land or in the holiday industry for minimal pay.[38]

#### Family and household structure in UK society

Further social change has occurred in the structure of the family. A significant development is the growing number of single-person households. There are various causes of this, discussed below.

Some 30 or 40 years ago a young person would have anticipated being single for a short time and remaining in the parental home, before marrying in their mid-twenties and settling down to a lifelong marriage and raising the average 2.4 children, being part of a traditional family unit. Society viewed being single as something undesirable that occurred as a result of misfortune and not an alternative that people chose. Being single was the result of being widowed or divorced because of an unfaithful spouse; indeed, both of these still occur today. The greater number of older people in the population and the higher life expectancy of both men and women now means that the number of older people living in single-person households will increase, usually due to the death of a spouse. The average lifetime is 79 years for a woman and 73 years for a man in the UK.

The UK has the highest divorce rate in the EU, with around 35 per cent of marriages ending in divorce, compared to a rate of around 10 per cent in the mid-1970s. Consequently, divorce is less shameful and more frequent than 30 or 40 years ago. Between 1970 and 1997 the increased divorce rate, coupled with a 50 per cent fall in the number of first-time marriages, added to the number of single-person households. The lower marriage rate means that more people cohabit, meaning that people in this category are quite likely to end up being single at some point in their adult life.

Further social changes have occurred with regard to the family. In 1996 married couples with children comprised 41 per cent of families, compared with 4 per cent of families composed of children living with cohabiting couples and 13 per cent of families consisting of lone parents with dependent children. The growth in lone-parent households results from the increase in divorce and fewer marriages. This is reflected in the statistic that in 1996 a third of all births were outside marriage, four times the 1971 figure. However, 80 per cent of the births outside marriage in 1996 were registered by both parents. Further social change in the family includes alterations in the role of parents. In 1985 both parents worked in half of all two-parent families with children; in comparison, the figure was 62 per cent a decade later.

Today fewer people believe that it is a wife and mother's job to look after the home than did so in 1986. However, mothers still spend four times longer than fathers on cooking and housework.[39] The growth in the number of women in the workplace and the resulting greater financial independence of women are due to several factors. First, women are no longer expected to marry and be totally financially dependent on a husband. Since the 1960s there has been improved and increased access to higher and further education, which has meant greater access to better jobs and careers for women. This, coupled with the introduction of legislation in the 1970s covering equality of women in the workplace in terms of equal opportunities and equal pay, has contributed to women's greater role in the workplace.

There are some mixed messages to be observed in the political reactions to social change. In contrast to the moral rhetoric of family values that it espoused at the hustings, the Conservative government in the late 1980s introduced independent taxation for married women for the first time, thereby recognising both the earning potential and the legal status of married women. Nevertheless, in an effort to promote the responsibility of fathers in providing for their children beyond separation and divorce, it also established the Child Support Agency, whose job was to track down and make pay the errant fathers of children living in one-parent families with their mothers.

The last category of single people is, of course, those who choose to be single. A 1996 Mintel survey predicted a 17 per cent increase in single-person households by the year 2000. The greatest rise is expected in well-off men and women who have never married and are aged under 35. Approximately two-thirds of single women and over half of single men are in the ABC1 economic group, with three-quarters of single men and two-thirds of single women in full-time employment.

Examples of businesses that have gained from the expanding number of single-person households include supermarkets and cinemas. Supermarkets have reported an increase in the sale of single-person ready-cooked meals and a considerable number of these will be purchased by single people who live alone. Cinema audiences more than doubled from 50m a year in the mid-1980s to 120m a year in the mid-1990s and this is in part attributed to single people in the population spending their disposable income. In 1996 the government estimated that the number of households in the UK would increase by 18.5 per cent by the year 2011, with the greatest rise being in single-person households. This is one factor that will influence companies in the construction industry and the type and size of houses that will be built.[40]

## The global sociocultural environment

Definitions of culture and how it affects individuals and national society are covered in detail in Chapter 4. However, here it is useful to consider some of the sociocultural issues that affect organisations once they leave their home environment and operate overseas.

**Exhibit 2.24 The global sociocultural environment**

| | Political | Economic | Sociocultural | Technological |
|---|---|---|---|---|
| Local | | | | |
| National | | | | |
| Global | | | • Cross-cultural issues<br>– Language<br>– Behaviour<br>– Culture shock | |

## Cross-cultural issues

It has already been mentioned that organisations have to consider political and economic circumstances in other countries when contemplating international business operations. Culture plays just as important a part in governing international business at the global level of the external environment.

### Language

The most obvious illustration of culture is language. Investing in accurate language assistance when operating in another cultural context is a vital but largely underestimated consideration for organisations assessing the potential costs of international operations. The cost of getting it wrong is often much greater, but is often ignored in short-term decision making. It is only necessary to consider how many misunderstandings occur between native speakers of English who originate from the UK, the USA and Australia to begin to appreciate the difficulties of translating and interpreting other languages. The UK lecturer who asks an American colleague to invigilate an examination will be met with a blank look, as the American will be expecting to proctor it. The English tourist wanting to buy flipflops to wear on the beach in Australia will not be understood by the shop assistant, who will in turn shock the customer by offering thongs instead.

The back catalogue of language mistranslations and the choice of unfortunate words for products is huge. Volkswagen's multi-passenger vehicle (MPV) suffered when it was introduced to the UK from being called Sharon, a name not associated with the profile of customer to whom it was expected to be sold. Proton, Malaysia's national car manufacturer, did its market research and decided not to introduce its basic model to the UK under its Malaysian name, as 'Saga' is the brand name of products targeted at senior citizens, to whom Proton did not wish to limit sales. The Vauxhall Nova, General Motors' 1980s mini hatchback, was branded Opel Corsa in the rest of Europe as 'Nova', which was meant to have connotations of new, actually translates as 'does not go' in many European languages. The Rolls-Royce Silver Mist had to be renamed for the German-speaking market because *Mist* in German is a colloquial word for excrement.

However, translation or the choice of words meaning other things in different languages is not the only skill required overseas. Interpreting – which is not translating the words, but rather saying the right thing in the target language – is a crucial skill. At a business meeting between British and Chinese businesspeople, when the British host says 'We hope you have enjoyed your stay in the UK', a direct translation will sound arrogant and rude, so a skilled interpreter will replace this phrase with the customary 'We are sorry we did not look after you properly', and the courtesy requirements of both sides are fulfilled.

### Behaviour

Other types of cross-cultural issues would relate to the consumption of alcohol in Muslim states and of beef in India, where the cow is sacred to Hindus (McDonald's had to substitute a Hindu-friendly version of the burger). How much physical contact or personal space people are customarily allowed is also problematic, as some cultures remain physically very distant from each other, while in others regular touching is commonplace. Again, whether or not physical contact is permissible between the sexes or between people of the same sex is an issue. For example, in many Middle and Far Eastern cultures, any touching between the sexes is unacceptable, while man-to-man handholding and bodily contact are quite normal. This understandably becomes a minefield of danger to the foreign executive. In Japan, blowing one's nose in public is quite taboo, while in China the public expectoration of waste is unsurprising.

Therefore, the behaviour of foreign executives, the design of products and services, and the labelling, packaging and advertising of goods and services must all be subject to intense scrutiny.

### Culture shock

The greater the distance between home and host culture, the more likely it is that the host culture will provide elements of everyday life that shock the individual travelling there. From language to food, from individual behaviour to collective customs, culture shock is a real and debilitating influence on the individual businessperson abroad. As it is based in experience, it is difficult to know how culture shock can be dealt with until it has been experienced, since there is still a huge difference between knowing something is going to happen (I have learnt that China is crowded, so I expect that when I go there I will experience a lot of pushing and shoving on the streets) and actually experiencing it happening (everywhere I go in China people touch or push me – I come from a culture where touching strangers is almost taboo, so I hate it!). Nevertheless, organisations can invest in pre-departure orientation programmes, training people for overseas postings through contact with natives, visits and access to expatriates who have already lived in that culture. It is essential to consider not only the expatriate executive, but also the relocation and comfort of family members as well.

## The technological elements of the external environment

Technology has an influence on all aspects of business from the very general to the very specific, at all different levels of the external environment, local, national and global.

### Communications technology

The advent of technology has made it easier for people to communicate with each other, whether they operate in the political, economic, social or general business arena at a local, national or global level. The technology that has made communication easier takes the form of mobile phones, fax machines, video conferencing, the Internet and the world wide web.

#### Mobile phones and faxes

Mobile phones have grown in popularity since the mid-1980s. Initially, mobile phones were large (therefore visible) and expensive. Hence in the 1980s they were very much the necessary executive accessory. The first mobile phones were also heavy and cumbersome to carry around all the time. The technology improved throughout the 1980s and 1990s and the size and styling of mobile phones improved to give small and slim models that slip easily into a jacket pocket or handbag. In line with improved technology and styling, the price of mobile phones fell and connection is possible for under £50, which compares very favourably with a cost of over several hundred pounds when they first came on the market. So many more people now carry mobile phones. The key benefit to organisations that issue personnel with mobile phones is that staff are contactable all the time (unless the mobile phone is switched off) and should be able to contact customers and clients without having to return to an office.

Fax machines are another development of the 1980s that became commonplace in the office of the 1990s, allowing letters or documents to be transmitted

**Exhibit 2.25**
**The technological external environment: communications technology**

| | Political | Economic | Sociocultural | Technological |
|---|---|---|---|---|
| Local | | | | • Communications technology<br>– mobile phones and faxes<br>– video conferencing<br>– Internet and world wide web |
| National | | | | |
| Global | | | | |

over a telephone line in a number of seconds or minutes. This compares favourably with the day or more it would take a letter or document to reach the same destination by post, hence making communication quicker.

If you look at Exhibit 2.3, the generic LoNGPEST grid, you will see that P, E and S are clearly divided into local, national and global factors. With technological factors this is much less clearcut and the Technological factor may apply at all levels (see also p. 10 in Chapter 1 for an explanation of this). Therefore, only that section of the Technological factor has been shown that is being discussed in the following text.

### Video conferencing

Video conferencing is becoming increasingly popular among businesses. Large companies such as British Petroleum (BP) have been using video conferencing since 1983 and have in-house studios in global locations. The greatest benefits of video conferencing are conferred when it is used by people in two or more places in different parts of the world. For example, two teams of people based on opposite sides of the Atlantic in the UK and the USA working on joint projects will need to meet regularly. The teams could meet by video conferencing rather than one team flying across the Atlantic for a meeting. Alternatively, a multinational that needs to have a meeting of senior managers, one manager based in Australia, one in Russia, one in Hong Kong and one in the UK, could all meet by video conferencing, assuming there are facilities in each location. SmithKline Beecham has over 30 video conferencing studios world-wide[41] and is able to carry out video conference meetings of the type described above. Organisations that do not have video conferencing studios are still able to participate if they book ahead and rent time in a video conferencing studio like the ones owned and run by Regus Management in London.[42]

The most common form of video conferencing is the dedicated studio and people in the organisation wanting to video conference have to go from their desks to the studio to meet with their colleagues in a studio in another location. The obvious future development of video conferencing is to take it out of the studio and into the office. This is likely to happen as the technology and equipment develop. Portable video conferencing equipment that can be wheeled about, in the same way as televisions and videos on stands can be moved about offices, is cheaper and more flexible than the dedicated studio system of video conferencing. The portable system allows video meetings to take place from the desktop or from a meeting room, which may be more convenient than going to a central studio location.

The other development is PC-based video conferencing, which is popular with heavy users of video conferencing. However, a drawback of PC-based systems is that they are not particularly user-friendly as a shared resource. Consequently, organisations may be disinclined to install the necessary video conferencing board and software in every PC in the organisation. Also the current quality of the audio and video in PC-based systems is not as good as it is on portable video conferencing systems.[43] The cost of a video conferencing studio system will vary from £18 000 to £100 000, compared to about £5500 for a portable system and under £700 for a PC-based system per computer.[44] Because the prices for portable systems have fallen and their operation and quality have improved, PC-based systems are unlikely to overtake portable systems in popularity until their price and quality similarly improve. The use of studio and portable video conferencing systems for internal communication in companies is the most likely continued use of video conferencing in the immediate future.

The extensive use of PC-based video conferencing systems for communication between different companies will not occur until there is a critical mass of users in different organisations all with the necessary technology in their PCs. Inter-company video conferencing will also require telephone companies to publish a video conferencing directory, hence allowing users to know who has a PC-based video conferencing unit.[45]

Be aware that if meetings have always been disorganised and unproductive in an organisation, then video conferencing will not solve those problems, it will merely automate them.[46] The key benefits of video conferencing include the following:

- Less time spent travelling to and from meetings, therefore reducing the cost and stress of travelling. The National Economic Research Council estimates that by 2007 video conferencing could replace 20 per cent of business travel.[47] A foreign business return trip can take from 8–55 hours depending on the distance involved: a door-to-door return trip from London to Frankfurt may take 8–10 hours travelling time; at the other end of the scale a return trip to Australia from the UK will take over 50 hours.
- The possibility of more people attending the meeting and that technical experts can be called into the meeting at relatively short notice if their knowledge and expertise are required.
- Eye contact and body language being seen, a clear advantage over phone conferencing.
- Enhancement of teamworking and communication among teams whose work is spread out across the globe.

### The Internet and the world wide web

The Internet is an array of inter-connected networks to which millions of computers around the world are attached. There are a large number of Internet sites that hold information, for example company sites or the *Financial Times* site. The world wide web (WWW) allows linking of Internet sites and research and retrieval tools, meaning that the WWW and the Internet are often seen as one and the same. Search engines such as Yahoo! and AltaVista allow searches to be carried out very easily and it is equally easy to repeat and refine any searches that have previously been carried out.[48]

Electronic mail, commonly known as e-mail, is a frequently used application of the world wide web. The use of e-mail allows messages to be sent to anyone who has an e-mail address, whether an individual at home or someone in a company. Sending a message by e-mail is quicker than posting a letter; cheaper than faxing a message; and on occasions is more convenient than using the telephone. E-mail also has the advantage that the same message can be sent simultaneously to a large or small group of people.

Companies can take up a presence on the Internet by setting up a website. This will give a company 24 hours a day, 365 days a year, worldwide exposure on the Internet. Companies are able to use their Internet presence to advertise their products and services and collect addresses and details from potential customers who visit the site, with a view to e-mailing or posting further information. The Internet is equally accessible by both large and small companies, although large companies may have more money to spend on designing a site.

It is suggested that when designing a website the following are considered:

- The site should indicate the type of business the company is in early on.
- Do not make the pages over-elaborate or extensive as this can result in a lengthy wait when downloading individual pages.
- Do not use text that is too small and difficult for people to read on-screen.

- Do decide whether the company wants to collect details of potential customers and/or create an awareness of products and services.
- Do plan for how the company is going to deal with responses from actual and potential customers who visit the site, as a large amount of e-mail may be generated by an Internet presence.[49]

A company that has set up an Internet site also has to consider how it is going to persuade people to visit the site. Many companies advertise in the more traditional media, such as the television and press, and include their Internet address in the advertisement. An additional method is to be included on search engines so that when users type in a keyword such as books or beer the relevant Internet sites are listed. It is best if a company or its Internet site is listed in the first ten, as potential visitors do not tend to scroll down the list and are therefore less likely to visit Internet sites further down the list. Companies are also able to advertise on search engines by paying to have their logo appear, although costs are high.[50]

There are other methods of encouraging people to visit a company's Internet site and recent examples include Shell and Barclaycard. In the UK Shell has developed and expanded its loyalty card scheme to allow points to be collected whenever a customer visits certain Internet sites. The Shell loyalty card is a SMART card and visitors to seven Internet sites, including the Commercial Union and Victoria Wine, will find SMART symbols there. SMART card holders who visit and browse through any of the seven sites can collect points and trade them in for goods from a Shell petrol station. Barclaycard has launched an Internet system that allows customers to check their account and pay their credit card bill using a debit card.[51] The use of the Internet in this manner is in its infancy and it will grow as the technology develops and as a greater number of people become linked up to the Internet in their own homes and though digital television.

## Organisations and the application of technology

In addition to the communications technology discussed above, organisations make use of other types of technology on a daily basis. For some organisations technology is increasingly the key to improved product or service delivery. Banking is such an industry and is looked at in detail in this respect later in this section.

### The personal computer

The manner in which organisations and the people in them operate has been radically altered over the last 20 years by the advent and development of the personal computer (PC). When first introduced to organisations PCs were stand-alone machines and few people in the organisation had a PC on their desk. At the end of the twentieth century virtually all organisations use PCs widely in carrying out their daily business. Accordingly, in many organisations everyone from the managing director to the most junior clerical

**Exhibit 2.26**
**The technological external environment: organisations and the application of technology**

| | Political | Economic | Sociocultural | Technological |
|---|---|---|---|---|
| Local | | | | • Organisations and the application of technology<br>– the personal computer<br>– the banking and financial services industry |
| National | | | | |
| Global | | | | |

assistant will have a PC on their desk. The other fundamental difference is that PCs are no longer always stand-alone machines, but are at the very least networked together within the organisation, hence facilitating internal communication. It is also extremely common for an organisation's computer networks to be linked to the outside world via the WWW, hence allowing communication and information retrieval, as described in the previous section on communications technology.

The most familiar types of software developed and used on a daily basis by organisations are wordprocessing and spreadsheet packages, which are commonly used for producing letters, reports and analysing data. The proper use of PCs and the appropriate software make the storage, alteration and manipulation of data and information easier and less time consuming.

### The banking and financial services industry

Banking and financial services is a good example of an industry where service delivery has been continually altered and modernised since the late 1970s. In banking the most obvious application of technology has occurred in the development of the cashpoint machine, now found on the high street, inside bank or building society branches, in supermarkets and in public buildings such as railway stations. The other main technological development has been in the use of central computer-based systems to hold customer details and account records. These applications of computer technology allow any customer with a passbook or cashpoint card to withdraw money from their bank or building society account anywhere in the country and even overseas. This contrasts to the 1970s, when withdrawing money from a bank account meant that you had to visit the branch at which the account was held before the bank shut at 3.30pm, but not during the 12.30–1.30pm lunch hour when the banks closed.

This greater reliance on technology to perform at least some routine tasks has been part of the reason that both banks and building societies have been able to use and train their staff to sell a much wider range of financial services. These include savings accounts; tax-free savings accounts such as TESSAs, PEPS and ISAs; pensions; insurance for cars, home buildings and contents, travel and health; a range of different types of mortgages; and credit cards.

The development of new ways of delivering banking and financial services is likely to continue well into the twenty-first century and the medium of delivering the full range of financial services is where the development will be.

There are various types of ways in which high-tech banking can be delivered. Several possible methods are being considered and tested today.

Online transactional banking is sometimes called electronic or online or home banking. Home banking allows customers to access their bank accounts via PCs or TVs in their own home. There are a number of different methods by which home banking can be delivered: private dial-up services; managed networks; the Internet; and TV-based services, which are all discussed below. The home-banking facilities available include:

- showing balances and statements on-screen;
- settling bills;
- moving money from one account to another;
- making arrangements to settle bills and move money;
- looking at current standing orders and direct debits;
- observing transactions, including the use of a search and sort facility;
- ordering new cheque books;
- transferring information into other software, for example, to be used by a personal financial adviser.[52]

#### PC private dial-up services

This form of home banking involves the bank or building society providing the customer with the necessary software. The customer loads the software on to a PC in the home, which allows direct access to the bank or building society via a modem link. This type of home banking is being tested or used by many of the high-street banking organisations, including NatWest, HSBC, Barclays, Bank of Scotland and the Nationwide Building Society.[53, 54]

### Managed networks

A network administered and run by a third party may be used to provide home banking. The use of Internet service provider CompuServe's network by the TSB to provide online banking is a good example.[55] However, TSB is now part of Lloyds TSB and provides home banking over the Internet.

### TV-based services

The use of TV to provide home-banking services is perhaps the area where the greatest potential exists, as it does not require there to be a PC in the home. Early TV banking systems used satellite TV to show account information and a touchtone phone allowed a limited range of banking activities and transactions to be carried out.[56] However, the advent of digital TV is expected to herald an expansion of TV-based home-banking services.

### Internet

The Royal Bank of Scotland and Barclays are two organisations that provide Internet banking services. One of the challenges faced by banks and building societies is that in relative terms only a small number of homes contain computers linked up to the Internet and only those people who are familiar and feel comfortable with the Internet are likely to be happy to bank via it. This is coupled with the likelihood that customers are liable to weigh up the potential ease of use of Internet banking services against concerns about how quickly these services can be accessed over the Internet and the chance of security breaches occurring. Finally, if a customer finds banking with one bank or building society over the Internet convenient, then the same customer will have equally easy access to competitors' financial services.[57] The high street banks and building societies are subject to further competition from Internet-only banks which do not have the high overheads associated with retaining high street premises.

In summary, banks and building societies operating PC private dial-up services and managed networks can control how quickly customers are able to access services and the security of the system. These are both issues that concern customers, hence being able to control them is clearly beneficial. This contrasts with satellite TV and Internet home-banking services, where making the system secure is viewed as more problematic. Also with Internet home banking access to the system may be slow. However, the introduction of TV-based systems will provide access to the mass market and Internet home banking will eventually attract a greater number of customers than the dialup and managed network systems.[58]

## The millennium bug

One of the most recent technological issues that organisations have had to face and deal with is the millennium bug, known as Y2K by the computer trade. This resulted because computer disk space was expensive in the 1970s and 1980s. To save disk space and hence money, computer programmers

**Exhibit 2.27**
**The millennium bug**

| | Political | Economic | Sociocultural | Technological |
|---|---|---|---|---|
| Local | | | | |
| National | | | | |
| Global | | | | • Millennium bug |

used two digits to represent the year part of a date, so 1998 became 98. The concern for many organisations was that the year 2000, represented by two digits, 00, could be read as 1900 by computer systems and the outcomes of this happening were unknown. However, staff at a Scottish bank were eager to know what the year 2000 would do to the bank's computer systems, so the clock on the mainframe computer was switched forward to a minute before the turn of the century. The bank's staff watched and waited, as initially the system continued to process financial records as normal. When the computer's clock reached the year 2000, the figures being produced by the computer made no obvious sense. Then bank's older staff recognised them as financial data expressed in pounds, shillings and pence, as you would expect in 1900.[59] It was thought that 80 per cent of all mainframe computers were likely to be affected by the millennium bug.[60] Difficulties caused by the millennium bug arose earlier than expected: in 1992 Mary Sendar, who lives in Winona, Minnesota, USA, was invited to join her local kindergarten class, along with other local children born in 88. She was 104 years old and born in 1888.[61]

The millennium bug problem was mainly viewed as a problem striking mainframe computers and the software running on them. However, electronic equipment containing embedded computer chips was also affected. Such equipment and devices include stand-alone devices controlling the operation of industrial devices such as pumps, turbines and conveyer belts that are not usually reprogrammable; and items that can be reprogrammed, such as radar, satellite systems, hospital life-support systems, traffic lights, power stations and weapons systems. To solve the year 2000 problem for programmable embedded chips involved physically locating, checking, testing and if necessary replacing each individual chip. The cost for each chip was small, but the total cost for any organisation checking embedded chips was huge, as there are many embedded chips in all sorts of systems.[62, 63]

The cost of resolving the millennium bug issue has been estimated at billions of pounds, £31bn for the UK and £309bn for Europe.[64] The government's response to the problem was to set up Taskforce 2000, whose remit was to make organisations aware of the issue. Taskforce 2000 was replaced by Action 2000 in September 1997, with a shift in focus to what businesses needed to do to be Year 2000 compliant.

In UK central government alone the cost of making equipment and systems year 2000 compliant was estimated at £393m. This estimate included the Highways Agency spending £14.3m, the Telecommunications agency spending a modest £333 220, the Home Office £20m, the Ministry of Defence a massive £200m and Social Security £45.6m.

The private-sector courier company DHL spent $25m over two years to make its computer systems year 2000 compliant. This involved rationalising the company's software and computer systems around the world, entailing checking, analysing, repairing and testing 20m lines of computer code, 1000 servers and 25 000 network users, a very extensive job.[65]

One of the main difficulties that organisations faced in their efforts to become Year 2000 compliant was the availability of computer programmers with the skills necessary to deal with the millennium bug. Many computer programs written and developed in the 1970s and 1980s for mainframe computers were written in computer languages that are no longer taught, such as Cobol and Assembler. Therefore programmers with skills in languages such as Cobol were able to find very lucrative employment in the freelance contract market for programmers, with salaries of up to £500 a day. Hence organisations were forced to find ways of encouraging programming staff with the appropriate skills to remain with the company rather than taking up contract freelance work. In the event the dawn of the year 2000 passed without major computer failure affecting business and public services.

## Conclusion

This chapter aims to provide detail on what constitutes the external environment of an organisation and discussion on how organisations are influenced by the constituents of the external environment. A clear understanding of the detail and discussion provided on elements of the external environment at the local, national and global levels should help in understanding external environmental analysis and the necessary information gathering that are both presented in Chapter 1.

ETHICAL ISSUES CASE STUDY 2.2 FT

# Sierra Leone head denies UK military backing

**By Jimmy Burns**

Robin Cook, foreign secretary, claimed a new and important ally in the arms-to-Sierra Leone affair when president Ahmed Tejan Kabbah said he had received no military backing from the UK government.

In a letter released yesterday, hours before Mr Cook faced a parliamentary grilling over the affair, President Kabbah of Sierra Leone said he had not asked or received from the UK government any military assistance either overtly or covertly.

The letter, sent to Tony Blair, the prime minister, also claimed the role played by the UK company Sandline in helping President Kabbah return to power had been exaggerated.

The letter was contradicted by Sandline, which yesterday released further details of what it alleged was a legitimate military involvement in Sierra Leone which had the backing of the UK government.

The company said its main role had been immediately before and during the attack by the west African peacekeeping forces, Ecomog, early in February this year which led to the restoration of the Kabbah presidency. It claimed its support for the Ecomog forces had included advising on military planning, intelligence gathering, logistics and communications.

Further questions about the nature of the UK presence in Sierra Leone were emerging last night as the government admitted for the first time that a senior British officer had fought alongside pro-Kabbah troops last June after being deployed to Sierra Leone as part of a three-man military advisory team.

The officer, Major Lincoln Jopp, was recommended by the MoD for a Military Cross which was awarded by the Queen. He was wounded while helping defend a position occupied by pro-Kabbah troops which were under attack from forces loyal to the rebel government.

The Defence Ministry would not comment last night on independent witness accounts made available to the Financial Times claiming that Maj Jopp was embroiled in the fighting alongside at least eight white mercenaries.

The MoD insisted the fact that Maj Jopp and the chief executive of Sandline, Lieutenant Colonel Tim Spicer, both belonged to the same Scots Guards regiment was 'purely coincidental'.

The MoD did not release Maj Jopp's name on Monday when it was asked to detail the extent of its military involvement in Sierra Leone. 'His name did not appear in the briefing we got from officials,' the ministry said.

Meanwhile Mr Cook's insistence yesterday that no intelligence about Sandline's activities had been passed either to officials or ministers was questioned by observers close to Sandline who suggested some degree of tracking of the company's activities by M16 and military intelligence.

*Source*: *Financial Times*, 13 May 1998. Reprinted with permission.

**Questions for ethical issues case study 2.2**

1 Should the UK government have given approval for Sandline's activities in Sierra Leone? Give reasons for your answer.

2 Was Sandline's behaviour proper in supplying expertise to Sierra Leone to help restore a democratically elected government:

(a) with UK government approval?

(b) without UK government approval?

Explain your answers.

3 Would the behaviour of the UK government have been proper if it had used the military intelligence services to trace Sandline's activities in the lead-up to the company's activities in Sierra Leone? Give reasons for your answers.

EXIT CASE STUDY 2.3 FT

# The growing business: coffee bean counters

**By Scheherazade Daneshkhu and Flona Lafferty**

When Sahar Hashemi, a lawyer, returned to the UK two years ago after a stint in New York and frequent forays into its expresso bars, she was shocked to find how difficult it was to get a decent cup of coffee.

She and her brother Bobby, an investment banker, threw in their jobs and set up Coffee Republic, one of a new breed of coffee bars that have proliferated in London during the past few years.

The two, who opened their first bar in South Molton Street in late 1995 with £20 000 of their own capital and a £75 000 small-company loan from the Department of Trade and Industry, now have six bars and plan to have 20 in a year's time.

Other groups, including the US-based Seattle Coffee Company, Whitbread's Costa Coffee, and Aroma, plan a rapid expansion during the next few years. All have rushed to capitalise on the yen for coffee suddenly acquired by a nation of tea drinkers. But is such growth sustainable, and how many casualties might there be?

All the operators insist the market is large and will continue to grow in London and other city centres. Ms Hashemi, who estimates that there are about 150 coffee bars in the UK, believes there is easily room for 10 times that number. 'The market is huge – it's a question of a handful of players going in and capturing it. No one player will be able to dominate – there'll be four or five with their own edge.'

Each claims to have its own competitive advantage – Costa is the biggest, Coffee Republic claims it has the best coffee, while Prêt à Manger says it can afford to compete on price because of its larger sandwich sales.

Finlay Scott, chief executive of Aroma, which has 15 outlets since opening in 1995, says: 'We are all in the same sector but we address different parts, so competition isn't a problem. Whereas Prêt focuses on food and Costa on coffee, Aroma sits in the middle.'

England's coffee bars have their roots in the 17th century when they became a meeting place for men to discuss business. Garraway's was the scene of rash speculation during the South Sea Bubble of 1720; Dryden held court at Will's, while Lloyds of London grew out of the coffee house where underwriters met.

The new bars, by contrast, draw their inspiration from the US with slick service and quick turnaround times. Many of the recent entrants point to the success of Starbucks, the big Seattle-based coffee company, as evidence of the UK's potential and the importance of developing a brand to support growth. Starbucks has 1300 outlets in the US and plans to have 700 more by 2000. There are more than 10 000 coffee bars in the US.

The expansion in premium coffee bars follows a steep rise in coffee drinking at home, a market worth more than £800m, of which real coffee – roast and ground – consumption accounts for 10 per cent. Tea accounts for 40 per cent of hot beverages drunk at home compared with more than 50 per cent for coffee.

The Real Coffee Association, an alliance of coffee manufacturers, says the growth has been accompanied by a trend towards drinking better quality coffee and more varieties. Bad office coffee is being rejected in favour of takeaway coffee, in spite of the higher prices demanded by coffee bars.

The Seattle Coffee Co, the US concern that opened its first site in Covent Garden 18 months ago, now has 30 more and plans to open a further 20 by the end of the year. Scott Svenson, its founder, says: 'The typical uninitiated customer who came into our first store would have asked for something familiar like a cappuccino. Now they'll experiment with different flavoured syrups and play around with strengths – it's all about getting people to discover the world of coffee.'

Aroma's Mr Scott says: 'We've moved away from the unbranded, fragmented, angry Italian of 20 years ago; our pubs are dying on their feet and we won't accept horrid coffee from a company vending machine.'

But the stampede into this growing market is creating business problems. Competition for sites is putting pressure on prices in high streets, near offices or in railway stations. Mr Svenson says each time an outlet becomes available, his company faces a fierce bidding war with rivals.

Sinclair Beecham, who co-founded Prêt à Manger in 1986, predicts trouble for companies that pay too much for leases. 'Some people are getting over-excited and being over-competitive when it comes to bidding for new outlets. They might find it tough in the future when they can't deliver the necessary profitability. The property market is very over-heated at the moment, which means it's suddenly going to fall very hard, leaving certain companies with a lot of debt.'

Justin Worsley, senior consultant at The Henley Centre, the consumer consultancy group, agrees that demand for coffee bars will grow as increased work loads and time pressures are causing Britons to develop American grazing habits. A change in social attitudes has also made eating in the street more acceptable. Indeed, carrying a cup of premium brand coffee is developing into a style badge, he says.

But however much operators protest their ability to differentiate, they are basically focused on selling a good cup of coffee, Mr Worsley argues. This means they are vulnerable to low-cost competitors. He believes the answer

**Exit case study 2.3 *continued***

to securing long-term growth lies in developing coffee bars into places to meet friends, as in the US, as an alternative to pubs.

'A lot of the market is concentrated on coffee. That in itself is a limitation on the growth because not everyone likes coffee,' says Mr Worsley. 'Just as pubs have got away from spirits and beer with the introduction of food, wine and alcopops and have become a community centre, the long-term growth of coffee bars is in the sociability market.'

As for whether coffee bar growth is linked to the economic cycle, Ms Hashemi believes they are recession-proof in spite of the high prices charged: 'In a recession you may not be able to afford the best house or the best car but you can afford the best cup of coffee. It's affordable luxury – a way to treat yourself.'

*Source*: *Financial Times*, 16 September 1997. Reprinted with permission.

**Questions for exit case study 2.3**

1 (a) Perform a LoNGPEST analysis of the external environment of Coffee Republic.
  (b) Clearly identify the influences that are most likely to affect the success or failure of the business.

2 How important is competition to the success or failure of the venture?

## Short-answer questions

1 Name three influences from the political external environment, one each from the local, national and global levels.

2 Name three influences from the economic external environment, one each from the local, national and global levels.

3 Name three influences from the sociocultural external environment, one each from the local, national and global levels.

4 Explain why the influence of technology on an organisation is considered differently from the influence of economic, political and sociocultural factors.

5 Indicate the biggest technological influences on organisations at the end of the twentieth century.

6 Define autocracy.

7 Describe a democracy.

8 List the characteristics of a free market.

9 List the characteristics of a planned economy.

10 List the characteristics of a mixed economy.

11 Explain the function of Chambers of Commerce in the business community.

12 Summarise the role of the IOD and CBI in business.

13 Explain the role of trade unions.

14 Explain the role of NATO in the external environment.

15 Describe the role of the Commonwealth in the external environment.

16 Illustrate the role of the Bank of England in the economy of the UK.

**17** Explain the role of the London stock market in the economy of the UK.

**18** Indicate the role of the WTO in global trade.

**19** Explain the role of NAFTA in North American trade.

**20** Explain the role of ASEAN.

**21** Explain the role of EFTA and how it differs from the EU.

**22** Indicate by example what you understand alliances and agreements to be.

**23** State three key changes in the demographics of the UK at the end of the twentieth century.

**24** Cite three key social changes occurring in the UK at the end of the twentieth century.

**25** Demonstrate by example why language, behaviour and culture shock are all cross-cultural issues.

## Assignment questions

1. Choose *one* of the following organisations: Institute of Directors (IOD), Confederation of British Industry (CBI), or Trades Union Congress (TUC). Identify and collect appropriate information and write an essay of 2000 words that demonstrates and evaluates the influence of your chosen organisation on the individual employee, the workplace and the economy.
2. Select two organisations from different industries or sectors and collect relevant and appropriate information. Compare and contrast the impact of technology on your chosen organisations over the next three years and present your findings in a 2000-word report.
3. Consider the position of a foreign student or employee who has travelled over 2000 miles to work or study in your home country. Write an essay of 2000 words that identifies and evaluates:
   - the global sociocultural issues of language, behaviour and culture shock with which the foreign student or employee would have to contend;
   - the likely impact of global sociocultural issues on his/her performance at college/university or in the workplace.
4. Identify and collect appropriate information and write an essay of 2000 words comparing and contrasting the roles of ASEAN and NAFTA in their home geographic regions.

## References

1 http://www.businesslink.co.uk/services/index.htm.
2 http://www.bbc.co.uk/politics97/analysis/rozenberg2.shtm1.
3 Ibid.
4 Ibid.
5 Ibid.

6 Manning, A (1997) 'If it's good enough for everyone else, it's good enough for us', *Independent on Sunday*, 11 May.

7 *China-Britain Trade Review*, quoting HM Customs and British Trade Commission (Hong Kong).

8 http://www.europa.eu.int/abc-en.htm.

9 Barber, L (1997) 'No turning back from brave new Europe', *Financial Times*, 17 July.

10 http://www.baltinfo.org/CBSS.htm.

11 http://www.priairienet.org/acas/96F23024.html; Ejime, P (1996) Panafrican News Agency (PANA), 'Reconciling Nigeria with the Commonwealth', 23 June.

12 http://wwwafricanews.org/west/nigeria/stories/19970821_feat1.html; Ejime, P (1997) Panafrican News Agency (PANA), 'Nigeria reassessing Commonwealth membership', 21 August.

13 http://www.thecommonwealth.org/about/general/general1.html.

14 http://www.nato.int/welcome/home.htm.

15 Buchan, D, Clark, B and White, D (1997) 'Nato expansion deal covers divide', *Financial Times*, 9 July.

16 *Daily Telegraph*, 28 February 1998.

17 http://www.efta.int/docs/EFTA/Gen . . . rmation/Informationsheet/221.html.

18 Ibid.

19 http://www.oecd.org/about/works.htm.

20 http://www.nafta-sec-alena.org/english/nafta/preamble.htm.

21 Crawford, L (1997) 'Hazardous trades bring pollution and health fears down Mexico way', *Financial Times*, 6 June.

22 Dombey, D and Dunne, N (1997) 'Peso crisis turbo-charges revolution in motor trade', *Financial Times*, 11 June.

23 Ibid.

24 Dunne, N (1997) 'Erie fends off economic slings and arrows', *Financial Times*, 27 June.

25 Bardacke, T (1997) 'Cambodia rebuffed by Asean', *Financial Times*, 11 July.

26 http://www.aseansec.org/history/asn_his2htm.

27 http://www.wto.org/htbin/htimage/wto/map.map?8,33.

28 Williams, F (1997) 'India import curbs row goes to WTO', *Financial Times*, 17 July.

29 Jonquieres, G de and Walker, T (1997) 'New dawn in the east', *Financial Times*, 3 March.

30 Nicholson-Lord, D (1995) '"Greys" take over from the young as big spenders', *Independent*, 27 January.

31 Braid, M (1995) 'Tomorrow belongs to them', *Independent*, 1 October.

32 Nicholson-Lord, op. cit.

33 Johnson, P (1990) 'Our ageing population – the implications for business and government', *Long Range Planning*, 23 (2), April.

34 Braid, op. cit.

35 Gwyther, M (1992) 'Britain bracing for the age bomb', *Independent on Sunday*, 29 March.

36 Braid, op. cit.

37 Timmins, N (1995) 'A powerful indictment of the eighties', *Independent*, 10 February.

38 Mills, H (1995) 'Shift in wealth divides rich South and poor North', *Independent*, 1 February.

39 Jury, L (1998) 'Britons remain true to family values', *Independent*, 7 August.

40 Cooper, G (1996) 'The single society', *Independent*, 20 March.
41 Johnstone, V (1998) 'Companies switch on to virtues of global vision', *Daily Telegraph*, 30 April.
42 Ibid.
43 Baxter, A (1997) 'Face to face across borders', *Financial Times*, 26 March.
44 Johnstone, op. cit.
45 Baxter, op. cit.
46 Johnstone, op. cit.
47 Ibid.
48 Shankar, B and Sharda, R (1997) 'Obtaining business intelligence on the Internet', *Long Range Planning*, 30 (1), February.
49 Bird, J (1996) 'Untangling the Web', *Management Today*, March.
50 Ibid.
51 Bowen, D (1998) 'www + flair = new.business', *Management Today*, May.
52 Daniel, E and Storey, C (1997) 'On-line banking: strategic and management challenges', *Long Range Planning*, 30 (6), December.
53 Kelly, S (1996) 'Virtual banking – banks gain net interest', *Computer*, 17 October, referenced in Daniel and Storey, op. cit.
54 Lavin, P (1996) 'Manage your money. How to do your banking electronically', *What Personal Computer*, July, referenced in Daniel and Storey, op. cit.
55 Kelly, op. cit.
56 George, C (1996) 'Armchair banking? Rest assured', *Daily Telegraph*, 25 June, referenced in Daniel and Storey, op. cit.
57 Daniel and Storey, op. cit.
58 Ibid.
59 Cane, A (1997) 'Countdown to calamity?', *Financial Times*, 31 May.
60 Taylor, P (1997) 'Year 2000: a $600bn headache', *Financial Times*, 8 January.
61 Ibid.
62 Cane, A (1997) 'Microchip inaction makes a date with disaster', *Financial Times*, 6 March.
63 Cane, op. cit. 31 May 1997.
64 Cane, A (1997) '"Millennium bomb" threat to international telecoms', *Financial Times*, 2 May.
65 Taylor, P (1998) 'Defusing the millennium bomb', *Financial Times*, 7 January.

## Further reading

The following books and articles all look at aspects of the external environment of organisations:

Baer, W S (1998) 'Will the Internet bring electronic services to the home?', *Business Strategy Review*, 1 (1), Spring.
Bird, J (1996) 'Switching on intranets', *Management Today*, December.
Bird, J (1998) 'The great telephony shake-up', *Management Today*, January.
Ghosh, S (1998) 'Making business sense of the Internet', *Harvard Business Review*, March/April.
Griffith, M W and Taylor, B (1996) 'The future for multimedia – the battle for world dominance', *Long Range Planning*, 29 (5), October.
Gwyther, M (1999) 'Sold to the man on the internet', *Management Today*, June.
Mitchell, A (1998) 'New model unions', *Management Today*, July.

Molina, A H (1997) 'Newspapers: the slow walk to multimedia', *Long Range Planning*, 30 (2), April.

Pickard, J (1999) 'Grey areas', *People Management*, 29 July.

Smith, D (1998) 'How single is the single market?', *Management Today*, January.

Thomas, R (1999) 'The world is your office', *Management Today*, July.

van de Vliet, A (1996) 'Whatever happened to leisure', *Management Today*, May.

Wallace, P (1999) 'Agequake', *Management Today*, March.

Worthington, I and Britton, C (2000) *The Business Environment*, 3rd edn, Chapters 3–6, Harlow: Financial Times Prentice Hall.

# 3 The competitive environment

Claire Capon and Andrew Disbury

## Organisational context model

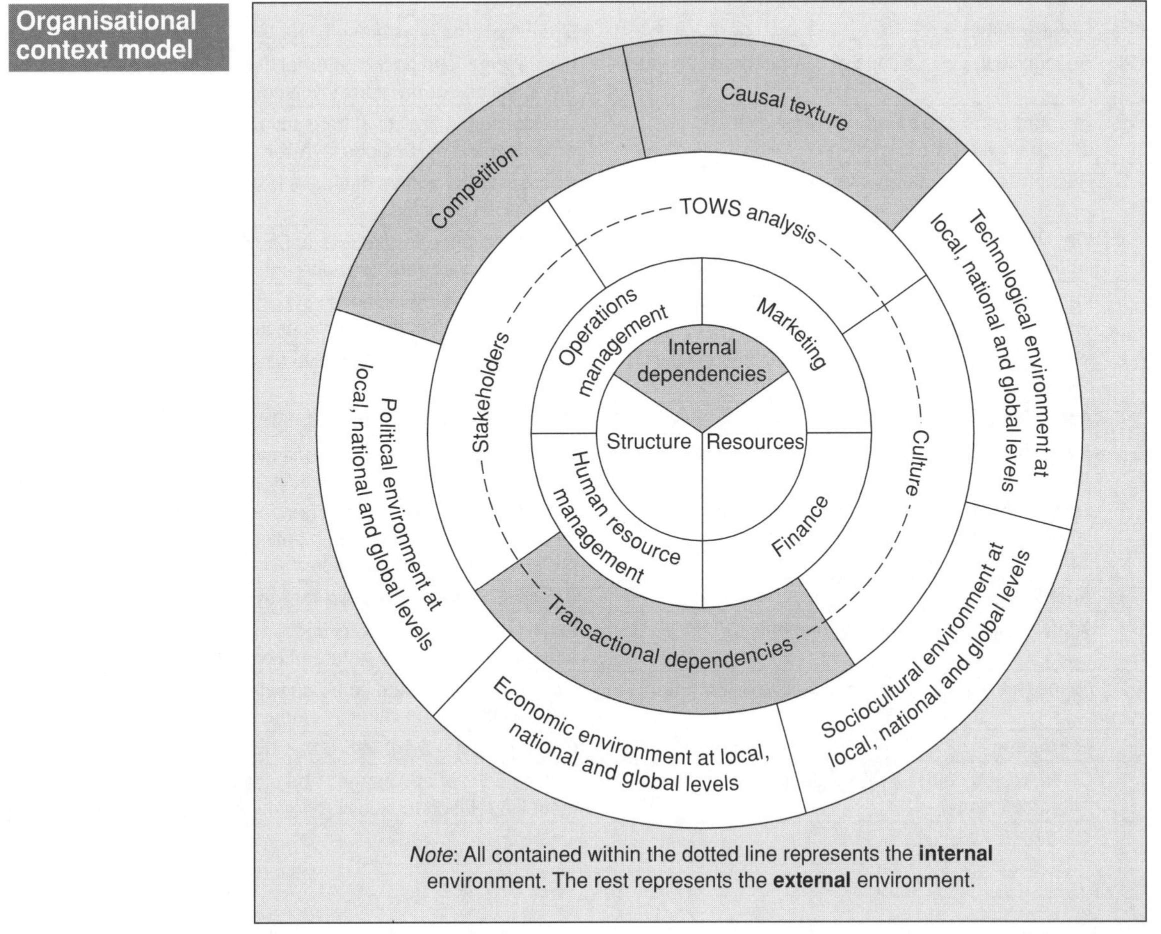

**Exhibit 3.1**

## Learning outcomes

While reading this chapter and engaging in the activities, bear in mind that there are certain specific outcomes you should be aiming to achieve as set out in Exhibit 3.2 (overleaf).

**Exhibit 3.2 Learning outcomes**

| | Knowledge | Check you have achieved this by |
|---|---|---|
| 1 | Define competition | recalling the elements affecting the nature of competitive rivalry and drawing Porter's five forces model of competition |
| 2 | Define internal dependencies | identifying the type of situation in which internal dependencies could occur |
| 3 | Define transactional dependencies | identifying the type of situation in which transactional dependencies could occur |
| 4 | Define causal texture | identifying the type of situation in which causal texture could occur |
| | **Comprehension** | **Check you have achieved this by** |
| 1 | Recognise the occurrence of internal dependencies | discussing the impact of internal dependencies on the organisations or divisions involved |
| 2 | Recognise the occurrence of transactional dependencies | discussing the impact of transactional dependencies on the organisations or divisions involved |
| 3 | Recognise the occurrence of causal texture | discussing the impact of causal texture on the organisations or divisions involved |
| 4 | Identify the role of the regulatory bodies for privatised industries | discussing the effectiveness of particular regulatory bodies in the light of current developments in the industry or sector |
| 5 | Identify the role of UK statutory bodies in the regulation of competition in UK | discussing the effectiveness of particular statutory bodies in the light of current developments in specific industry sectors |
| 6 | Identify different ways in which organisations can compete | explaining different competitive strategies |
| | **Application** | **Check you have achieved this by** |
| 1 | Apply Porter's five forces of competition to particular industry sectors | illustrating the influence of substitutes, threats, buyers and suppliers on particular organisations in given sectors |
| 2 | Employ Porter's generic competitive strategies to a range of products or services provided by an organisation | illustrating the application of generic competitive strategies by explaining the relationship between costs, price and profits for products or services |
| | **Analysis** | **Check you have achieved this by** |
| 1 | Analyse the competitive environment | examining various industry sectors and come up with conclusions about the nature of competitive rivalry observed |
| 2 | Assess an organisation's competitive environment | looking at options and come up with recommendations to improve the organisations' competitive position |
| | **Evaluation** | **Check you have achieved this by** |
| 1 | Compare and contrast PEST analysis and Porter's five forces model | making a list of advantages and disadvantages of each model in relation to studying organisations in context |

ENTRY CASE STUDY 3.1

FT

# Return to the wild frontier

## Wrangler's European challenge to Levi's jeans bucks the campaign trend its rival has set

**By Alison Smith**

Jason Maddox and Trey King are used to riding tough animals as they compete on the rodeo circuit in Arkansas. Now Wrangler's hopes of breaking through to mount a serious challenge to Levi's in the European jeans market are riding on them.

The two rodeo riders are the stars of a television advertising campaign which breaks in the UK today and in Germany next week. 'We believe this is an opportunity to break out of the pack chasing Levi's and become a challenger for the number one position,' says David Smith, marketing director for Wrangler in Europe. Wrangler has earmarked £15m for spending on the European campaign this year.

Substantial sales are at stake. In the UK market alone, more than 39m pairs of jeans at a value of more than £900m were sold in the 12 months to the middle of last year. Levi's has more than 20 per cent of the branded market, while Wrangler has 8–9 per cent. The gap between the two brands is narrower in Germany, where each has a much smaller share of the market. In Ireland the two are broadly neck and neck.

The new Wrangler ads contrast sharply with the current Levi's campaign, already showing across Europe. Levi's ad, created by Bartle Bogle Hegarty, shows a man falling off a boat in a storm, and then escaping from three mermaids, who try to take his jeans off but fail because they fit too closely.

The down-to-earth Wrangler's ads from Abbott Mead Vickers show the rodeo riders in action, and talking about the extensive injuries they have received. The style is documentary, and the message is that Wranglers are worn by 99 out of 100 rodeo riders.

'A mistake a lot of brands have made is to forget their western roots and float off into something else. We are the authentic western jeans, and we have to take that positioning and make it aspirational and exciting,' says Smith.

Wrangler's strategy of moving outside the advertising genre of boy meets girl to chart topping soundtrack, created by Levi's since the mid-1980s, is seen by other advertising agencies as sensible for a number two brand. 'Unless you do something absolutely different, you will be seen as a "me too" Levi's ad,' says one. Levi's advertising is so dominant, however, that creating a significantly different approach is very hard, he adds.

Beyond that general difficulty, there lies a particular issue for Wrangler. The emphasis of its advertising is that real western values are epitomised by the riders who make their living by risking their lives in the arena. Yet they are taking that message to young European consumers, whose view of the west is likely to have been formed by Hollywood and American television. The gritty tone of the campaign may well not be what they expect to associate with the west.

Nigel Marsh, board account director at AMV, sees this as an advantage. 'If it is not what they expected, it will stick in the mind.'

Tom Blackett, deputy chairman of Interbrand, a brand consultancy, is not convinced about the campaign. 'It may appeal to the laddish tendency but I'm not certain it's an enduring image.'

Moreover, there is the risk inherent in any ad campaign based on real life that it ceases to be an unvarnished reality as soon as it appears in an advertisement. Wrangler is already talking about personal appearances by the two rodeo riders in the films. But the more the campaign makes them stars, the more it undermines their strength as real-life characters.

Wrangler has a strong start to the campaign, running with a mix of films instead of just one film as is more often the case. But it is by no means clear it will have an easy ride.

*Source*: *Financial Times*, 1 April 1997. Reprinted with permission.

## Introduction

Now that we have defined and analysed the external environment in Chapters 1 and 2, this chapter deals with how organisations relate to elements of their environment, and are affected directly by the actions of these elements towards them. This chapter will also consider how actions that take place between elements of the environment could, although not directed towards the organisation, nevertheless have secondary impacts on the organisation's ability to achieve its aims and objectives. Competition is a key element of the external environment and is examined later in this chapter.

The organisational context model at the start of this chapter illustrates that the external environment of any given organisation is a complex and dynamic arena. Therefore, to understand the reasons underpinning organisational success or failure, it is necessary to examine and understand how organisations relate to the other organisations and to the individuals who comprise their specific environment. In industrialised nations, competition is an ever more important part of any organisation's external environment. Therefore the study of the competitive environment is key to understanding how and why organisations interact with the key players surrounding them. When studying an organisation, or when working in one, it is tempting but misleading to see the organisation as a stand-alone unit, a body of people and tasks existing in isolation of any other place.

## Environmental linkages

Emery and Trist[1] suggest that understanding the environment of organisations depends on a firm grasp of the linkages within an organisation's environment. They identify three types of linkage: from within to within; from outside in and inside out; outside linkages.

**From within to within linkages**

From within to within linkages are internal dependencies inside the organisation. These internal dependencies can be co-operative or confrontational in nature. A co-operative inter-dependency would be the marketing and production department agreeing to work closely together on developing a new product. A confrontational inter-dependency would occur if two divisions in an organisation were to disagree on the resources allocated to their divisions; *see* Exhibit 3.3, link number 1. This type of confrontational inter-dependency can usually be resolved by an appropriate third party, a more senior employee, for example a supervisor, senior manager or director, depending on the type of confrontation.

**From outside in and inside out linkages**

From outside in and inside out linkages are called transactional dependencies. Transactional dependencies are links in and out of the organisation. In any organisation there are a number of transactions that take place between organisations and elements of the external environments; *see* Exhibit 3.3, link number 2. These transactions can be simple or complex, frequent or rare, consistent or one-off. Depending on the kind of transaction, the level of dependence between the organisation and the environmental element will change. Obvious transactional dependencies are those between an organisation and its suppliers, buyers and competitors. The organisation manages such transactional dependencies by negotiation with the other parties involved.

**Exhibit 3.3 Environmental linkages**

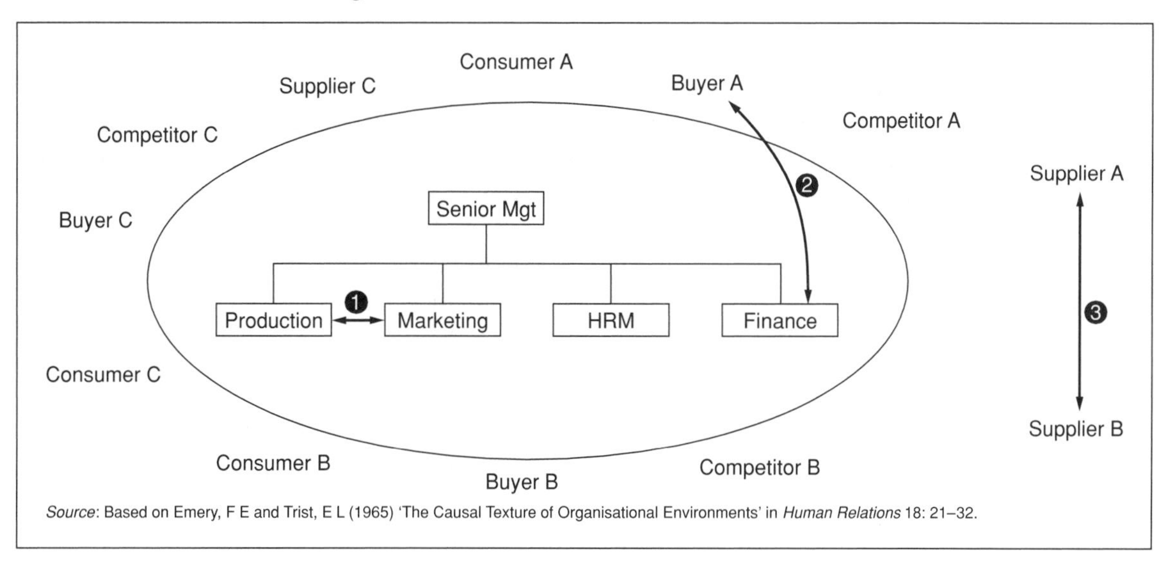


*Source*: Based on Emery, F E and Trist, E L (1965) 'The Causal Texture of Organisational Environments' in *Human Relations* 18: 21–32.

If a manufacturer, such as Nissan, operates a just-in-time production system (*see* Chapter 7), then the manufacturer is not keeping stocks of raw materials or component parts. Therefore the manufacturer is hugely dependent on suppliers' abilities to provide the resource inputs in an accurate and timely fashion. If the right piece does not appear in the correct place on the assembly line at the proper time, then the entire production process grinds to a halt. Thus there is a frequent and consistent transactional dependency between the manufacturer and the supplier, with the balance of the power resting with the supplier. Nevertheless, a collaborative, rather than combative, relationship is usually sought in such situations.

Complete independence is neither possible nor desirable for an organisation: not possible since organisations cannot provide all the resources or components they need internally, and not desirable since they must have customers to be in business, and *de facto* these customers are external to the organisation. Too much dependence on suppliers or customers, or other elements of the external environment, is also not desirable, as it leaves the organisation vulnerable. Thus senior managers need to attempt to tip the balance of power as much as possible in favour of their organisation to achieve at least inter-dependence and at most the upper hand.

**Outside linkages**

Outside linkages are referred to as causal textures. Causal textures are interdependencies outside the organisation in the external environment, which can have an effect on the organisation; *see* Exhibit 3.3, link number 3. The causal texture in which an organisation operates is called that because there is a cause-and-effect relationship between what elements of the external environment do and how this affects third-party organisations. The simplest example of causal texture would be two suppliers agreeing to fix prices to the detriment of retail buyers. This causes the cost of supplies to go up for buyers, who in turn increase prices charged to members of the public, which results in lower sales and reduced profits.

This type of linkage or causal texture is difficult for organisations to uncover and manage effectively, so responses to such linkages may be slow and apprehensive. This contributes to an organisation's doubt and uncertainty in managing the external environment. A very clear example of an organisation affected by causal texture is Clan Douglas. This is a manufacturer of luxury cashmere sweaters, based in the Borders area of Scotland, whose biggest market is the USA.[2] In March 1999 the US government began a 'banana row' with the EU, with Britain being the most affected European country. The row centred around Britain's favouring of bananas produced by small farmers on Caribbean islands that are former British colonies, as opposed to bananas produced by US companies on plantations in South America. The USA viewed this as unfair trade and therefore imposed a 100 per cent import tax on certain luxury goods imported by US companies from Britain and Europe. These goods included cashmere sweaters, which sell for $350, and fountain pens made by Mont Blanc and Watermans, which retail at $300.[3] The doubling of the price of these goods to over $600 would effectively kill the US retail market and hence put affected manufacturers out of business. Hence a company like Clan Douglas in the Borders would be severely affected by such outside linkages or causal texture. Such companies have absolutely no influence over the banana war, nor could they have been reasonably expected to have anticipated the disagreement.

## The competitive environment

The behaviour of competitors in the external environment, be they local, national or global, very often affects organisations. Therefore it is necessary to understand and analyse the nature and role of competition. Competition is readily recognisable in the private sector, and is a mainstay of the free-market philosophy, as explained in Chapter 2. The analysis of competition is also applicable to the study of public-sector organisations. In the UK in particular, the rhetoric of the free market and competition have been introduced into the public sector through wholesale privatisation or the introduction of quasi-markets into public-sector services. This was clearly demonstrated by the privatisation of public utilities such as gas and electricity, which are still monitored by a regulator.

When examining an organisation's external environment, competition is considered an important element of that and is therefore analysed in addition to the general external environment, which is examined by LoNGPEST analysis. The importance of competitors and competition is further illustrated and reinforced in the entry case study, 'Return to the wild frontier', and the effort that Wrangler went to in 1997 to project a very different image from its competitors, Levi.

## The regulation of competition in the UK

The UK economy is a mixed economy, which means that there is some influence and regulation of the competitive behaviour of firms. Some of the regulation of competition is by bodies that may intervene if there is a likelihood of anti-competitive behaviour occurring. These bodies are the Office of Fair Trading, the Monopolies and Mergers Commission and the Restrictive Practices Court.

### Office of Fair Trading

The Office of Fair Trading (OFT) is a non-ministerial department of the government, established in 1973, headed by the Director General of Fair Trading. The OFT has the power to investigate companies suspected of carrying out anti-competitive behaviour such as discriminatory pricing, predatory pricing or refusing to supply non-appointed distributors. Under the 1980 Competition Act, a company found to be behaving in an anti-competitive manner can be referred by the OFT to the Monopolies and Mergers Commission for further investigation. The principal tasks of the OFT are:

- to identify and put right trading practices which are against the consumer's interests;
- to regulate the provision of consumer credit;
- to act directly on the activities of industry and commerce by investigating and remedying anti-competitive practices and abuses of market power, and bringing about market structures which encourage competition.[4]

### Monopolies and Mergers Commission

The Monopolies and Mergers Commission (MMC) was established in 1948 and looks at the way monopoly power and unfair competition are exercised in specific cases that are referred to it by the Office of Fair Trading or an appropriate Minister of State. Such an example is shown in the ethical issues case study for this chapter, 'MMC to probe ice cream market', which illustrates the debate over access to ice-cream freezer cabinets in shops and the MMC's role in trying to resolve the situation. The MMC may also be asked to investigate proposed mergers where £30m of assets are involved or if a market share greater than 25 per cent would result.

### Restrictive Practices Court

The Restrictive Practices Court was established in 1956 and controls practices that are presumed to be against the public interest. Restrictive practices can relate to the price of goods, conditions of supply, qualities or descriptions, processes or areas and persons supplied. A case in which the Restrictive Practices Court has recently ruled is that of over-the-counter medicines. The Office of Fair Trading launched a four-year investigation into the price-fixing agreement for over-the-counter medicines such as aspirin, vitamins and cough and cold remedies. In the 1970s this agreement guaranteed small chemists shops a reasonable profit and living by virtue of an assured profit margin on over-the-counter medicines. However, in the mid-1990s 40 per cent of the market for over-the-counter medicines was held by large supermarket chains. The Office of Fair Trading investigation began when one such supermarket chain, Asda, challenged the protected position of these medicines. In March 1999 the Restrictive Practices Court announced that it would allow the Office of Fair Trading to launch a full-scale hearing into the 30-year-old price-fixing agreement. The matter is unlikely to be fully resolved until 2001.

## Regulatory bodies for privatised utilities

The privatised utilities in the UK, which immediately after privatisation were not subject to intense competition, are each monitored by a regulator. Regulators exist for telecommunications, gas, electricity, water and rail transport. Their responsibility is to encourage competition and see that customers are not unfairly exploited where there may only be one supplier of a service, such as the supply of water to domestic premises.

### Office of Telecommunications

The Office of Telecommunications (OFTEL) was established by the Telecommunications Acts 1984, the year that British Telecom was privatised. Like the Office of Fair Trading, OFTEL is a non-ministerial government department and is headed by a Director General who is appointed by the Secretary of State for Trade and Industry. OFTEL regulates the prices charged by telecommunications providers, including line rental and call charges. It also provides licences and conditions of operation for new entrants to the telecommunications industry, which includes licensing new types of service and regulation of the equipment market. The key functions of OFTEL are:

- to ensure that licensees comply with their licence conditions;
- to initiate the modification of the licence conditions either by agreement with the licensee or, failing that, by reference to the Monopolies and Mergers Commission (MMC) together with the Director General, of Fair Trading to enforce competition legislation – under both the Fair Trading Act 1973 and the Competition Act 1980 – in relation to telecommunications;
- to advise the Secretary of State for Trade and Industry on telecommunications matters and the granting of new licences;
- to obtain information and arrange for publication where this would help users;
- to consider complaints and enquiries about telecommunications services or apparatus.[5]

### Office of Gas Supply

The Office of Gas Supply (OFGAS) was established under the Gas Act 1986 to regulate British Gas, which was a privatised monopoly. The role of the regulator was to set prices of gas, monitor service levels, safeguard the interests of the consumer and encourage competition in the supply of gas to industrial markets. In 1995 an amended Gas Act was introduced giving OFGAS responsibility for managing competition in the domestic market and issuing licences to companies wishing to transport and supply gas.[6]

**Office of Electricity Regulation**

The Office of Electricity Regulation (OFFER) is a non-ministerial government department established under the Electricity Act 1989 at the time when the regional electricity boards were privatised. OFFER regulates the price of electricity charged by electricity generators, such as PowerGen and National Power, and the price that regional electricity companies charge both domestic and industrial users. The main duties of the regulator are to:

- secure all reasonable demands for electricity are met;
- secure that licence holders are able to finance their licensed activities;
- promote competition in the generation and supply of electricity;
- protect the interests of electricity customers in respect to price charged, continuity of supply and the quality of services provided;
- promote efficiency and economy of the part of licencees in supplying and transmitting electricity.[7]

**Office of Water Supply**

The Office of Water Supply (OFWAT) is a government department led by the Director General of Water Services. OFWAT regulates water supply and pricing to domestic and industrial customers. OFWAT checks that prices for different types of customers – metered or unmetered, large or small, urban or rural – are fair. Generally the prices charged by water companies should reflect the cost of supplying clean water and getting rid of dirty and draining water from homes and premises.[8]

The duties of the Director General of Water Services are laid down in the Water Industry Act 1991. These duties cover ensuring that:

- the functions of a water and sewerage company, as specified in the Act, are properly carried out;
- companies are able to finance their functions, in particular by securing a reasonable rate of return on their capital;
- no undue preference is shown and there is no undue discrimination in the way companies fix and recover charges, and that rural customers are protected;
- other aspects of customers' interests are protected – including quality of services and benefits from the sale of land transferred to the companies at privatisation, or acquired since then.[9]

OFWAT works closely with two other water regulatory bodies: the Environment Agency, which implements water quality standards in inland waters, estuarial and coastal waters; and the Drinking Water Inspectorate, which regulates standards for drinking water.[10]

**Office of Passenger Rail Franchising**

The Office of Passenger Rail Franchising (OPRAF) was established under the Railways Act 1993 to award franchises to the 25 train operating companies on the railway network in Great Britain. OPRAF oversees and manages the franchises, with the accent on safeguarding passengers' interests. The Franchising Director acts according to Objectives, Instructions and Guidance (OIG) which are given to him/her by the Secretary of State for Transport. The latest OIG were issued in November 1997 and outline the franchising director's key objectives as:

- to increase the number of passengers travelling by rail;
- to manage the franchise agreements to promote the interests of passengers;
- to secure a progressive improvement in the quality of services available to passengers.[11]

**Office of the Rail Regulator**

The Office of the Rail Regulator (ORR) was also established under the Railways Act 1993. The Rail Regulator is independent of ministerial control. The main purposes of the Rail Regulator are to:

- issue licences to operators of trains, networks and stations;
- enforce competition law in connection with railway service;
- protect the interests of users of railway services, including disabled people;
- promote the use and development of the national railway network for freight and passenger;
- provide through ticketing;
- sponsor a network of Rail Users' Consultative Committees (RUCCs), representing passengers' interests.[12]

## Assessing the nature of competition

Porter[13] presents a model for examining competition in an industry or sector, *see* Exhibit 3.4. He argues that five basic forces drive competition in an industry: competitive rivalry; threat of new entrants; threat of substitute products or services; bargaining power of buyers; and bargaining power of suppliers.

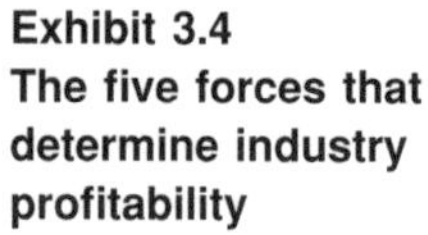

**Exhibit 3.4**
**The five forces that determine industry profitability**

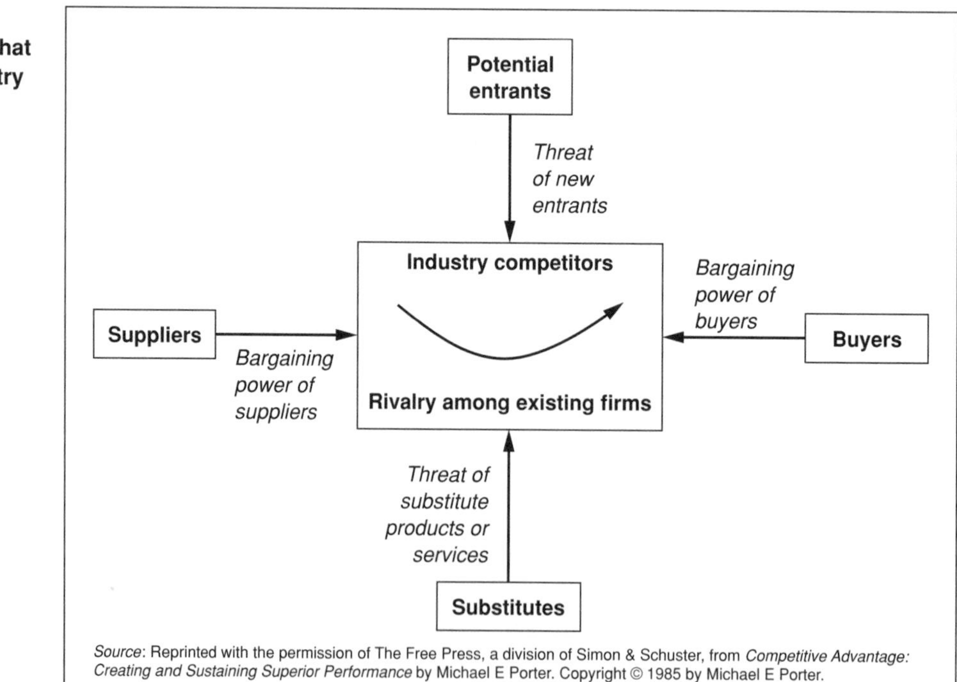


*Source*: Reprinted with the permission of The Free Press, a division of Simon & Schuster, from *Competitive Advantage: Creating and Sustaining Superior Performance* by Michael E Porter. Copyright © 1985 by Michael E Porter.

These five forces need to be examined and understood if the nature of competition in an industry or sector is to be fully appreciated.

### Competitive rivalry

The first of the five forces examines the nature of competitive rivalry within a particular sector or industry. There are a number of factors affecting how fierce competitive rivalry is in an industry or sector and consequently how difficult the market is for organisations operating there. The key questions to examine concern who the present and potential competitors are, and how intensive the competition is between them. In some industries there are numerous competitor companies, all of a similar size and capacity, all holding comparable market shares and all seeking to dominate the industry. There may be no dominant company or companies within the sector and little to distinguish between the brands and products that are available to the customer. The market itself may be established or mature, with little prospect of major innovation or design surprises. In such an industry or sector, the intensity of competitive rivalry will be very high, as mature companies have to battle to retain market share, sustain differentiation and maintain their customer base.

The entry case study to this chapter discusses the competitive rivalry between jeans manufacturers Levi and Wrangler. Both are established companies seeking sustainable differentiation of their product, with the aim of maintaining market share and tempting customers away from the competitor's jeans. In the case of jeans, the role of advertising on television and in the cinema is key to promoting the appropriate image to attract jeans-wearing customers. The different approaches of Levi and Wrangler to image and advertising are discussed in the entry case study.

The supermarket industry in the UK is good illustration of extremely fierce competitive rivalry. Tesco is dominant in terms of market share, but other players all have high profiles and similar market presence. For many customers there is little real difference between the big supermarket chains, with Tesco, Sainsbury's and Safeway all offering very similar ranges of products and services. Supermarkets compete ruthlessly and do so by offering an ever wider range of differing services, including those complementary to their core business. Supermarkets have opened banks to provide financial services, in addition to a range of ancillary services that are now standard, including dry cleaning, cafés, photo processing, clothes stores, recycling collection points and pharmacies.

### Threat of new entrants

The second of Porter's five forces is the threat of new entrants to an industry or sector. It is necessary to identify which companies are likely to be able to enter the market in order to compete with existing operators, and to recognise the other markets in which these potential new entrants currently operate. The threat of new entrants will be greatest if an industry is attractive enough to entice them. The attractiveness of an industry depends on there being a sufficient customer base to support the new entrant's business along with existing organisations. High potential profits and low set-up costs make an industry attractive. This gives a combination of low financial risk combined with high potential returns, which is an attractive combination.

In the exit case study for this chapter, the move by Superdrug into the business of retailing up-market perfume and aftershave was undertaken in part as an opportunity for Superdrug to make good profits. The standard profit margin made by an authorised retailer on a bottle of up-market perfume or aftershave is 40 per cent. Hence Superdrug was able to cut its profit margin to, say, 20 per cent and still make an attractive profit. This, combined with the relatively low set-up costs needed to exploit the opportunity, the cost of the perfume and securing supplies, meant it was too good an opportunity for Superdrug to miss.

In contrast, an industry for which markets are mature, set-up costs high and profit margins slim will be unattractive and face a low threat of new entrants. The UK telecommunications industry is a good example of an industry where the threat of new entrants is low. At privatisation, the government decided that British Telecom should have competition. However, it was recognised that new telecoms companies would not have the financial or technological capabilities to be able to establish their own telecommunications networks across the country in order to be able to compete with BT. Thus Mercury, the first new entrant to telecoms, was allowed by law to route its telephone services via the BT network. It was envisaged that gradually competition would grow and BT's dominance would be challenged. Cable operators, providing television and telephone through one wire, have begun to lay their own network across the UK, but progress is slow. New entrants need considerable financial resources in order to be able to stay in the new market long enough to gain a small foothold in BT's competitive environment. Mercury is owned by global player Cable & Wireless, which has helped it to achieve such a foothold.

### Threat of substitute products

When considering the nature of competition, the alternative products or services available to be purchased require consideration. A substitute product or service provides the same function as the good for which it is a replacement. Straightforward examples of substitute products include tea as a substitute for coffee, or cans and cartons as substitute forms of packaging that have largely replaced glass bottles as containers for milk and soft drinks. In relation to the entry case study, trousers made from material other than denim would be a direct substitute for jeans. Substitute products or services will be a threat if customers perceive that the alternatives perform a similar or equally good function. The threat from substitutes will be greater if the alternatives provide better value for money. This can be achieved by the substitute's being equally good and cheaper or by its being equal on price, but offering a better product or more added value.

The consumer frenzy that is Christmas shopping provides a vivid example of the exercise of choice and decision making between alternatives or substitute products or services. For the man who is buying a gift for a wife or girlfriend, the initial decision between substitutes is perhaps a choice of a bottle of eau de toilette, a compact disc, a new sweater, or a piece of jewellery. If the customer decides on a compact disc, there are further decisions to be made. Which band or artist should have recorded the compact disc? Should it be their new Christmas album? Should it be an older album? Should it be a compilation of greatest hits? Where should it be brought: in a music superstore like HMV or Virgin, in WH Smith, by mail order from Britannia Music or online from Amazon.co.uk? The function of the merchandise purchased is to provide a good Christmas present, hence all of the possibilities mentioned are substitutes for each other and so pose a threat.

### Bargaining power of buyers

Earlier in this chapter it was mentioned that organisations have varying degrees of dependence on their customers and suppliers. Porter's five forces of competition model refers to this as the 'bargaining power of buyers and suppliers'. There are two types of buyer, the industrial or commercial buyer who purchases goods on a large scale on behalf of the organisation for whom s/he works, and the individual consumer.

The bargaining power of a commercial buyer depends on a number of factors. If, for example, the threat of substitutes is strong, then a number of choices exist and the buyer will shop around to find the best deal and most suitable choice. The bargaining power of the buyer is also strengthened if alternative sources for the supply of a product exist. In this situation the buyer will have the upper hand when negotiating supply and price. In the UK the supermarket sector is a good example of a group of organisations that have high bargaining power as buyers. In purchasing food to sell to the general public the supermarkets purchase in bulk and there are many substitutes and alternative sources of supply for them to capitalise on. Hence they can drive a hard bargain in terms of price and product.

In buying fruit and vegetables, supermarkets require suppliers of products like apples or tomatoes to grow a particular variety, supply fruit of uniform appearance and of a pre-determined size. This allows for attractive in-store displays to tempt customers to buy the produce, which is aided by the uniform appearance of the fruit and vegetables, which also limits the amount of handling and rummaging though the goods by customers. If the size of a particular fruit or vegetable is important to the supermarket, then that will be specified in the contract with the grower, along with the variety and delivery date. For round fruit and vegetables size is specified as the circumference in millimetres. For example, in February 1999 Marks & Spencer sold pre-packaged South African plums, labelled as being of the Harry Pickstone variety, size 50/55mm. It should be noted that powerful buyers like supermarkets often work closely with their suppliers in developing new products and the systems for producing them.

The individual consumer is usually much less powerful as a buyer compared with large organisation. The bargaining power of an individual buyer is influenced by factors similar to those for commercial buyers. The strong threat of substitutes and the number of choices available will allow the individual consumer to shop around for the best deal. The city centre office worker who goes out to buy his/her take-away lunch everyday has a number of choices. They can buy a sandwich, drink and packet of crisps from a city centre store like Marks & Spencer or Boots; a bakery which is one of a chain of local bakeries; or a sandwich shop like Deli France. Other substitutes are available for the individual consumer to consider, such as a burger, fries and drink from a fast-food outlet like McDonald's or Burger King. Individual consumers are free to exercise choice but have no real power as individuals to negotiate over the price they pay for their lunch.

However, if individual consumers choose to act in unison they may be able to exercise power. For example, in the mid-1990s when the health scare over British beef and BSE erupted, McDonald's faced the prospect of large numbers of its customers acting together and refusing to buy and eat its hamburgers. Therefore it switched from supplies of British beef to supplies of Dutch beef.

**Bargaining power of suppliers**

The other side of the transactional relationship is the power exerted by suppliers. There are a number of different cases when the bargaining power of suppliers is high. In industries or sectors where there are few possible suppliers, then they will be able to exert a good deal of influence on the organisations to which they supply raw materials, components or finished goods for retail. In the supply of highly specialist technology, of highly prized or rare materials where the quantity is low and price is high, the supplier is more powerful as it controls something that is greatly sought after. Thus oil-producing nations have the ability to bring the world to its knees since the most modern industrialised nations are entirely dependent on the supply of crude or refined oil. In some cases, suppliers may not be entirely satisfactory and alternatives may be available, but the cost of switching from one supplier to another is too high in the short term to be affordable, even if, in the long term, the savings would be greater. Where there are plentiful suppliers who can not easily be substituted, supplier power is low.

The exit case study to this chapter provides an excellent illustration of suppliers with low bargaining power. The suppliers of fragrances to French perfume houses such as Yves Saint Laurent and Chanel are family-based firms located in the Grasse area of France. These firms are not paid for the research and product development work they undertake. They only receive payment if they win a contract with a fragrance house, which is issued once the fragrance house is satisfied with the fragrance that has been developed for them. The developers of the fragrances will be competing against other similar firms and they stand a one in ten chance of being successful and winning a contract with a perfume house like Chanel. Hence the bargaining power of the fragrance developers as suppliers to the fragrance houses is very low indeed.

## Guidelines for assessing competition

In assessing the competitive environment faced by an organisation in a particular industry or sector, it can be useful to consider the following areas in relation to the five forces driving competition.

**Competitive rivalry**

Identify present and potential competitors in the industry or sector. Assess the intensity of competition between the different organisations. Is this likely to change?

**Threat of new entrants**

Does a threat of new entry into the industry exist? From which organisations does it arise? Identify the industry in which potential entrants currently operate. Evaluate the likelihood of new entrants coming into the industry.

**Threat of substitute products and services**

Identify alternative products and services. In what industry are present and potential substitute products located? Assess the likely impact of substitute products and services on the organisation and industry being analysed.

**Bargaining power of buyers**

Name the buyers of the organisation's products and services. Identify and evaluate any sources of power the buyers have with regard to the organisation being analysed.

**Bargaining power of suppliers**

Name the suppliers of the organisation's key resources and inputs. Identify and evaluate any sources of power the suppliers have with regard to the organisation being analysed.

## Competitive strategies and competitive advantage

In order to operate successfully in an industry or sector where substantial competition is created by the five forces of competition, organisations need to adopt a competitive strategy. They may choose to follow one of Porter's competitive strategies to gain competitive advantage; *see* Exhibit 3.5. Competitive advantage is gained when an organisation achieves a position in an industry, due to cost or differentiation factors, which allows it to make above-average or superior profits. Supermarket chain Iceland's home delivery service, which allows customers to order food shopping from home and have goods delivered direct to their door, is a source of differentiation that gives the company a competitive advantage over other food retailers, although this advantage is being eroded as its competitors follow suit.

Porter[14] suggests that there are two decisions that organisations have to make to arrive at a suitable type of competitive strategy. The first is to decide if the basis of competition is to be based on cost or added value. The second is to decide if a broad target market (mass market) or a narrow target market (niche

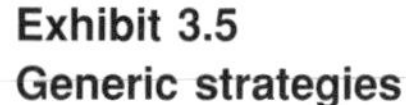

**Exhibit 3.5**
**Generic strategies**

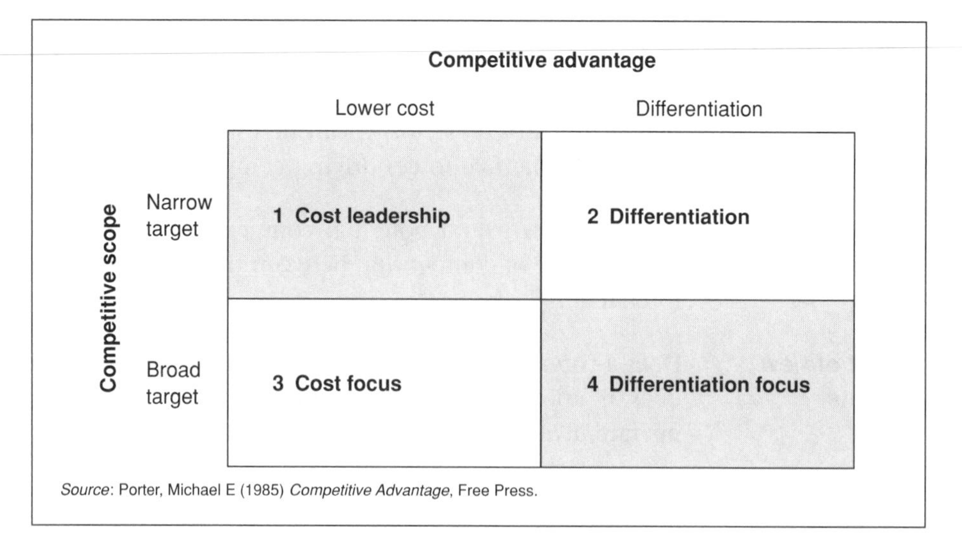


*Source*: Porter, Michael E (1985) *Competitive Advantage*, Free Press.

**Exhibit 3.6**
**Examples of generic strategies**

Competitive advantage

| Competitive scope | Lower cost | Differentiation |
|---|---|---|
| Broad target | **Cost leadership**<br>• Milk Tray<br>• Ford Fiesta<br>• Standard-class rail ticket from London to Glasgow | **Differentiation**<br>• Thorntons Continental Selection<br>• Volkswagen Golf<br>• First-class rail ticket from London to Glasgow |
| Narrow target | **Cost focus**<br>• Safeway own-label mint thins<br>• Skoda Felicia<br>• Coach ticket from London to Glasgow | **Differentiation focus**<br>• Imported Belgian handmade truffles<br>• Audi A3<br>• Chartered private jet from London to Glasgow |

market) is to be served. Porter suggests that combinations of price and market type give rise to the following competitive strategies: cost leadership, selling a standard product to a mass market; differentiation, selling an added-value product to a mass market; cost focus, selling a low-cost product to a niche market; differentiation focus, selling an added-value product to a niche market (*see* Exhibits 3.5 and 3.6).

## Cost leadership

If an organisation serves or aims to serve a broad target or mass market, to be operationally efficient it will supply standard products or services to many consumers. If an organisation is following a cost leadership strategy, then it will be seeking to be the lowest-cost producer in its industry or sector to supply a mass market. A successful cost leader will have achieved its position while offering products and services of a quality comparable to those offered by direct

**Exhibit 3.7**
**Cost, profit and price relationships for cost-based competitive strategies**

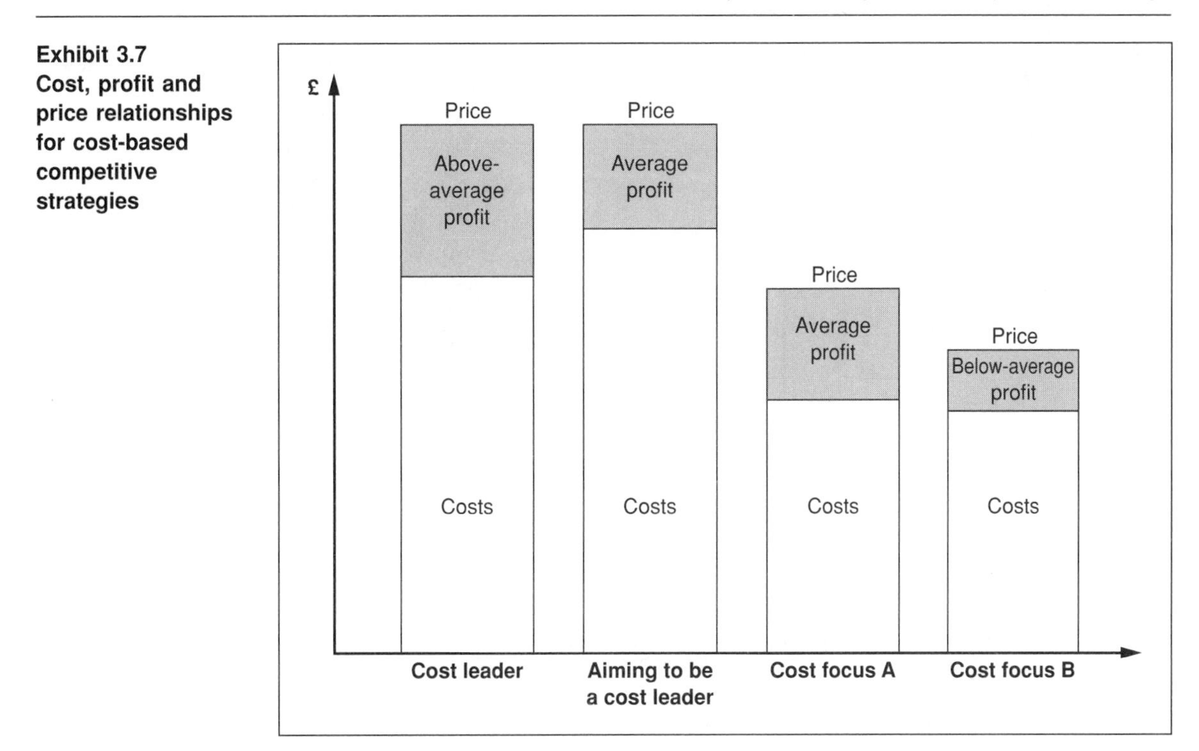


competitors. Hence in a mass market there are likely to be a number of key players, and the competitive rivalry could be fierce. However, because of the lack of specialism or high technology, costs for most organisations following a cost leadership strategy are likely to be average, as are profits.

The successful cost leader who has achieved below-average costs, but is selling at an average price, will make greater or superior profits; *see* Exhibit 3.7. It is difficult for any competitor serving a mass market with standard products and services to charge above-average prices, as the competitive rivalry in the mass market prevents individual competitors from raising the price of standard products significantly. Exhibit 3.7 shows examples of organisations and products that follow a cost leadership strategy.

**Differentiation**

An organisation following a differentiation strategy perceives that it is still able to serve a broad target market, but by providing a product or service that is different and better; *see* Exhibit 3.5. A product or service is different and better due to its added value. Added value arises via the addition to the product or service of extra features, or from the better quality of the product or service. The customer needs to be prepared to pay extra for additional features or quality; if this is not the case, then the organisation has wasted money in providing the added value. In this situation any superior profit that may have been made as a result of the added value is lost.

In an added-value or differentiated product costs should be average in areas that do not add value and extra costs incurred only for added value, therefore

**Exhibit 3.8**
**Cost, profit and price relationships for differentiation-based competitive strategies**

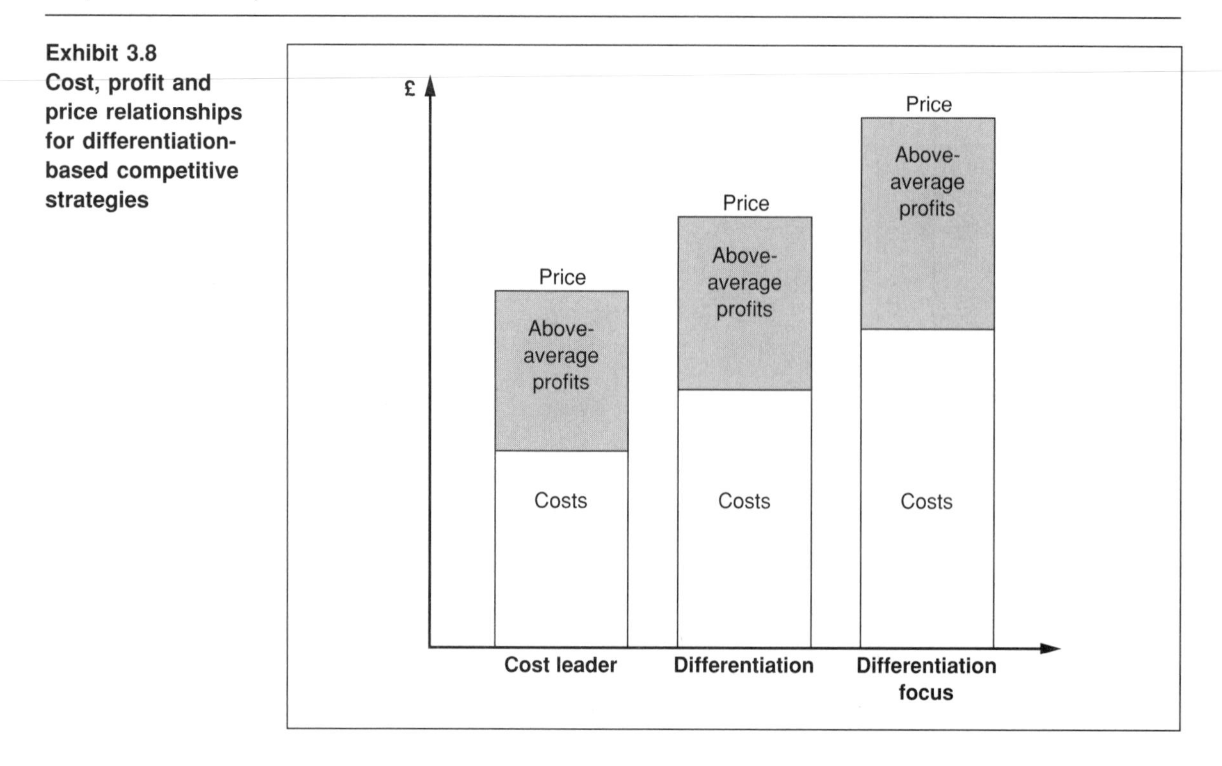


overall costs are higher than average. This allows a higher or premium price to be charged, giving rise to superior profits; *see* Exhibit 3.8. This can be seen in department stores or supermarkets that have higher than average prices but still gain market share because their customers are willing to pay extra for goods they consider to be better.

Added value is not only located in the perceived quality of goods or services. People are prepared to pay extra for packaging and labelling, the image or reputation of the brand or the lifestyle choices accompanying the use and purchase of certain goods; *see* Exhibit 3.6. This point is also illustrated by the fiercely pursued advertising strategies used by Wrangler and Levi, both anxious for their jeans to have the greater appeal and preferred image in the eye of the customer. This is important, as both companies produce basically the same product, good-quality casual wear denim trousers. Without the image created around the jeans, they would not attract a higher than average price. Customers are willing to pay more for real or perceived added value, as outlined in Exhibit 3.8.

### Focus strategies

An organisation employing a focus strategy will centre its efforts on a number of niche market sectors and serve only them to the exclusion of other broad market segments; *see* Exhibits 3.5 and 3.6. The successful application of a focus strategy rests on there being clear and significant differences between the segments focused on and the other market segments in the industry and the market segments that are focused on being poorly attended to by the competitors serving the broad target market.

### Cost focus

The cost focus strategy is followed by organisations aiming for a narrow target market, where customers are very price sensitive; *see* Exhibit 3.5. Organisations following a cost focus strategy will seek to deliver low cost and thus low-priced products and services to the market. In order to make profits, costs must be maintained at a minimum through using the lowest-priced raw materials, manufacturing processes, packaging, delivery and advertising. The cost focus strategy allows for average or below-average profits to be made, depending on how well costs are controlled and the level at which prices are set; *see* Exhibit 3.7.

### Differentiation focus

A differentiation focus strategy is used by organisations wishing to serve a narrow target market where consumers are prepared to spend a great deal of money in order to acquire luxury, top-of-the-range goods or services; *see* Exhibit 3.5. The relationship between costs, prices and profits is shown in Exhibit 3.8. Examples of products that follow a differentiation focus strategy are shown in Exhibit 3.6.

## Choosing a competitive strategy for competitive advantage

The achievement of competitive advantage and hence superior profits is central to the strategy of any organisation. For competitive advantage to be achieved successfully, a company must be clear about which type of competitive strategy is being followed: cost leadership, differentiation, cost focus or differentiation focus. Porter argues that an organisation that tries to follow a combination of competitive strategies, i.e. cost leadership and differentiation, will only achieve average or below-average performance. Hence competitive advantage is not achieved, which Porter calls being 'stuck in the middle'.[15]

However, it is clear from looking at some organisations that it is possible to follow all of Porter's generic competitive strategies and still be successful. For example, British Airways pursues all of Porter's generic strategies by offering a range of different airline tickets: first class, business class, economy class, and low cost via Go, its budget subsidiary. The supermarket chain Tesco also follows a variety of Porter's generic competitive strategies. In February 1999 the Tesco store on Abbeydale Road in Sheffield displayed a trolley of a typical selection of weekly groceries, just inside the main entrance, with the slogan, 'All this for £14.52'. The groceries in the trolley all came from Tesco's value line of products, with their distinctive blue-and-white striped packaging. Clearly a competitive strategy devised to tempt the price-sensitive customer into the store, this was, in Porter's parlance, a cost focus strategy. In contrast, Tesco also follows a differentiation focus strategy and sells a very up-market range of pre-prepared foods, called 'Finest', which is clearly designed to appeal to a customer prepared to pay a great deal of money for a meal. Hence it is possible for organisations to follow both cost- and differentiation-based competitive strategies.

An alternative view of competitive strategies is presented by Gilbert and Strebel.[16] They argue that there are two constituents of competitive advantage: lower delivered cost and higher perceived value. However, unlike Porter, Gilbert and Strebel go on to argue that lower delivered cost and higher perceived value can be used together to give a company a superior position in an industry or sector. The essence of their argument is that to achieve competitive advantage a firm must strive to give the highest perceived value for the lowest delivered cost. Examples of organisations that strive to offer such a combination are IKEA and the John Lewis Partnership. IKEA, a chain of Swedish furniture stores, prides itself on offering 'good design at best ever prices'. It achieves this by seeking out and using good design, efficient suppliers, and innovative and rationalised global distribution systems.[17] The commitment of the John Lewis Partnership's retail stores to offering best quality at lowest prices is summarised in its own slogan, 'never knowingly undersold'. John Lewis offers a refund, if a customer finds the same item cheaper elsewhere.

Competitive advantage is achieved by creating 'disequilibrium' between the perceived value of the product and the asking price by increasing the perceived value or reducing the asking price. This alters the terms of competition and could drive competitors out of the market or influence them to offer more perceived value for the same price or the same value for less money. However, Gilbert and Strebel go on to argue that the number of competitive formulas is small, not large. They give two possible reasons for this, the first being that there is an 'internal logic to each business system' that dictates the possible combinations of perceived value and delivered cost that must exist for the whole business system. This number of combinations is clearly finite. The second somewhat obvious reason is that there are only two basic generic competitive strategies, high perceived added value and low delivered cost. Depending on a company's industry position and circumstances, there can only exist variations around these two themes, which must therefore limit the number of competitive formulas.

Further to this, Gilbert and Strebel proceed to suggest that strategic advantage is obtained by the implementation of generic competitive moves in a sequence, such that the implementation of one prepares for the implementation of another, which should of course result in high perceived value for low delivered cost.

## Conclusion

This chapter examined the influence that competition has on an organisation, via environmental linkages, regulation of competition and the five forces of competition. It also considered the choices that organisations may make with regard to competing in the external environment. Two different perspectives on this were discussed, Porter's generic strategies and the work of Gilbert and Strebel.

ETHICAL ISSUES CASE STUDY 3.2

FT

# MMC to probe ice cream market

## Watchdog's third investigation of industry in six years will examine problems over supply

**By John Willman**

The Monopolies and Mergers Commission is to make its third probe in six years into the UK's ice cream industry with a comprehensive investigation of the market for wrapped products sold for immediate consumption.

The market is dominated by Birds Eye Wall's, the frozen food division of Unilever, which has almost 70 per cent of sales with a range of products including Magnum, Solero and Cornetto.

Wall's has a dedicated distribution network and has placed almost 70 000 freezer cabinets in shops exclusively for its products. Competitors say this in effect excludes them from the market, particularly in small shops with room for only one cabinet.

Ordering the inquiry yesterday, John Bridgeman, director general of fair trading, said there were 'continuing problems with competition in the supply of impulse ice cream'.

The new inquiry was recommended by the MMC in its last report on the industry, on the wholesaling practices of Wall's. In July, the commission condemned the use of exclusive distributors as anti-competitive because it excluded rivals from access to retail outlets. It also said there was a case for a comprehensive look at the whole industry, including the practice of freezer exclusivity, which it cleared after an investigation which began in 1993.

Freezer exclusivity has been under attack elsewhere in the European Union, and in March the European Commission ruled that Unilever's Irish subsidiary had abused its dominant position by excluding rivals' products from its cabinets. Unilever, the world's biggest maker of ice cream, is appealing against the ruling to the European Court.

Mr Bridgeman said yesterday the market had not developed as anticipated when the MMC gave freezer exclusivity a 'clean bill of health' in 1994.

Birds Eye Wall's said it was disappointed at the decision and saw little point in another inquiry. 'Wall's has achieved its position in the ice cream market through innovation, market forces and fair competition,' said Iain Ferguson, chairman. Investors did not seem upset, however, as Unilever shares rose $12\frac{1}{2}$p to close at $657\frac{1}{2}$p.

Mars, the US confectionery group which has been campaigning for a full inquiry since 1993, welcomed the referral. It said products such as Mars and Snickers ice cream bars outsold Wall's in grocery multiples where they were usually sold on equal terms.

The new investigation was also welcomed by Treats, a former Wall's unit that makes ice cream under its own name.

The MMC has nine months to complete its investigation and will report to the trade and industry secretary.

*Source*: *Financial Times*, 23 December 1998. Reprinted with permission.

**Questions for ethical issues case study 3.2**

1 Consider the information in the case study. Do you think the use of the MMC is an effective way to police anti-competitive behaviour? Give reasons for and against the use of the MMC in this context.

2 'Wall's has achieved its position in the ice cream market through innovation, market forces and fair competition.' Do you feel this is a legitimate statement? Give reasons for your answer.

3 If the MMC findings are that exclusivity of freezer cabinets should not occur, what will be the likely effect on:

(a) Walls/Unilever?

(b) Mars?

EXIT CASE STUDY 3.3

# The fragrance industry

**By Claire Capon**

The fragrance industry is one of France's largest and most well-known glamour image industries, along with champagne and fashion. Consequently France tops the table of perfume users, with French women splashing out an average annual spend of £36 on perfume, more than double that of her British counterpart.

The fine fragrances sold by companies like Chanel, Yves Saint Laurent and Givenchy are likely to begin life in Grasse in Provence in southern France. It is here that 30 member companies of the 70-strong Syndicat National des Fabricants de Produits Aromatiques are based. These companies specialise in raw material extraction, perfumery compounds and food flavourings. Some of them are owned by large multinationals such as Rhône-Poulenc or Bayer and others are family run and retain their independence. The companies extract essential oils from raw ingredients like jasmine, with the oils fitting into one of seven fragrance families: citrus, flower, fougere, chypre, wood, amber or leather.[18] The companies in Grasse develop fragrances for the perfume houses by blending the oils together. However, they undertake this product development work for free and only get paid if they win a contract with a fragrance house. The chances of winning a contract are one in ten.

A fragrance house will spend very large sums of money to develop and launch a perfume on the marketplace. Dior spent £40m on developing its Dolce Vita fragrance in the mid-1990s, with a £2.5m advertising campaign in the UK alone in its first year on the market.

The world's bestselling women's perfume is Chanel No 5, with retail sales of $130m in 1996. It is estimated that the sales of perfume and aftershave worldwide will top $21bn by the year 2000.[19] Therefore it is not surprising that relationships between the French government and France's fragrance industry are good, with the fragrance industry proving effective at lobbying the government on its own behalf. The French government is in turn effective at lobbying and influencing the European parliament and associated bodies on behalf of French industries, including the fragrance industry. As a result, the French fragrance industry is in a strong position and some of the large fragrance houses, like Givenchy and Yves Saint Laurent, have selective distribution arrangements that have been approved by the European Commission. The fragrance industry claims it sells images of glamour, passion, fantasy and exclusivity, which have taken the established fragrance houses 40 years to create and hence require selective distribution to protect the image that is being sold and to ensure that it remains up-market and exclusive.

These selective distribution agreements are implemented by the fragrance houses, who decide which retailers will stock and sell their brands of perfume and aftershave. A fragrance house will assess a retailer on a series of criteria covering staff training, lighting, ambience, location, size of windows, shelf arrangements and the quality of floor coverings. These are all considered to contribute to the image of the fine fragrance in the mind of the consumer.

In the UK up-market perfume and aftershave are typically sold by retailers such as Boots, House of Fraser, John Lewis and Debenhams. Boots is the market leader with 30–35 per cent market share. Fine fragrances are defined as perfume, eau de parfum, eau de toilette, eau de Cologne and aftershave lotion retailing at more than £15 per 50ml bottle.

However, in 1992 retailers such as Superdrug, Poundstrecher and What Everyone Wants started to sell fine perfume and aftershave like Chanel No 5, Aramis and Giorgio at lower prices. This was viewed as a threat to the profitability of both the fragrance houses and the usual high-street retailers. The costs and profit associated with the production of a bottle of perfume and aftershave costing £30 in the high street breaks down as follows: retailer – £12; tax – £5.13; fragrance house – £10.59 (to cover advertising, marketing, packaging and profits); perfume and aftershave – £2.28. The practice of department stores offering luxury gifts free when a purchase of a particular brand of perfume or aftershave is made also reinforces the existence of high profit margins for the retailers.

The major fragrance houses refused to supply retailers such as Superdrug on the basis that their shops, when assessed, failed to meet the criteria discussed above. The shops were thus defined as down-market and unsuitable for the retailing of fine perfume and aftershave. However, Superdrug was making considerable investment in modernising its stores and installing special fragrance counters with dedicated staff, and it continued to gain supplies of fine perfume and aftershave from the grey market; not illegal, just unauthorised. Supplies of an up-market fragrance are usually leaked on to the grey market at one of three points in the distribution chain. A manufacturer that has been commissioned to produce perfume or aftershave for a company like Chanel or Dior may over-produce and dispose of the surplus on the grey market, hence allowing an unapproved retailer or middleman to buy directly from the manufacturer. Over-production is difficult to police, particularly if there are several manufacturing sites spread across the world. Alternatively, the approved wholesaler may order too much stock and sell the surplus to an unapproved retailer or a grey market trader. The third possibility is that an approved retailer over-orders and is prepared to sell the

**Exit case study 3.3** ***continued***

surplus on the grey market. This allows companies like Superdrug or the supermarkets to acquire fine fragrances at a cut price and therefore to sell them at a price undercutting the approved retailers.

The dispute between Superdrug and the fragrance houses centred on the selective distribution arrangements agreed with the European Commission. Therefore in late 1992 Superdrug complained to the Office of Fair Trading that it should be allowed to sell fine fragrances at discounted prices and as such should be supplied through the normal channels via the fragrance houses and their distributors. Superdrug submitted a 200-page report in support of its case, which outlined the alleged discrimination that it felt it was facing. This alleged discrimination related to two issues: the refusal of fragrance houses to supply Superdrug with perfume and aftershave, and the refusal of major newspapers (*Sunday Times* and *Observer*) and magazines (*Vogue, Marie Claire* and *Woman's Journal*) to carry Superdrug's advertisements for fear of losing lucrative advertising revenues from the fragrance houses.

The Office of Fair Trading, concerned that the selective distribution arrangements used by the fragrance houses might have been resulting in lack of competition and high retail prices, decided that there was a case to answer and referred it to the Monopolies and Mergers Commission for a full six- to nine-month investigation. The Monopolies and Mergers Commission issued its report in November 1993 and stated that a complex monopoly existed in the fine fragrance market, but that it was not acting against public interest.

The power of manufacturers of fine fragrances was reinforced in 1996 when the European Commission extended until 1999 the regulations allowing selective distribution of certain goods, including fine fragrances.[20] In December 1996 the European Court of First Instance also ruled in favour of Yves Saint Laurent and Givenchy against the French supermarket chain Leclerc. Leclerc argued that the criteria used by the fragrance houses for selecting outlets for its fine fragrances excluded all Leclerc stores and discriminated against new methods of distribution, i.e. multi-product stores. The court ruled that Yves Saint Laurent and Givenchy needed to be able to preserve the consumers perception of the product as one with the 'aura of luxury and exclusivity'.[21]

**Questions for exit case study 3.3**

1 By applying Porter's five forces of competition model, assess the nature of competition in:
- the fine fragrance industry (Chanel, Givenchy, Dior);
- the retailing of fine fragrance industry (Superdrug, Boots).

2 Using your general knowledge, the fine fragrances discussed in the case study and substitute fragrance products identified in answer to Question 1, name ten fragrance products. Assess and justify which of Porter's competitive strategies each product follows.

3 If the selective distribution agreement for fine fragrances is abolished, what will be the likely effect on companies such as Chanel, Yves Saint Laurent, Dior and Givenchy?

## Short-answer questions

1 Define an internal dependency.

2 Define a transactional dependency.

3 Define casual texture.

4 Name three bodies regulating competition in the UK.

5 Name three bodies regulating privatised utilities in the UK.

6 List Porter's five forces of competition.

7 Specify three factors making an industry attractive to new entrants.

**8** List three factors making substitute products or services a threat.

**9** Specify three factors strengthening a buyer's power.

**10** List three factors strengthening a supplier's power.

**11** Name Porter's generic competitive strategies.

**12** Define competitive advantage.

**13** According to Porter, can competitive advantage be achieved if more than one of his competitive strategies is followed?

**14** In your opinion, can competitive advantage be achieved if more than one of Porter's competitive strategies is followed?

## Assignment questions

**1** Choose a private-sector industry that delivers a service to members of the public and analyse the competition in the industry by use of Porter's five force of competition. Select three companies in your chosen industry and discuss the competitive strategies they follow. Present your findings in a 2000-word report.

**2** Consider the universities and colleges in the town or city nearest to where you live. Assess the nature of competition between these colleges and universities by use of Porter's five force of competition. Determine if the university or college that you attend is influenced or affected by Emery and Trist's linkages. Present your findings in a 2000-word report.

**3** Research the deregulation of the domestic fuel market for gas and electricity in the mid to late 1990s. Write a report that:

- identifies the changes in the companies providing domestic fuel;
- evaluates the effect of these changes on competition in the domestic fuel market.

## References

1 Emery, F E and Trist, E L (1965) 'The causal texture of organisational environments', *Human Relations*, 18: 21–32.
2 BBC1, *Breakfast News*, 2 March 1999.
3 Edgecliffe-Johnson, A (1999) 'NY trade in luxuries becomes a victim', *Financial Times*, 6/7 March.
4 http://www.oft.gov.uk/html/about/strct.htm.
5 http://www.oftel.gov.uk/about/oftguide.htm.
6 http://www.ofgas.gov.uk/guide/gbot.htm.
7 http://www.open.gov.uk/offer/about.htm.
8 http://roof.ccta.gov.uk/ofwat/rolereg.htm.
9 Ibid.
10 Ibid.
11 http://www.opraf.gov.uk/about/index.htm.
12 http://www.rail-reg.gov.uk/about.htm.

13 Porter, M E (1985) *Competitive Advantage*, New York: Free Press.
14 Ibid.
15 Ibid.
16 Gilbert, D and Strebel, P (1988) 'Developing competitive advantage', in Quinn, J B, Mintzberg, H and James, R M (eds) *The Strategy Process*, Harlow: Prentice Hall.
17 IKEA catalogue 98, Inter IKEA Systems BV, 1997.
18 Harrington, C (1996) 'Heavenly scents', *Independent*, 20 July.
19 Rawsthorn, A (1997) 'Fragrant rival fails to outsell No 5', *Financial Times*, 25 January.
20 Pickard, S (1997) 'Message in a bottle', *Financial Times*, 11 February.
21 Tucker, E and Oram, R (1996) 'EU delays end of exclusive agreement', *Financial Times*, 15 October.

## Further reading

Bell, E (1998) 'Winner takes all', *Management Today*, September.

Bennett, R (1999) *Corporate Strategy*, 2nd edn, Chapter 3, Harlow: Financial Times Pitman Publishing.

Brown, K (1999) 'Competition policy could find it tougher to cut prices than win friendly headlines', *Financial Times*, 12 March.

Cole, G A (1994) *Strategic Management*, Chapter 4, London: DP Publications.

Collis, D J and Montgomery, C A (1998) 'Creating corporate advantage', *Harvard Business Review*, May/June.

Johnson, G and Scholes, K (1999) *Exploring Corporate Strategy*, 5th edn, Chapter 3, Harlow: Prentice Hall.

Jones, I W and Pollitt, M G (1998) 'Ethical and unethical competition: establishing the rules of engagement', *Long Range Planning*, 31 (5), October.

Lynch, R (2000) *Corporate Strategy*, 2nd edn, Chapter 5, Harlow: Financial Times Prentice Hall.

Lynn, M (1999) 'Can nice guys finish first?', *Management Today*, February.

Moore, K (1998) 'Power struggle', *Management Today*, February.

Porter, M E (1979) 'How competitive forces shape strategy', *Harvard Business Review*, March/April.

Richardson, B and Richardson, R (1992) *Business Planning*, 2nd edn, Chapter 5, London: Pitman Publishing.

Thompson, J L (1997) *Strategic Management: Awareness and Change*, 3rd edn, Chapter 9, London: International Thomson Business Press.

Thurlby, B (1998) 'Competitive forces are also subject to change', *Management Decision*, 36 (1).

Trueman, M and Jobber, D (1998) 'Competing through design', *Long Range Planning*, 31 (4), August.

Wheatley, M (1998) 'Seven secrets of effective supply chains', *Management Today*, June.

Worthington, I and Britton, C (2000) *The Business Environment*, 3rd edn, Chapter 12, Harlow: Financial Times Prentice Hall.

# 4 Inside organisations

Claire Capon with Andrew Disbury

## Organisational context model

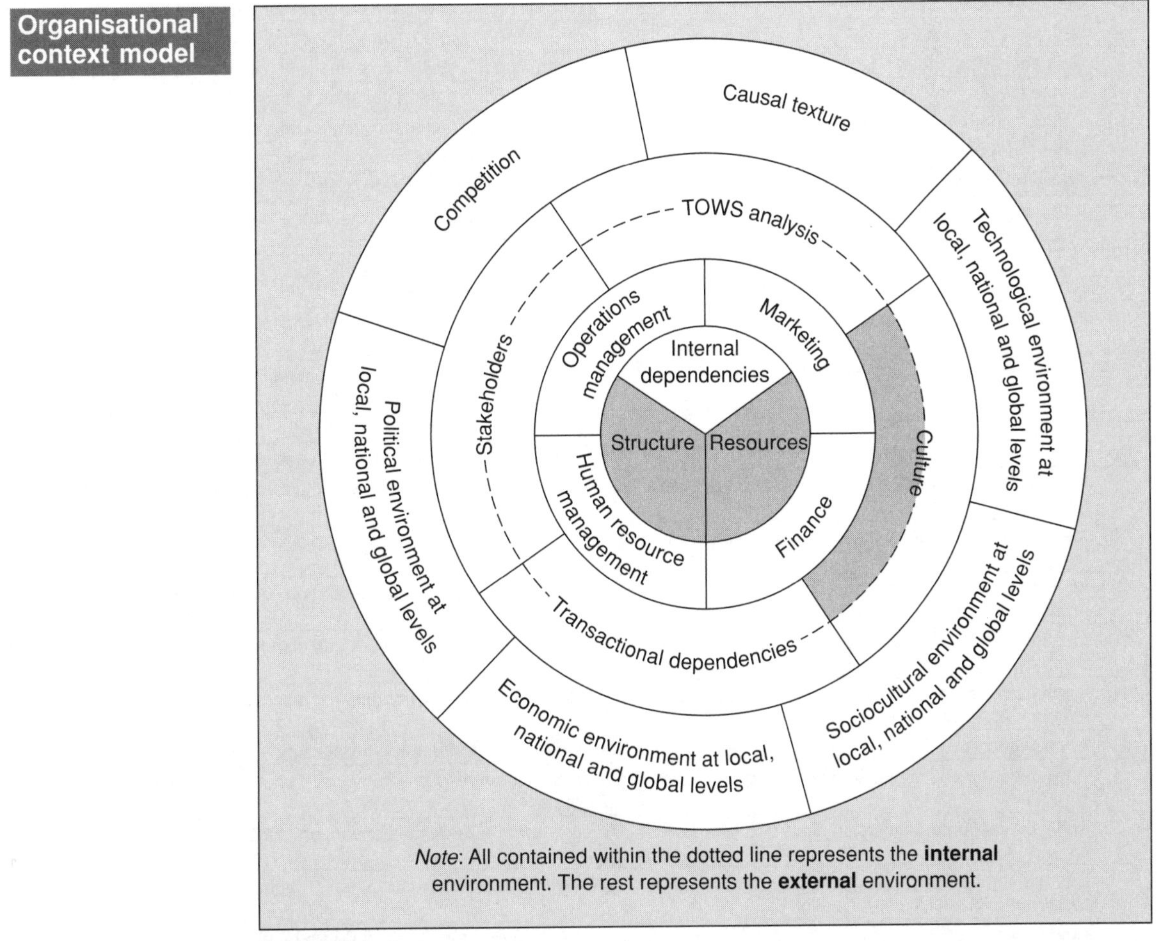

**Exhibit 4.1**

## Learning outcomes

While reading this chapter and engaging in the activities, bear in mind that there are certain specific learning outcomes you should be aiming to achieve as set out in Exhibit 4.2 (overleaf).

**Exhibit 4.2 Learning outcomes**

| | Knowledge | Check you have achieved this by |
|---|---|---|
| 1 | Define the term organisation | stating in one sentence what an organisation could be said to be |
| 2 | List the elements of the internal environment | reviewing the generic elements of the internal environment from memory |
| 3 | List the categories of resource inputs | reviewing generic resource inputs |
| 4 | Define the simple structure | sketching the simple structure and listing its main features |
| 5 | Define the functional structure | sketching the functional structure and listing its main features |
| 6 | Define the divisional structure | sketching the divisional structure and listing its main features |
| 7 | Define the holding company structure | sketching the holding company structure and listing its main features |
| 8 | Define the matrix structure | sketching the matrix structure and listing its main features |
| 9 | Define the power culture | sketching the power culture and listing its main features |
| 10 | Define the role culture | sketching the role culture and listing its main features |
| 11 | Define the task culture | sketching the task culture and listing its main features |
| 12 | Define the person culture | sketching the person culture and listing its main features |
| 13 | Define a centralised organisation | listing the characteristics of a centralised organisation |
| 14 | Define a decentralised organisation | listing the characteristics of a decentralised organisation |
| | **Comprehension** | **Check you have achieved this by** |
| 1 | Recognise the organisational resource conversion process | explaining the role of inputs, processes and outputs in the resource conversion process |
| 2 | Recognise specific human resource inputs | discussing the role of specific human resource inputs in the resource conversion process |
| 3 | Recognise specific tangible resource inputs | discussing the role of specific tangible resource inputs in the resource conversion process |
| 4 | Recognise specific intangible resource inputs | discussing the role of specific intangible resource inputs in the resource conversion process |
| 5 | Recognise organic growth | explaining how organic growth occurs |
| 6 | Recognise external growth | explaining how external growth occurs |
| 7 | Identify the type of organisation for which the simple structure is suitable | discussing the advantages and disadvantages of the simple structure |
| 8 | Identify the type of organisation for which the functional structure is suitable | discussing the advantages and disadvantages of the functional structure |
| 9 | Identify the type of organisation for which the divisional structure is suitable | discussing the advantages and disadvantages of the divisional structure |
| 10 | Identify the type of organisation for which the holding company structure is suitable | discussing the advantages and disadvantages of the holding company structure |
| 11 | Identify the type of organisation for which the matrix structure is suitable | discussing the advantages and disadvantages of the matrix structure |
| | **Application** | **Check you have achieved this by** |
| 1 | Apply the resource conversion chart to organisations generally | producing a resource conversion chart for a hypothetical organisation |
| 2 | Apply the resource conversion chart to a specific organisation | producing a resource conversion chart for a real organisation |
| 3 | Illustrate the application of generic organisational structures to a specific organisation | sketch the structure for a real organisation |
| | **Analysis** | **Check you have achieved this by** |
| 1 | Differentiate between different organisational structures | comparing and contrasting different organisational structures |
| 2 | Differentiate between different organisational cultures | comparing and contrasting different organisational cultures |
| | **Integration** | **Check you have achieved this by** |
| 1 | Create your own organisational structure for an organisation | proposing an argument in support of your organisational structure |
| | **Evaluation** | **Check you have achieved this by** |
| 1 | Choose an organisational structure for specific organisations | comparing and evaluating different possible structures |
| 2 | Choose which organisational culture dominates in a specific organisation | comparing and evaluating different possible cultures |

ENTRY CASE STUDY 4.1

FT

# Wm Morrison plans to grow in the south

**By Peggy Hollinger**

Wm Morrison, the Bradford-based supermarket group, is moving south in an expansion programme which it claims will create 2250 jobs next year.

Its first two superstores in the south of England would be opened in 1998 as part of a 'rapid expansion' £200m investment programme, which will also include three further stores in the north. The new jobs are in addition to 3200 in this year's opening programme.

News of the expansion programme came as the group announced a 7 per cent increase in pre-tax profits to £135.8m for the year to February 2, on sales 3.6 per cent higher at £2.1bn.

Mr Ken Morrison, chairman, said the company had achieved record sales although it had opened no stores in the period. Furthermore, 'these sales have been made while competition has remained fierce,' he said.

Current sales were running 1.6 per cent ahead of last year, excluding petrol sales. Supermarket like-for-like sales were 0.4 per cent ahead, with the strongest growth coming from non-food products.

Morrison reported improved margins, in spite of higher costs. This was partly due to an increase in discounts from suppliers due to Morrison achieving sales targets on certain products.

The final dividend is 1.375p, for a total 21 per cent higher at 1.7p, payable from earnings per share that rose from 10.67p to 11.01p.

*Source*: *Financial Times*, 21 March 1997. Reprinted with permission.

## Introduction

The purpose of this chapter is to explain the internal organisational elements of the organisational context model (*see* Exhibit 4.1). The chapter examines:

- the organisational resource conversion process;
- the legal restrictions on private-sector businesses in the UK;
- typical generic structures that organisations tend to adopt and their relationship to organisational size;
- issues of organisational culture;
- the four common functional areas of organisations.

Organisational weaknesses and strengths are dealt with in Chapter 10.

## Organisational resource conversion

One characteristic that all organisations share, irrespective of sector or activity, is the fact that they take in resources, do something to them, then deliver outputs with a higher value than the inputs had, as a direct result of the processes to which they were subjected by the organisation. This is organisational resource conversion, and all organisations engage in it to a greater or lesser extent. This is as true of public-sector organisations as it is of private-sector companies, and can be observed in the service sector as well as in manufacturing. A consumer appliance factory buys steel, passes it through its machines and the hands of its workers, and sends out refrigerators that it sells

**Exhibit 4.3 Organisational resource conversion chart**

| | Resource inputs | Processes | Outputs |
|---|---|---|---|
| **Human** | • Owners/ shareholders<br>• Managers<br>• Employees<br>• Part-timers<br>• Contractors | • Goal setting<br>• Decision making<br>• Planning products and services<br>• Managing functions, (including HRM)<br>• Assembling parts<br>• Manufacturing goods<br>• Dealing with customers | • Job satisfaction or dissatisfaction<br>• Salaries and wages<br>• Bonuses<br>• Satisfied or dissatisfied customers |
| **Tangibles** | • Money (loans, overdrafts, profits, private capital)<br>• Buildings<br>• Machines and equipment<br>• Raw materials<br>• Components<br>• Energy (gas, water, electricity)<br>• Market research data | • Assembly<br>• Manufacture<br>• Service delivery<br>• Supply<br>• Quality control<br>• Accounting<br>• Distribution<br>• Formal communication systems<br>• Formal information systems | • Products<br>• Services<br>• Waste materials<br>• Waste energy<br>• Effluent<br>• Profit or loss |
| **Intangibles** | • Systems<br>• Design<br>• Information<br>• Innovation | • Informal communication<br>• Culture<br>• Corporate memory<br>• Informal information flow | • Professionalism<br>• Happiness<br>• Image and reputation<br>• Innovation |

for more than the cost of the steel and labour. A university recruits school or college leavers, submits them to several years' teaching, learning and assessments, and passes out graduates who command higher salaries than they could have done without their higher education experience. Further examples are given later in this section.

In examining the resource conversion process, the organisation can be depicted (*see* Exhibit 4.3) as a chart containing a sequential list of inputs, conversion processes and consequent outputs. There are also three kinds of inputs, processes and outputs: human, tangible and intangible. These are discussed in detail later in this chapter.

A generic organisational resource conversion chart delineates those activities that all organisations share. The generic chart can be adapted and applied to individual organisations according to their particular context. Following this generic chart, the model is applied to four fictional organisational examples.

The chart for any specific organisation contains those elements that are generically applicable to any organisation, such as buildings, as well as elements that apply specifically to individual organisations' particular sector, activities or other contextual factors, e.g. a bakery needs flour and yeast as its raw materials in order to bake bread. Therefore, when using the organisational resource conversion chart to examine closely what any given organisation actually does, an exact model of its resource conversion process must contain both generic inputs, processes and outputs as well as the specific elements, which are those inputs, processes and outputs found only in the individual organisational context.

It must also be recognised that while inputs, processes and outputs can be shown separately in the conversion model, they should not be considered as independent of each other. In the same way that organisations cannot be considered as independent of their external contexts, clear transactional dependencies can be identified between the various cause-and-effect stages of the resource conversion process, i.e. there can be no outputs if there are no conversion processes, and the processes cannot operate without inputs. Initially, however, each stage of the process is considered separately here.

As previously mentioned, resource inputs can be grouped into three categories: human resources (the people in the organisation); tangible resources (e.g. machines and money); and intangible resources (e.g. information). Both finance and human resources are covered in detail in this book (Chapters 8 and 9 respectively) as internal functional departments of the organisation, each interacting independently and together with elements of the external organisational context. Here, however, we consider the money and people that these departments use as financial and human inputs to the organisational resource conversion process.

Human resource inputs are obviously key to organisational success, as without people there can be no organisational activity. Despite hierarchical definitions of role and responsibility within organisations, which differentiate employer and employee, manager and worker, and functionaries in different departments, people all make fundamental contributions to the organisation's activity through their efforts. These efforts must be as effective and efficient as possible in order for organisational operational costs to be minimised and outputs and profits maximised. Human resources, once input to the organisation, obviously contribute to the conversion processes themselves. The human resource inputs to the conversion process rely on people at all levels of the organisation, whatever their specific role or skills level. Without the right people in the right roles there is little or no possibility of the organisation being able to achieve its goals. Employers or managers have specific roles and responsibilities, which usually include recruiting and utilising the other human resources that the organisation acquires. Employees also have specified roles and responsibilities, which usually become more and more defined according to how low down the hierarchy the human resource input is made.

Organisations have to consider what human resource inputs they need to begin, maintain and develop their operations, and how many people are necessary in order to achieve a critical mass for operational efficiency and effectiveness. People in organisations then need to be managed on an ongoing basis in order to ensure that they have the relevant skills at the appropriate level to fulfil the responsibilities their roles require. The numbers of human resources input to the organisation then need to be monitored in order to respond to the changing needs of the operations, i.e. is it growing, shrinking or maintaining its size as a reaction to changes in the marketplace? Human resource management, then, is concerned with getting people into the organisation, making sure they can do the jobs they need to do while they are there, and planning for getting them out of the organisation when necessary. Human resource management as a functional area is considered in Chapter 9.

Perhaps the most important tangible resource input is money. As in the case of human resources, without money the organisation is also totally inoperable. Whether in the private or public sector, all organisations must raise money in order to survive. Financial inputs come from private sources such as capital investment, bank loans, budget allocations from head office or reallocation of retained profits. Public sources of financial inputs include grants or allocated budgets from different levels of the state, including local government, or from supra-national bodies such as the European Union. Financial inputs contribute to the organisational resource conversion process in three main ways.

First, financial resources fund the acquisition of all other inputs. Money is needed to fund the human resources via their salaries or wages. How much the organisation can afford to pay its employees affects its ability to recruit and retain the people it needs to accomplish its tasks. Motivation is also directly linked to the amount of financial resource devoted to rewarding the efforts of its human resources. In the case of a manufacturing or assembly operation, money purchases the raw materials or component parts necessary to be able to manufacture the organisation's portfolio of products. There is also the machinery that assists in the manufacturing, any spare parts that might be needed to repair or maintain the machines and the energy needed to run these machines, all of which could be classed as inputs to the resource conversion process. In the service sector, money still funds all other inputs in terms of the service design, physical location of the service delivery and paying for the operation of customer service departments. It also funds the purchasing of external market information, which informs the internal operations of the organisation.

Secondly, financial resources fund the conversion processes themselves, not only by having paid for the inputs the organisation needs, but also by providing the money the conversion processes themselves need. A factory that has purchased its machines needs a healthy cashflow to be able to run them on a daily basis, pay weekly wages or monthly salaries and provide the money to fund all aspects of operations management, marketing, accounting and quality control that enables it to function. In the public-sector example of a hospital, money must be found to fund the daily activities of routine surgery, accident and emergency departments, organ transplants and the space for patients to recuperate.

Finally, financial resources are needed in the output stage, in order to fund the ultimate delivery and distribution of the organisation's services or products to its customers in the marketplace, at the appropriate time or location. There is also advertising of available products and services and dissemination of information to develop or maintain the organisation's external image, which all depend on having sufficient money to do them effectively. Even the disposal of waste products has a financial implication for the organisation.

The next tangible resources to be considered here are raw materials or component parts. These are obviously necessary for organisations to be able to make the final products or to provide the services they offer. Key issues when considering raw materials or component parts are their cost and quality, as organisations seek to maximise their own profits by minimising the cost of their inputs, without compromising the quality of their product or service through using inferior materials purchased cheaply.

Organisations also all need premises or buildings, and must have sufficient equipment in those premises for manufacturing, communications and health and safety requirements. There are issues here in relation to size of premises, their location – are they near to population centres, motorway networks, sea ports – and their appropriateness for the required function – are there costs of conversion or is it a purpose-built greenfield site?

Slightly less obvious tangible inputs are the resources purchased from the utility companies, e.g. water (including sewage and effluent treatment), gas and electricity, as well as local authority services such as transport infrastructure and refuse collection. These are inevitably key, as organisations cannot function without the input of energy to fuel their machines and light and heat their offices. In addition, roads providing access to the organisation for cars and lorries, convenient bus routes, bus stops and other connections for public transport as well as the regular collection and disposal of its waste are vital to ensure the operation can function efficiently and effectively.

Finally, there are intangible inputs to organisations, which consist of information and design. Information is the essential tool enabling the organisation to understand, analyse and react to activities in its external environment. The key information inputs that organisations use for this purpose concern the marketplace and competitors. In terms of market information, the organisation needs to know how big the market is, what potential customers are likely to need or want, and how it could provide products or services that meet those needs. In terms of competitor information, the organisation needs to monitor what products or services competitor organisations provide, how much are they charging for these and what share of the market they currently serve. Marketing is considered in more detail in Chapter 6.

Information about the external political and economic environment is equally key for organisations, especially at times of impending significant change such as a general election or national budget, as such information must be used to attempt to predict or anticipate changes in the external environment, hence allowing the organisation to adapt to anything that is likely to have a major effect on it. How well this resource input of information is gathered, analysed and disseminated within the organisation – sending the right information to the right department at the right time – is both a symptom and a cause of operational effectiveness. A good information flow demonstrates that an organisation understands and analyses the external environment and ensures that it continues to be able to do so.

The design of an organisation's systems and processes can also be considered as an input to its resource conversion process. The efficient and effective design of the organisation, from its internal communication systems, its accounting procedures, the factory layout or the positioning and decoration of its reception area, are all manifestations of its interaction with the external environment. The particular conversion processes that add value to the resource inputs vary according to the organisation and the business it is in. Perhaps the most easily understandable resource conversion process is manufacturing, where raw materials such as steel are turned physically into products such as automobiles. Some materials may be subassembled elsewhere, and these components may later be assembled by another organisation as part of its resource conversion process. In the service sector, planning and executing the delivery of a service would also count as a resource conversion process, as this is also concerned with adding value to inputs in order to produce outputs.

In order to contextualise the resource conversion process, four different fictional contexts are given below, showing organisational resource conversion charts for public- and private-sector contexts, in manufacturing and services, and at varying levels of the external context.

**Little Mester Ltd, surgical steel instrument maker**

The case of a self-employed steel instrument maker (or 'Little Mester') in Sheffield shows a very different resource conversion process chart from the generic model. The context level is largely local.

**Exhibit 4.4 Organisational resource conversion chart: Little Mester Ltd, surgical steel instrument maker**

| | Resource inputs | Processes | Outputs |
|---|---|---|---|
| **Human** | • The owner/manufacturer ('Little Mester')<br>• Suppliers | • All aspects of the business, including planning, manufacturing and selling done by the owner<br>• Providing raw metal and small components | • Job satisfaction<br>• Development of the toolmaker's craft<br>• Wages and profits for the business<br>• Satisfied customers |
| **Tangibles** | • Loans, mortgages, overdrafts, sales income, profits<br>• Workshop and home<br>• Machinery<br>• Raw metals<br>• Component parts<br>• Energy | • Supplying metal and components<br>• Manufacture and assembly of tools<br>• Quality control (probably informal – throwing away mistakes)<br>• Accounting<br>• Delivering products to hospitals<br>• Letters and adverts | • Surgical tools<br>• Waste metal (maybe re-input as recycled raw metal)<br>• Profit or losses |
| **Intangibles** | • Very informal systems for monitoring customer needs and reacting to them | • Everything is in the owner's head and hands | • Professionalism<br>• Happiness<br>• Image and reputation<br>• Specialisation |

**Natpower Ltd, electricity generator**

The case of 'Natpower' concerns a private-sector service provider, in this case an electricity generator, in a predominantly national context. Its particular resource conversion chart also contains generically applicable elements as well as context-specific items.

**Exhibit 4.5 Organisational resource conversion chart: Natpower Ltd, electricity generator**

| | Resource inputs | Processes | Outputs |
|---|---|---|---|
| **Human** | • Owners/ shareholders<br>• Managers<br>• Full- and part-time workers in electricity generation<br>• Contractors supplying services (e.g. repairs) | • Goal setting<br>• Decision making<br>• Planning for meeting electricity needs<br>• Managing functions<br>• Generating electricity from gas or coal<br>• Dealing with electricity supply companies | • Job satisfaction or dissatisfaction<br>• Salaries and wages<br>• Bonuses<br>• Satisfied or dissatisfied customers |
| **Tangibles** | • Income, profits<br>• Electricity generation plants<br>• Generators and the supply grid<br>• Coal or gas to convert and to power the plant<br>• Data to plan electricity needs<br>• Additional energy from overseas<br>• Legal regulation | • Generating electricity<br>• Quality control<br>• Accounting<br>• Supplying electricity via the grid to electricity supply companies<br>• Formal internal communication systems<br>• Formal internal information systems | • Electricity<br>• Waste (pollution) |
| **Intangibles** | • Systems<br>• Design<br>• Information<br>• Innovation | • Informal communication<br>• Culture<br>• Corporate memory<br>• Informal information flow | • Professionalism<br>• Happiness<br>• Image and reputation<br>• Innovation<br>• Safety |

**China National State Steel Corporation**

Although economic reforms in the late 1980s and early 1990s have led to the establishment of both a new private sector and newly privatised industries in China, much of the large-scale means of production remain under state control. Thus the example of the China National State Steel Corporation allows resource conversion at the global level of the external environment to be examined in the context of a public-sector manufacturing environment. Again, there are generic and specific elements in its resource conversion process chart. A huge, state-owned steel corporation in a country governed by a Communist Party has responsibilities far beyond those of private companies in capitalist countries. The organisation not only has to fulfil the demands of the government's plan for production, but also has to provide for all the social welfare needs of its employees and their families, including healthcare, schooling, daily provisions and housing. The environment level would be national, although there are global implications for competition and the international political arena.

**Exhibit 4.6 Organisational resource conversion chart: China National State Steel Corporation**

| | Resource inputs | Processes | Outputs |
|---|---|---|---|
| **Human** | • Managers<br>• Employees<br>• Families | • Liaising with government departments regarding the five-year economic plan<br>• Operational decision making<br>• Planning products<br>• Managing functions<br>• Manufacturing steel and goods<br>• Dealing with other organisations which use the steel | • Meeting the needs of employees and their families in terms of wages and social welfare needs |
| **Tangibles** | • State funding, sales income<br>• Steel plants all over China<br>• Machines and equipment<br>• Raw materials<br>• Components<br>• Energy (generated on site)<br>• Five-year plan | • Assembly<br>• Manufacture<br>• Supply<br>• Quality control<br>• Budget management<br>• Distribution<br>• Chinese Communist Party (CCP) organisation<br>• Formal communication systems, including propaganda radio<br>• Formal information systems (CCP) | • Fulfilling the five-year plan<br>• Products<br>• Social services (housing, hospitals, schools, shops)<br>• Waste materials<br>• Waste energy<br>• Effluent and pollution<br>• Profit or loss |
| **Intangibles** | • Planning, manufacturing and control systems<br>• Product design<br>• State and Party information | • Informal communication between workers and families<br>• Culture<br>• Corporate memory<br>• Informal information flows | • Basic standard of living for all<br>• Party control of workers and families |

**Superbuys**

A large, privately owned supermarket's resource conversion chart would, again, be fairly standard. The context would be global and national: global since many products are imported and there are branches overseas; and national as its largest customer base is in the UK.

## The legal structures of business

**The sole trader**

A sole trader business is owned and administered by one person who is personally liable for all the debts of the business, this being the primary disadvantage of this type of legal structure.[1] The sole trader is the owner of the business, having raised the capital for it from personal funds. The sole trader has direct control over the business and takes all the decisions relating to products and services, customers and markets, staff employed and future development. The principal advantage of the sole trader legal structure is that all profits appertain to the owner.[2]

**Exhibit 4.7 Organisational resource conversion chart: Superbuys**

| | Resource inputs | Processes | Outputs |
|---|---|---|---|
| **Human** | • Owning family<br>• Managers<br>• Shop workers<br>• Warehouse workers<br>• Truck drivers<br>• Suppliers of branded and own-brand products | • Goal setting<br>• Decision making (e.g. loyalty cards, banking)<br>• Planning products, services and new stores<br>• Managing functions<br>• Manufacturing own-brand goods<br>• Dealing with customers | • Job satisfaction or dissatisfaction<br>• Salaries and wages<br>• Bonuses<br>• Satisfied customers |
| **Tangibles** | • Capital, investment<br>• Stores, warehouses<br>• Equipment (checkouts, computers, shelving, lighting)<br>• Imports and domestic goods<br>• Energy<br>• Market research data | • Manufacture<br>• Shelf stacking<br>• Supply and distribution<br>• Quality control<br>• Accounting<br>• Formal communication systems<br>• Formal information systems<br>• Enhanced services (loyalty cards, banking, cafés, petrol stations) | • Products<br>• Services<br>• Waste materials<br>• Waste energy<br>• Effluent<br>• Profit or loss |
| **Intangibles** | • Systems<br>• Design<br>• Information<br>• Innovation | • Informal communication<br>• Culture<br>• Corporate memory<br>• Informal information flow | • Professionalism<br>• Happiness<br>• Image and reputation<br>• Innovation |

## Partnership

A partnership is where two or more individuals join together in a business venture with each partner having unlimited liability.[3] Therefore, should the partnership result in debt or bankruptcy, each partner can be held liable for the full amount owed by the partnership. Creditors can either sue each partner in turn or all of the partners jointly for as many of the partners' assets as will pay off their debts. The other option is for the partnership to have limited liability, but limited partnerships are uncommon because only some of the partners can have limited liability, so therefore limited liability is usually established by creating a limited company instead of a partnership.[4]

## The limited company

A limited company exists as a legal entity in itself, separately from its owners or managers. Therefore, liability for debts is limited to the amount of issued share capital, whether the shares have been sold or not.[5] The shareholders' personal assets cannot be claimed for the payment of business debts to creditors. The creation of a limited company requires the lodging of various documents with the Registrar of Companies, which must include a Memorandum of Association and Articles of Association. If all documentation is satisfactory, then a certificate of incorporation is issued, bringing the company into existence as a legal entity.[6]

## Organisational structures

The structures that organisations adopt are usually aligned to one of the five generic organisational structures. These are: the simple structure; the functional structure; the divisional structure; the holding company structure; and the matrix structure.

The simple and functional organisational structures are centralised structures where all the important and long-term decisions are taken by top management. Top management will determine the rules and procedures that closely govern and direct the jobs and tasks of managers further down the organisation, who are responsible for the departments, products/services and markets on a day-to-day basis.

The divisional, holding and matrix organisational structures are decentralised, containing divisions/subsidiaries/project teams that have a significant amount of decision-making power and responsibility of their own. Co-operation and co-ordination between the divisions/subsidiaries/project teams and the board of directors are crucial if the spreading of power and responsibility throughout the structure is to work for the organisation as a whole.

### The simple structure

A company that adopts a simple structure (*see* Exhibit 4.8) is likely to be a small business in the private sector or one in the very early stages of its growth and development. The simple structure is centralised, with all short-, medium- and long-term power and decision-making responsibility resting with the managing director, who is also likely to be the owner of the business. The managing director/owner controls and oversees all aspects of the company's operations. Therefore, the simple structure is suitable for a small business in the early stages of growth and development, allowing the managing director/owner to have control over the future growth and development of the business, as s/he has a financial stake in the business, along with expertise relating to the product or service sold by the business and the markets to which it sells.

The simple structure becomes less suitable as the business grows in size. The managing director/owner finds it more and more difficult to control and oversee the greater number of tasks and activities undertaken by a larger and growing business. However, the likelihood also exists that the managing director/owner has the skills, knowledge and abilities to run a small business, but may be lacking some of those necessary to run a larger and growing business. This situation usually requires the business to restructure if it is to

**Exhibit 4.8 The simple structure**

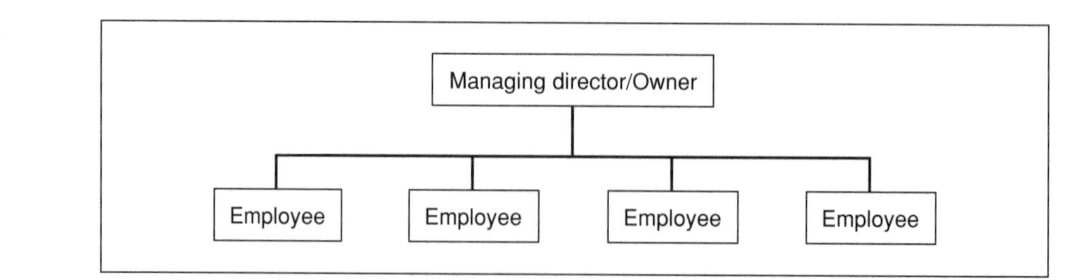

survive its growth and increase further in size. Simple structure businesses that grow in size commonly develop functional structures.

**The functional structure**

The functional structure (*see* Exhibit 4.9) is rigid and centralised with efficient management control systems, and is common both in companies that have outgrown the simple structure and in well-established public-sector organisations. Such organisations are medium sized and have a limited range of related products and services delivered to clearly defined and clearly segmented markets.

**Exhibit 4.9 The functional structure**

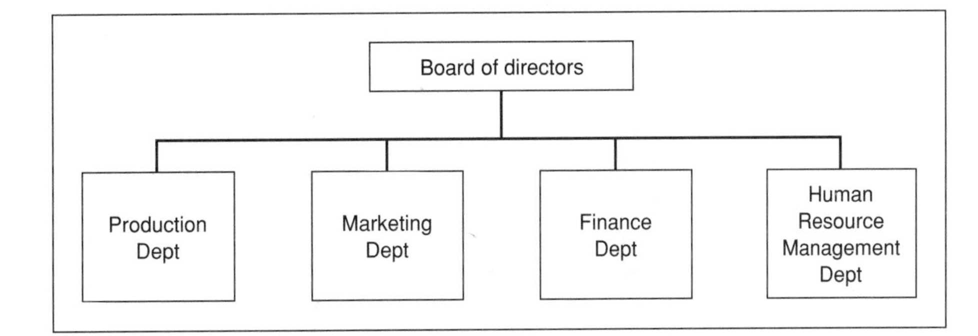


The functional structure also sees the introduction of specialist functional managers who head the different departments, e.g. marketing manager, operations manager, finance manager and human resource manager. In the case of growing private companies, these new managers provide the specialist skills, knowledge and abilities that may have been missing under the simple structure.

The managing director or board of directors will be in close contact with the new departmental heads who run the different departments on a day-to-day basis, hence the lines of communication and information flow within the functional structure are short and vertical. The structuring of the organisation around the different functions or tasks that have to be carried out by its employees results in job roles that are clearly defined and understood by everyone in the organisation. Short-term decision-making power and responsibility tend to rest with the departmental heads, who have to work together with the board of directors to ensure that what is happening at an operational level also reflects and feeds back into the long-term and medium-term decision-making process. Long- and medium-term decision-making power and responsibility, however, rest very much with the board of directors.

One of the reasons that organisations move on from the functional structure is that the organisation starts to diversify its product or service range and no longer has a limited range of related products or services for a clearly defined and clearly segmented market. An organisation that is outgrowing the functional structure is likely to have developed a wider range of products or services that are not so closely related and that sell to more diverse markets. The growth that has led to the need to restructure could have resulted from selling to markets that are more geographically diverse than was the case previously, e.g. the organisation has engaged in international business where before it dealt purely domestically. Another cause of growth could be selling to a wider customer base with more varied needs and wants than the traditional market segment.

When this type of situation arises, the functional structure becomes very stretched and cannot cope well with the increased diversity. The production and marketing managers now have to deal with a diverse range of products and services in a diverse marketplace, but in the functional structure there is only one marketing department and one production department to do everything. They are used to dealing with a limited range of closely related products or services and do not necessarily have the resources to service wider ranges or market segments. Thus, in essence, the functional structure has to change in order to cope with growth and diversification. It is the centralised and rigid nature of the functional structure that prevents it from adopting the decentralised and flexible practices needed to deal with more diverse markets and product ranges.

The rigid and centralised nature of the functional structure may also make difficult to operate activities requiring cross-functional teams. Since the organisation is structured in the form of vertical and hierarchical functions, its different departments have clear views on their operational responsibilities and find it difficult to act outside their perceived remit. A classic example of the need for cross-functional teams is product development, which requires inputs from the marketing, operations management and finance departments in order to research what customers want, develop products or services accordingly, cost their delivery and determine their price. This may be difficult to make work within the confines of a vertical and hierarchical functional structure, with its clearly defined and understood job roles.

## The divisional structure

A company that has adopted a divisional structure (*see* Exhibit 4.10) contains separate divisions based around individual product lines or services, e.g. a motor manufacturer may have a car division, a truck and van division, and a passenger service vehicle (bus) division. Alternatively, divisions can be based on the geographic areas of the markets served, e.g. Europe, Asia and North America. One organisation may contain some divisions based on product or service lines and others geographically allocated, e.g. cars, buses, trucks and overseas sales. However, one single division would not usually be based on both product and location, so it is rare to find divisions called 'cars North America' or 'buses Asia'. The divisional structure is decentralised and as such a company with a divisional structure usually offers a wide and diverse range of products and services compared to a company operating with the more rigid and centralised functional structure. The key benefit of a wide and diverse portfolio is the ability to spread either the general profitability of each division or the burden of the poor performance of one of the divisions by subsidising it from the profits of the others.

Management of this diversity requires the divisional heads or managers to have short-term and medium-term decision-making power and responsibility

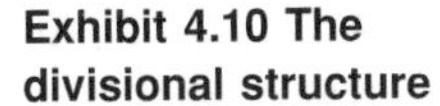

**Exhibit 4.10 The divisional structure**

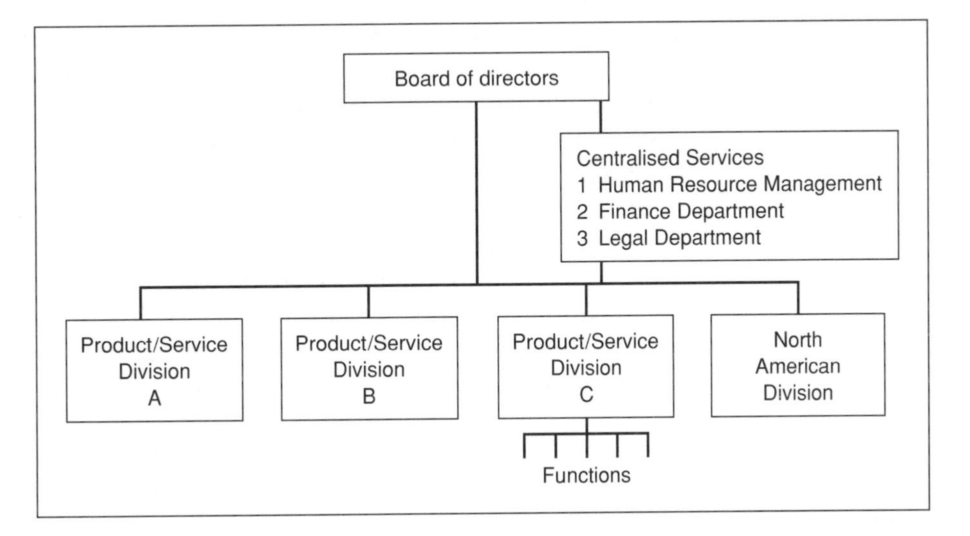


for the division they manage. This allows the managing director and his/her board of directors to concentrate on long-term planning for the organisation as a whole. However, good communication and working relationships must exist between the divisional heads or managers and the board of directors, as short-, medium- and long-term decision making must all relate to each other and link together if the whole company is to move forward in the same coherent direction.

The divisions in a divisionalised company will be cost centres or profit centres in their own right, having to manage budgets, meet budgetary targets and operate within budgetary constraints. Divisions must also satisfy performance criteria relating to profitability and asset use, with profit margins and return on assets likely to be the key measures applied to individual divisions. The company will aggregate the financial information on each division's performance to produce the overall annual company report and accounts. The separate divisions within the same organisation have an internal structure of their own. A common structure for individual divisions to adopt is the functional structure, because it is suitable for the size and range of activities of a discrete division under the larger divisionalised organisation.

The potential difficulties with the divisional structure relate to the allocation of resources, overall co-ordination of activities and the cost of running separate divisions. The existence of an element of competition between separate divisions and a very limited resource base may lead to conflict between the divisions, as each separate division vies for the best possible allocation of resources and wants to perform well in the eyes of head office. The diverse range of divisions in a company may also make company-wide co-ordination demanding. This may be exacerbated if there is duplication of key activities in each division, which makes the cost of running individual divisions high. For example, the presence of an HRM function in each division would be expensive duplication. The activities associated with the HRM function are equally applicable to all people who work for the company regardless of the division in which they work and, as such, centralisation of HRM activities makes economic and practical sense. With a centralised HRM function for all divisions, one HRM policy for the whole company can be developed and maintained, and every central and divisional employee recruited and measured against the central HRM policy. A different HRM function in each division risks resulting not only in wasteful duplication of effort, but also in different HR practices being adopted, thus diluting corporate culture and corporate control. Other centralised services could include the finance department and there may also be a discrete legal department to interact specifically with the various politico-legal issues at the various levels of the external context.

**The holding company structure**

The holding company structure (*see* Exhibit 4.11) is usually found in large industrial conglomerates with a parent company acting mainly as an investment company acquiring and divesting smaller subsidiary companies. A company operating as a holding company will usually have a small corporate headquarters from which the parent company will conduct business. This means that central overheads will be low because of the economies of scale that this company-wide co-ordination achieves. The finance and legal sections are part of the parent company and their purpose is to provide the expertise needed centrally in the acquisition and divestment of subsidiary companies.

The subsidiary companies continue to trade under their own names, with the parent company either wholly owning its subsidiaries or acting as a majority shareholder in them. Subsidiary companies will operate fairly independently of the head office, with all decision-making power and responsibility for their own performance resting with their management. Industrial conglomerates adopting the holding company structure are therefore very decentralised. However, the control systems implemented by head office will tend to centre on the subsidiary companies meeting tight financial targets with regard to profit forecasts, profit margins and return on assets, or risking swift divestment by the holding company.

**Exhibit 4.11 The holding company structure**

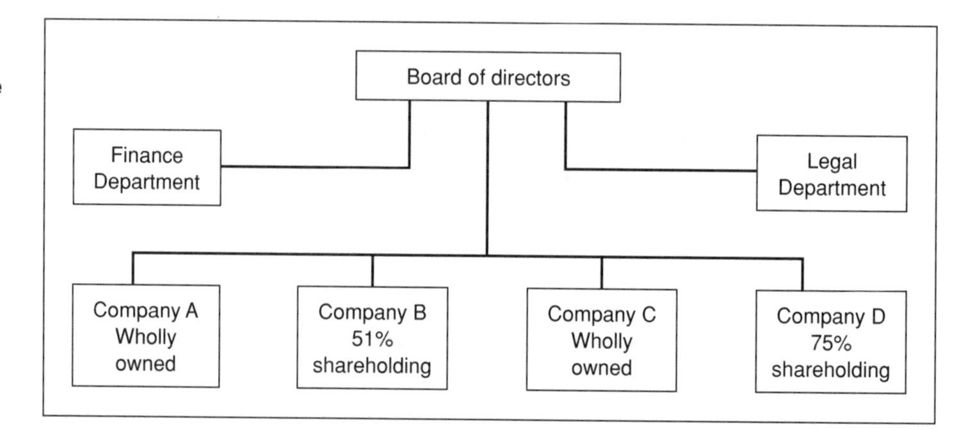

The ownership of a large number of subsidiary companies in a variety of different industries spreads the risk and profit for the parent company as a whole. The use of subsidiary companies to obtain diversity may ease divestment, especially in the light of poor performance, as that performance can be viewed as being ring-fenced in one or more companies and therefore contained.

The main potential disadvantage of the holding company structure relates to the subsidiary companies, which may view themselves as continuously up for sale. This type of situation invokes a high degree of uncertainty and the likelihood of change can be difficult for the subsidiary companies to manage on a permanent basis. The other potential disadvantage relates to the general fact that diversity is more difficult to co-ordinate and manage overall than simplicity, and so the holding company management faces a more complex task.

Hanson plc was a diversified conglomerate built up over 30 years by James Hanson and Gordon White. It acquired a variety of subsidiary companies over the years covering the areas of chemicals, tobacco, energy, building materials and building equipment. Given the variety of its acquisitions, Hanson became a holding company. Following the under-performance of the group in the early 1990s, when it lost 35 per cent of its stock-market value, the decision was taken in 1996[7] to demerge the Hanson group. The demerger resulted in four separately listed companies: Hanson, Imperial Tobacco, Millennium Chemicals and US Industries. Each has a turnover of more than £2bn and shareholders received shares in each of the resulting four companies[8] after the demerger.

## The matrix structure

The matrix structure attempts to merge the benefits of decentralisation with co-ordination across all areas of the business. Matrix structures are often used in organisations where there are two distinct and important areas of operation needing to be managed and co-ordinated in order to deliver the full product or service range. Matrix structures are often found in large multinational companies, educational establishments and small sophisticated service companies.

In a large multinational company (*see* Exhibit 4.12), the two arms of the matrix structure represent the product or services areas and the geographic areas in which the company operates. The product or service arm is responsible for the production of the product or delivery of the service, while each geographic arm of the matrix is responsible for the advertising, marketing, sales and distribution of those products or services to the end users in the geographic area for which they are responsible. The geographic arm becomes the customer of the product or service arm, as they purchase the product or service they need from them before selling these on to their geographically defined customer base.

The matrix structure in a university business school (*see* Exhibit 4.13) also has two arms, one responsible for delivering the products, higher education courses, to their internal customer, the other arm of the matrix. This second arm manages course administration and is therefore responsible for delivering the product on to the end users, the students. The internal customers, the course administrators, can be organised in terms of the type of external customer they

**Exhibit 4.12 The matrix structure: multinational company**

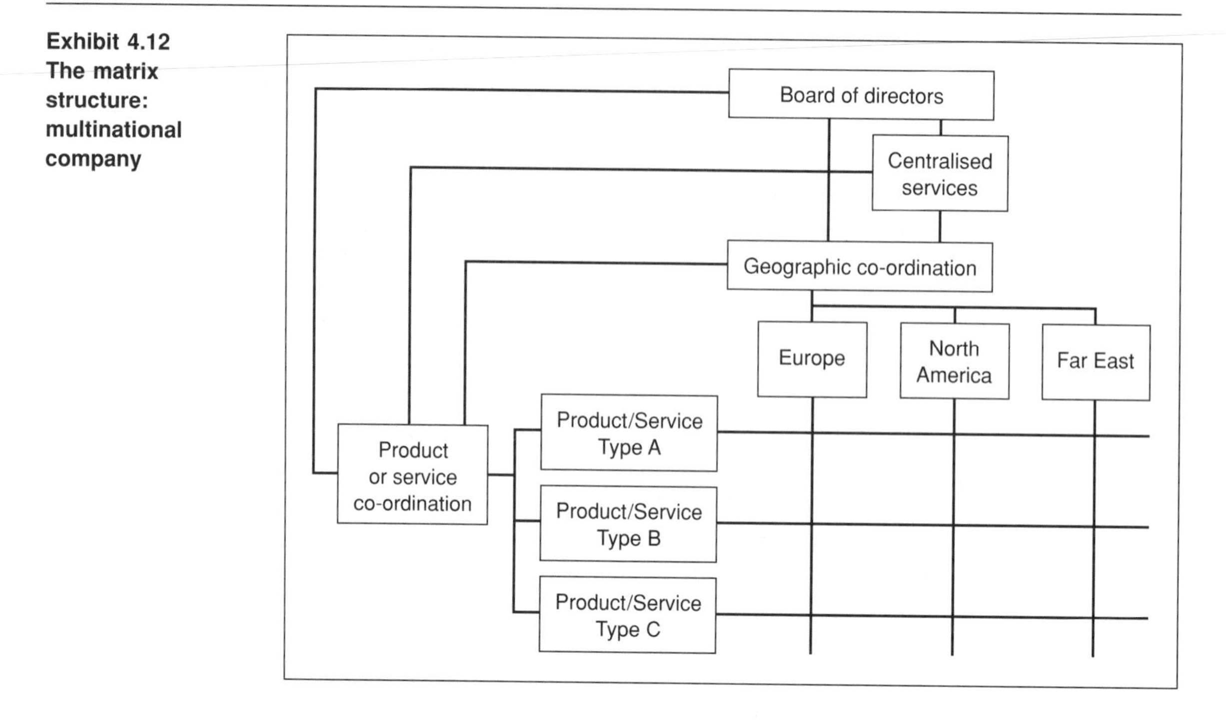


serve: full-time or part-time students; undergraduates or postgraduates; funded or fee-paying students. Whatever their provenance, each student grouping is enrolled on one of a number of courses or products, e.g. BA (Hons) Business Studies, HND Public Policy and Management, Master of Business Administration or MSc in Human Resource Management. Within the matrix structure, in order for this arm of the matrix to be able to run its courses and satisfy its external customers, it needs the services of the other arm of the matrix, which represents the staff, usually organised in groups that reflect their academic expertise and the subject they teach. This arm of the matrix is responsible for supplying tutors to classes and for all staff development issues to ensure that individual tutors are competent enough in terms of academic expertise and teaching methodology to deliver the services their internal customers require of them. Hence communication and co-ordination between the two arms of the matrix should centre on subject groups and the course leaders reaching agreement over who will teach which subject to which classes on which courses, teaching being the main activity occurring in the matrix. The direct contact between people from both arms of the matrix allows decisions to be made by the staff at the sharp end with direct responsibility for running courses and teaching students, which should avoid hierarchical bureaucracy. Decentralisation of responsibility for decision making to people from both arms of the matrix structure should increase the motivation of the staff involved, provided that job tasks and responsibilities are clear. A lack of clarity in people's roles, responsibilities and accountability is a potential disadvantage of the matrix structure.

**Exhibit 4.13 The matrix structure: educational establishment – university business school**

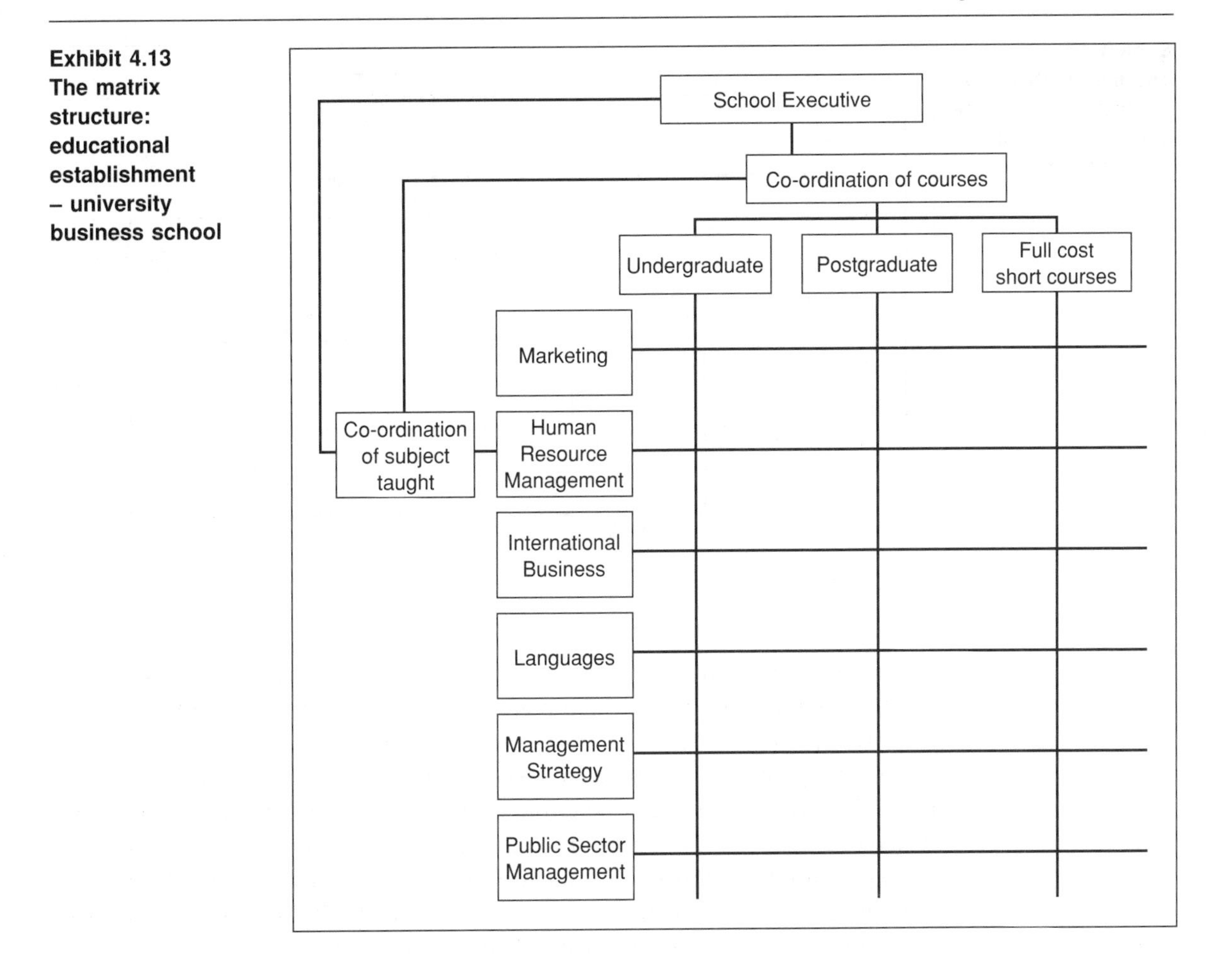


The matrix structure for a small sophisticated service company will contain professional expertise groupings on one arm of the matrix and may contain geographic groupings on the other arm. A design consultancy (*see* Exhibit 4.14) would operate by having offices in various different regions of Britain to handle the accounts and initial enquiries from clients, as well as having design staff working out of that office. The nature of the project from the client would determine if it could be handled by a team of designers available from the local design staff or if a team of designers from across the company's different offices would be required.

Bringing together a team to work on a specific project and then disbanding it once the project is complete, only to bring together another team for the next project, is a key feature of the matrix structure and is able to occur due its decentralisation and flexibility.

Whatever the context, whether multinational, private sector, public sector, manufacturing or service, the success of the matrix structure depends heavily on communication and co-ordination between the two arms of the matrix for any one product or service in any one market. This requires teams consisting

**Exhibit 4.14**
**The matrix structure: sophisticated service company – design consultancy**

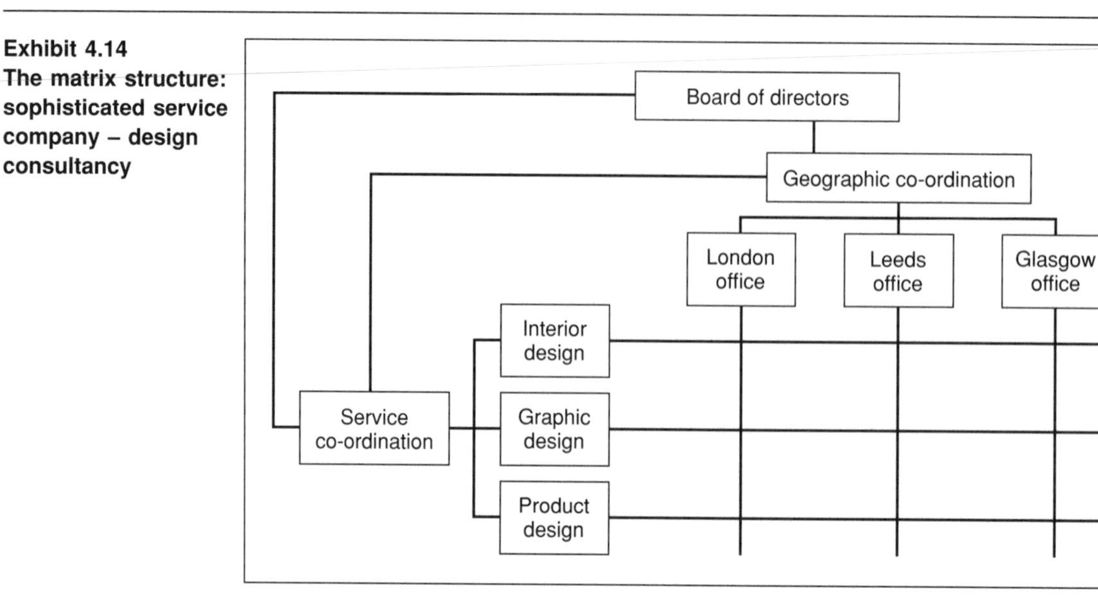


of people from both arms of the matrix to work together well at the points where they meet in the matrix and engage in their internal customer relationship. The quality of the end user or external customer experience depends heavily on the quality of this internal customer relationship between arms of the matrix. Therefore, the main potential difficulty with implementing the matrix structure occurs when people from both arms of the matrix fail to work together and co-ordination and communication between the two breaks down, thus impeding decision-making processes and adversely affecting the experience of the external customer or end user of the product or service.

## Changing organisational structure and size

Organisations are likely to change their structure over time, either by planned and deliberate restructuring or through emergent growth. That growth may result from the investment of cash, e.g. from a bank loan, from a share or rights issue, or from the reinvestment of profits in order to expand the business in some way. Expansion may occur via organic or external growth. Organic growth occurs when money is invested into the direct expansion of the business from scratch, e.g. buying and fitting a number of new high-street outlets, building new manufacturing premises or expanding the range of products or services better to serve current markets or to take advantage of new market opportunities. The main supermarket chains in Britain, Sainsbury, Tesco and Safeway, are examples of companies that expanded via organic growth throughout the 1980s and early 1990s with rapid building of new superstores.[9]

External growth of a business may take a variety of forms, all of which involve association with another business, such as acquisition, merger or joint venture.

Acquisition occurs when one company agrees to purchase another. The hostile acquisition of one company by another is often referred to as a takeover. Another type of agreement is a merger, where two companies agree to join together to form one company, for example Royal Insurance and Sun Alliance merged to form Royal Sun Alliance.

Problems with acquisitions arise if the price paid for the acquisition was higher than market value and/or if the expected future payoffs and benefits never materialise. Further difficulties include problems with selling any unwanted parts of the post-acquisition company and problems with integrating two different corporate cultures. However, acquisitions and mergers can engender benefits, including rapid increases in market share and presence, company size, and available expertise.

The opposite of acquisition is divestment and the opposite of merger is demerger. Divestment occurs when a company closes down a section of its operations, sells off the assets and makes the staff redundant, or a section of the business is allowed to continue operating but is sold to another company. Demerger occurs when a company decides to split into two or more companies. This has increased in popularity in recent years, with City institutions and investors preferring to invest in companies with a portfolio of activities that are clearly connected and focused. Therefore, a number of industrial companies[10] have demerged in recent years, including Courtaulds (Courtaulds and Courtauld Textiles), Racal (Racal Electronic and Vodaphone) and ICI (ICI and Zeneca). Zeneca assumed responsibility for the high value-added products such as pharmaceuticals, agrochemicals, seeds and speciality chemicals and ICI was to concentrate on the more mature markets of commodity chemicals, paints and explosives. The announcement of ICI's demerger in July 1992 saw the company's share price rise by 76p to £11.71.[11] In early 1997 Zeneca was valued at £16.3bn, three times the value of its former parent company. The ICI commodity chemicals company underwent further restructuring in 1997.[12]

Strategic alliances are contractual agreements between two companies, sometimes involving each company having a minority shareholding in the other. They do not usually result in the formation of a separate company.[13] The benefits of strategic alliances include the lower expense involved than for acquisition, each company keeping its own identity, access to each other's market knowledge, and their usefulness if acquisitions are impractical, i.e. government forbids a particular acquisition on legal and/or competitive grounds.[14]

A joint venture is a closer form of contractual agreement than a strategic alliance and results in the formation of a company whose shares are owned equally by the parent companies.[15] The benefits accruing for joint ventures are the same as those for strategic alliances. However, because of the structural ties that a joint venture involves, they often benefit from more motivated managers and the necessarily more intimate relationship between joint venture partners. These all contribute to the joint venture's success. Joint ventures can also be a more effective way of locking competitors out of the marketplace through enhanced collaboration.[16]

The potential disadvantages of joint ventures and strategic alliances are lack of control, limited motivation and differences in management style for any of the parties involved in either joint ventures or strategic alliances.

## Organisational culture and behaviour

The study of culture is always a fascinating but complex topic. It is important, since organisations are necessarily filled with people who bring with them the culture they have acquired in society. Culture exists in both the external and internal contexts. The external context of culture is dealt with in Chapter 2, as the sociocultural context or the 'S' of LoNGPEST analysis in the model, and in Chapter 5. Here we look at internal organisational culture and examine the implications of the 'chicken and egg' link with organisational structure. There are also clear links between organisational culture and efficient and effective operation of the four functional areas within the organisation. At all stages of the study of organisational context, the links between culture in society and culture in organisations should be observed so that conclusions can be drawn for the management of the people who work within the organisation.

In order to examine organisational culture in a specific context, we will use the generic models developed by Handy[17] from his observation of and research in organisations, which led to the isolation of four essential internal cultural models.

### The power culture

Small, entrepreneurial organisations, where the owner works with few employees, are likely to exhibit the power culture. This organisation is a club of intimates, where the colleagues or employees have been chosen by the owner/manager for their similarity to him-/herself. The centre of power, and all crucial decision making, is the owner/manager, who either is in personal charge of every aspect of the work or can trust colleagues and employees to do things instinctively the way the owner/manager would have done him-/herself. This is depicted by the model (*see* Exhibit 4.15), which resembles a spider at the centre of its web.

The formal structure that most closely echoes the power culture is the simple structure, discussed earlier in this chapter.

The choice of 'cloned' employees can be deliberate or subconscious on the part of the entrepreneur. Whether explicitly sought or not, an internal culture develops that is intimate and comfortable for those on the inside. This creates further issues for anyone different who tries to enter the organisation, and this is possibly the culture that is least open to equal opportunities issues, as the club members do not wish to admit new members not in their likeness. The power culture is exciting because of the risks involved in its operation, as colleagues at the centre of power make decisions in an unauthorised but implicitly supportive environment. They operate the way they think the

**Exhibit 4.15**
**The power culture**

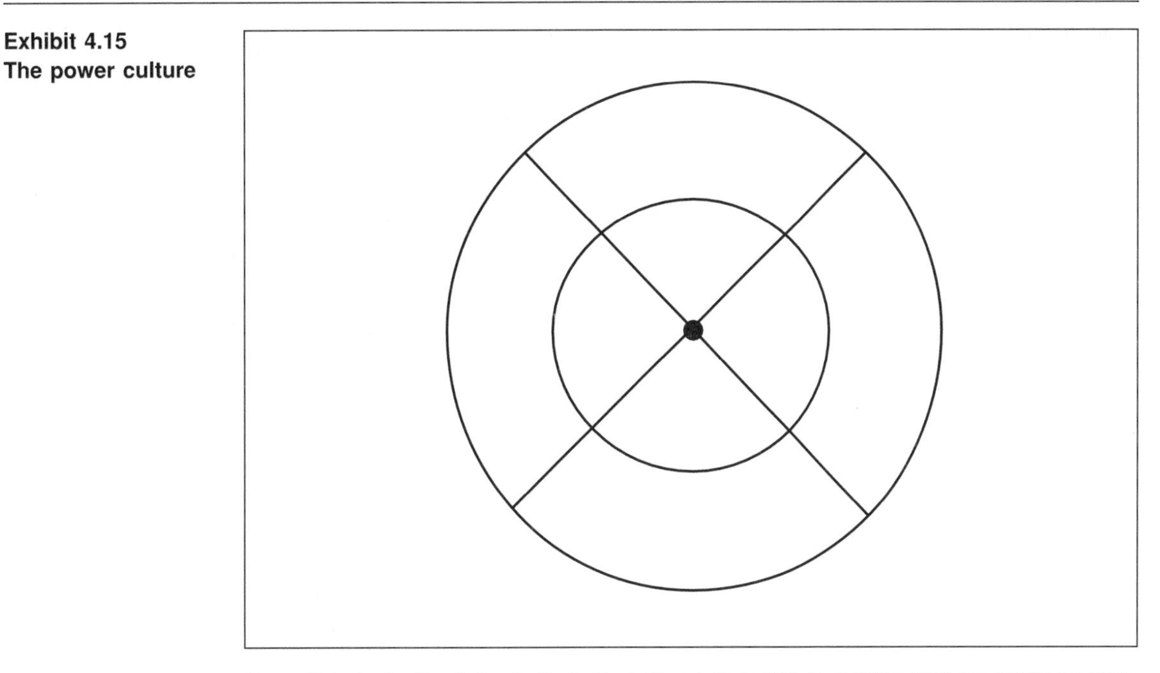

*Source*: *Understanding Organizations* by Charles Handy (Penguin Books 1976, Fourth Edition 1993). Copyright Charles Handy 1976, 1981, 1985, 1993. Reproduced by kind permission of Penguin Books Ltd.

entrepreneur should, are rewarded and congratulated when they are correct, but risk censure when they inadvertently make a mistake.

Without the leader, the power centre is lost and the club can break down. Should the leader become ill or die, the organisation grieves and can only recover from its loss with difficulty. This illustrates a danger of the club culture, in that the organisation becomes too reliant on the originator and entrepreneur who is all too literally the heart and soul of the organisation.

When the organisation grows, the club becomes too big for all the members to retain their former intimacy. Just as the structure changes with growth, developing from simple to functional structure, the culture also alters. The next of Handy's cultures is the role culture.

### The role culture

The role culture mirrors the functional and divisional structures and is evident within more mature and larger organisations with departments, divisions and different geographic areas. The symptoms of the role culture are based on everyone in the organisation having a specific job title and description and knowing what it is they are expected to do in their contribution to the organisational mission. Role culture organisations are functional, bureaucratic and highly systematised, with clearly documented, routine procedures and well-organised and efficient operations. Because of their size and the routine nature of their operations, these organisations develop into solid and predictable institutions, who operate the way they do because they have always done things that way. These cultures find an increase in the rate and speed of change a great threat, and are thus not adaptive to change.

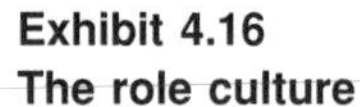

**Exhibit 4.16**
**The role culture**

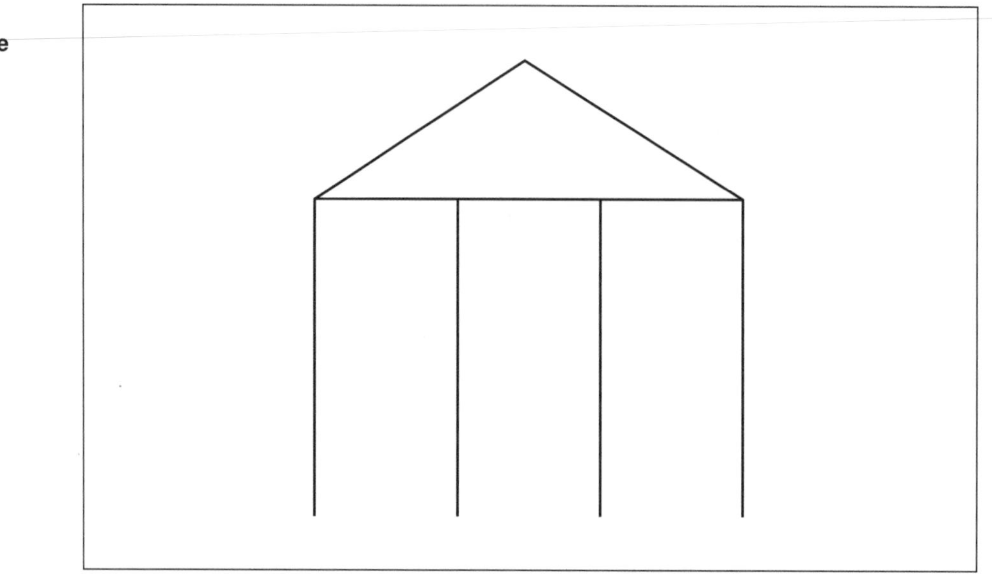

*Source*: *Understanding Organizations* by Charles Handy (Penguin Books 1976, Fourth Edition 1993). Copyright Charles Handy 1976, 1981, 1985, 1993. Reproduced by kind permission of Penguin Books Ltd.

### The task culture

Handy's task culture is more flexible and is often displayed in organisations that frequently undertake work for a variety of customers in a variety of fields. The task culture has close connections with the matrix structure.

Organisations with a task culture undertake very specific problem-solving or troubleshooting tasks as projects for internal or external clients, usually on a consultancy basis. The culture is extremely team oriented, since each task or project requires a fresh team to be constituted containing the required skills and knowledge that will enable the project to be completed successfully. Such a culture is highly flexible, but also expensive. This troubleshooting culture is often brought into an organisation to solve problems that others have found intractable. Members of the project team will exhibit their skills and competence through an extravagant use of resources, as they are used to being able to command anything that they need to get the job done.

### The person culture

The fourth of Handy's organisational cultures is the person culture, where a set of professionals agree to collaborate to perform a specific service. These people could be self-employed, or at least would have little notion of being employees of the organisation in the traditional sense. Rather, they would grant the organisation the benefit of their services for which they accept monetary gratitude. They may even avoid the word 'organisation', preferring, as Handy states, terms such as 'practice' to describe the collective activities in which they engage. The person culture centres on the particular professional skills that the individuals possess and without which the organisation could not operate. The individuals consider themselves to be highly valued, unique, creative and ultimately unaccountable. Examples of these person culture professionals would be academics, doctors, solicitors and management consultants.

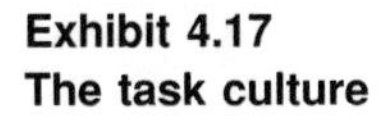

**Exhibit 4.17**
**The task culture**

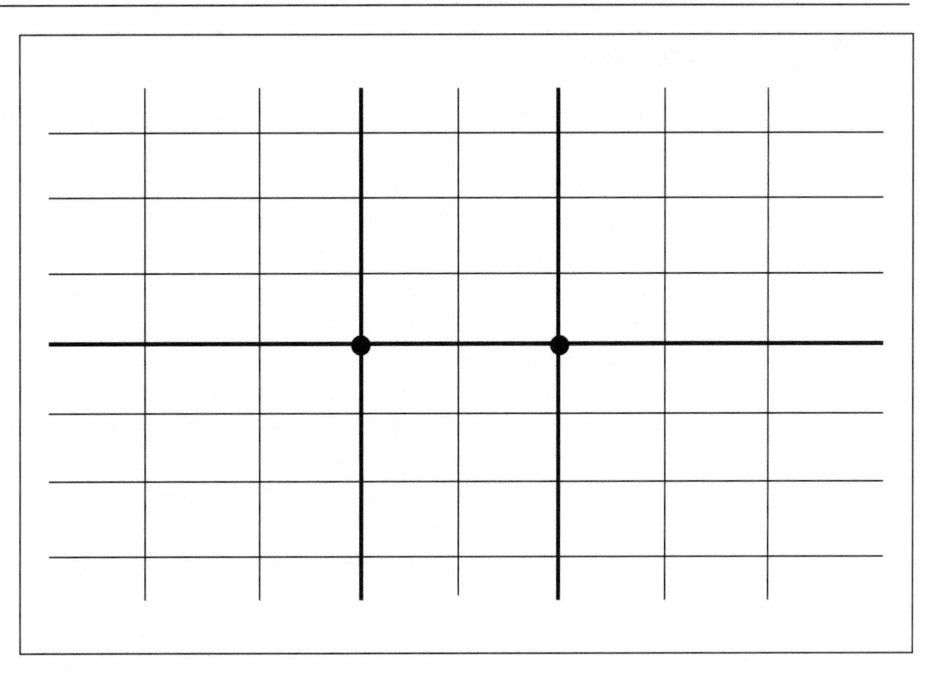

*Source*: *Understanding Organizations* by Charles Handy (Penguin Books 1976, Fourth Edition 1993). Copyright Charles Handy 1976, 1981, 1985, 1993. Reproduced by kind permission of Penguin Books Ltd.

**Exhibit 4.18**
**The person culture**

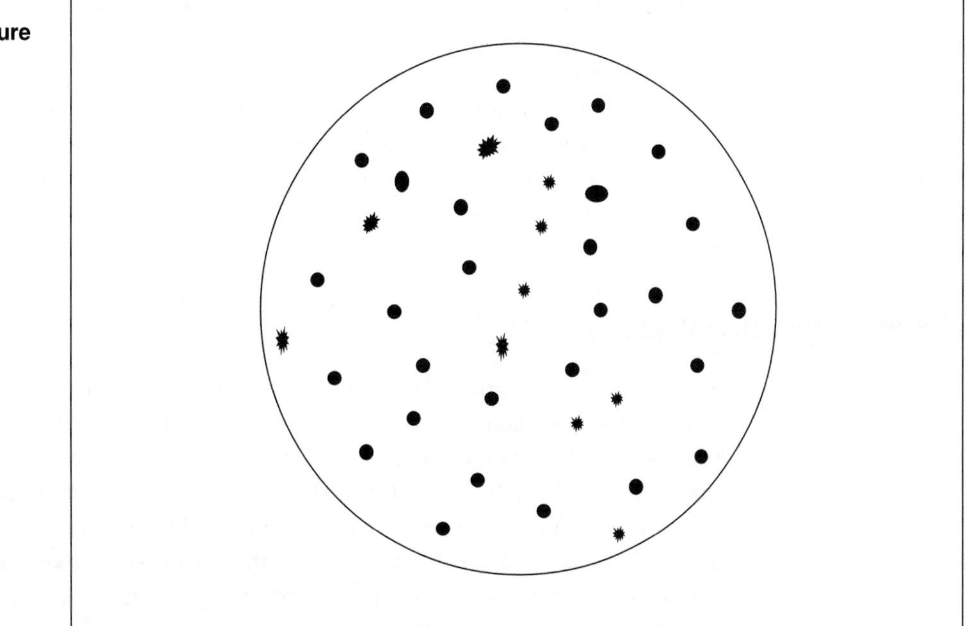

*Source*: *Understanding Organizations* by Charles Handy (Penguin Books 1976, Fourth Edition 1993). Copyright Charles Handy 1976, 1981, 1985, 1993. Reproduced by kind permission of Penguin Books Ltd.

## Culture and structure

There is undoubtedly a close link between culture internally and other elements of the internal environment as well as elements of the external context. The size and structure of the organisation dictate the ways in which the people within it are able to operate. A small, entrepreneurial organisation with a simple structure will inevitably see and hear the leader regularly, and so have close contact with the original whom they are designed to clone. A larger organisation with clearly defined departments and jobs will exhibit the role culture, so new recruits will be expected to learn their particular job and tasks quickly for two reasons: to enable them to fit into their position or role; and so they can transmit the required behaviour to any new people they meet connected with the organisation. The task culture has a cause–effect link with the matrix structure, as both are linked with the provision of goods and services in a task or project-based environment. The person culture is, perhaps, less tightly knit to one of the particular structures, but the service offered by an organisation employing such individualistic stars is definitely linked to the behaviour of the people delivering that service.

As we will see in Chapter 5, culture in society is generated by us and it also controls us. Whatever their provenance, sociocultural impacts on the human resource inputs to the resource conversion process are key to that organisation's operational abilities. Organisations strive for harmony between the work they have to do, the structure they adopt to make the work possible and the culture of the human resources they employ so their goals can be achieved. Organisations can be multicultural internally, and may need to be both global in their outlook on business, operating in many markets simultaneously, as well as local in their sensitivity to the needs and wants of employees and customers in a particular environment. Everyone lives culture, but only the clever are able to manage it.

## Internal functional areas

The four main internal functional areas in any organisation are marketing, operations management, finance and human resource management. These represent the principal activities that organisations perform to produce a product or service that can be sold for a profit and to manage appropriately the fundamental resources of finance and employees. Each internal functional area is looked at in greater detail later in this book: marketing in Chapter 6; operations management in Chapter 7; finance in Chapter 8; and human resource management in Chapter 9.

The marketing function is concerned with making sure that the organisation interacts appropriately with important external stakeholders such as customers and competitors. A marketing-focused organisation seeks to understand the needs and wants of customers and strives to satisfy those needs and wants via the development of suitable products or services. The meeting of customer needs and wants also includes the distribution of the product or service to customers and the provision of any after-sales service required. Additional services such as delivery or maintenance may be charged for separately.

Organisations seek to attract customers by advertising and promoting the products or services they offer. Advertising and promoting products or services informs customers or potential customers of how the organisation may be able to meet their needs and wants.

In seeking to attract customers and satisfy their needs and wants, organisations monitor, anticipate and react to competitors' strategies and marketing activities. This consideration of competitors is significant, as organisations usually seek to outdo competitors in terms of attracting greater market share and providing customers with better-quality or better-value products or services for the price charged. This requires the marketing function to liaise with other appropriate internal functions, such as operations management.

The operations management function is concerned with producing a product or provision of service delivery. It includes involvement in any product or service development or updating. Information indicating a need for product/service updating or redevelopment should arise from the marketplace and the marketing function. Hence, liaison between the marketing and operations management functions is crucial.

The activities of the production or service function also encompass the planning, sequencing and scheduling of resources needed to produce the product or deliver the service; ensuring the availability of goods or service as required by the customer; and liaising with the finance function to ensure that payment is requested and made for the goods or services delivered by the agreed method, cash or credit.

The finance function is involved in managing all aspects of an organisation's money including raising capital finance, generating sales revenue, meeting the expenses required to run the business, making investment decisions and generating and organising the information required for the annual report and accounts.

The human resource management function comprises all activities associated with recruiting, training, developing and rewarding the organisation's employees. It is responsible for overseeing the implementation of the human resource management policy on an organisation-wide basis. The precise detail of the type of employee, training, development and rewards may vary for different positions in separate sections of the organisation. Therefore, the implementation of the organisation's human resource management policy requires the involvement of both human resource management specialists and managers from the section of the organisation where the employees work.

## Conclusion

This chapter presented an overview of the internal environment of an organisation and the key issues that need to be managed. The resources required and used by organisations were examined, along with how organisations may be structured to allow efficient and effective management of resources. Also included was a look at the role of organisational culture and the key activities of marketing, operations management, human resource management and finance.

## ETHICAL ISSUES CASE STUDY 4.2

FT

# Women who make a difference

### Female-run companies are different, but not necessarily in the ways you thought

Female entrepreneurs are more autocratic in their management style than men, operate with fewer levels of authority and survive using hardly any written rules, with most information stored in their head.

These are some of the early findings of a study by Syeda-Masooda Mukhtar, fellow in small and medium enterprise management at Manchester Business School.

'I admit I was taken aback myself,' says Mukhtar, who had tended to subscribe to the stereotype of the consensual, collaborative female manager.

'These results are quite critical since there is a debate among policymakers as to whether to treat female businesses differently.' She reckons her study is proving that there are indeed gender differences, which may in turn suggest the need for adjustments in the formulation of public policy, for instance, on skills training.

When quizzed about management styles, nearly 63 per cent of women said they practised no delegation of authority in running their businesses. Only 48 per cent of the men interviewed said that was the case. These differences in management style held good across all sizes of companies surveyed, with the differences at their most pronounced among the few women running the (relatively) larger businesses.

Most of the women had no formal documented quality procedures, preferring informal techniques, such as memory power. The study claims to show that this does not make them worse managers; they appear to pay just as much attention to matters such as budgetary control or setting strategy objectives.

Mukhtar started her research because there was a dearth of data on the subject of women running their own businesses, even though their numbers had clearly increased. In the course of her research, one government official told her it was regarded as sexist to gather separate information about female-run companies.

Among the few pointers on the extent of the trend was that the number of women who are self-employed had more than doubled compared with a 54 per cent increase among men, according to the Department of Trade and Industry's Labour Force survey.

Mukhtar's study covers 5710 companies and is a random sample of small and medium-sized enterprises across all sectors and regions. All are members of the Federation of Small Businesses, the Forum of Private Business or the Rural Development Commission, and 6 per cent are run by women.

The female-led enterprises do conform to some stereotypes – notably as regards their size. Average annual turnover is just £35 000, with only 3 per cent passing £1m. Male-run businesses in the sample, by contrast, boast an average turnover of £350 000.

Making comparisons within the same sectors, women tend to have a narrower product range, are much less likely to form limited companies, and employ more part-time staff.

This does not necessarily imply that women are less successful or less entrepreneurial, Mukhtar reckons. They simply exhibit distinct characteristics and sources of motivation. They seem to go for stable, lower growth companies as a deliberate policy – which they run with a rod of iron. Men tend to achieve higher growth, but at the cost of a more volatile performance.

While the average life of the businesses in the sample is 10 years for men and three for women, Mukhtar claims this is largely because the women started more recently. When taken by age, the companies' survival rates are quite similar – and some 2 per cent in the female category had been around for more than 30 years.

Mukhtar suggests that a preoccupation with 'growth' businesses – among banks and others – as the sole criterion of success may have been overdone. 'I think banks have a very narrow definition as to what constitutes a successful business,' she says.

If male and female entrepreneurs tend not to share attributes, they should perhaps not be judged by the same yardsticks, the study suggests.

Then again, one of the things Mukhtar cannot gauge is which gender proves to be the more truthful when it comes to taking part in a survey.

**Questions for ethical issues case study 4.2**

1. According to the survey, in what ways are female-run businesses different from male-run businesses?
2. Do you agree with the government official who said it was 'sexist to gather separate information about female-run companies'?
3. Working together with other students, list the personal qualities you consider to be typically female or typically male. How is each list likely to affect the operation of the four internal functional areas?
4. Why do you think that in female-run companies 'Average annual turnover is just £35 000, with only 3 per cent passing £1m while male-run businesses in the sample boast an average turnover of £350 000'?

## EXIT CASE STUDY 4.3

FT

# Evolution plays its part in Hanson's big bang

**By Ross Tieman**

Hanson shareholders will meet in London today to approve the final stage of an extraordinary corporate restructuring.

By Monday, when trading begins in new Hanson stock and Energy Group shares, a £10bn-a-year conglomerate will have been split into five quoted companies, and £2.6bn of disposals completed.

Yet no blueprint ever existed for Hanson's fragmentation. It came about, says chief executive Mr Derek Bonham, 'by evolution, not revolution'. That alone makes it a remarkable story.

When Mr Bonham was elevated from finance director to the role of chief executive in April 1992, Hanson's destiny as an acquisitive conglomerate seemed unquestionable.

Over the best part of 30 years its chairman, Lord Hanson, and his partner, Lord White, had become renowned as fearless takeover barons, capturing, digesting and divesting a rich diet of companies on both sides of the Atlantic.

Mr Bonham was Lord Hanson's protégé, apparently destined to succeed him as chairman. Few imagined he would reverse the strategy of three decades. Yet that is precisely what he did.

Mr Bonham was determined that Hanson, unlike so many of its erstwhile targets, should not end up as a kind of industrial museum of unrelated businesses.

Beginning with a review of the company he had served for 21 years, Mr Bonham and his colleagues soon found themselves focusing on the characteristics needed to achieve premier market positions. Their conclusions were inescapable. Hanson needed a radical overhaul.

'We recognised that one couldn't just keep buying businesses,' says Bonham. 'You had to create growth.'

Bonham and his lieutenants began a massive clear-out. By the end of 1993, they had sold most of the smaller businesses whose success or failure was concealed by Hanson's size.

But not all. Shedding the diverse portfolio of small US businesses was proving slow. Returns on a bulk trade sale looked inadequate; a demerger impossible because of the tax implications. Bonham was stale-mated until a bright spark figured that the Inland Revenue might buy the idea of a dual-resident business. It did, but it was not until May 1995 that Hanson succeeded in demerging US Industries, clearing out 34 businesses and a raft of debt at a stroke.

Disposal proceeds were used to buy Eastern, the UK electricity distributor, and reinforce Hanson's strong points: tobacco, chemicals, coal mining and building products.

But the logic of the USI demerger had acquired a life of its own. Markets and fund managers had become more sophisticated; institutions more reluctant to invest in ill-assorted businesses that lacked synergy. In mid-January 1996 the Hanson board, still chaired by the founder, approved Bonham's response: a four-way demerger of the remaining group.

When it was announced, on January 30, shares in Hanson responded vigorously, reaching a new peak of 211½p. But enthusiasm proved short-lived. Markets for titanium dioxide, the main product of the now-demerged Millennium Chemicals, collapsed. New US accounting rules required massive asset write-downs and provisions. Conglomerates became the Cinderella stocks of the stock market.

'This last year has been one of the most brutal years of my life – and I have never shirked very hard work,' Bonham says. 'You are trying to run the businesses as well as trying to drive the demerger process. We have made people work harder than I think they ever realised they could.'

Partly this is because Hanson directors and divisional chiefs remained in control, limiting the role of advisers

**Exit case study 4.3 *continued***

– a decision Bonham believes was crucial to its timely completion.

But shareholders have yet to reap a benefit.

Analysts contrast Hanson's strategy with that adopted by a rather newer chief executive, Mr Ian Strachan, at BTR. While switching, like Hanson, to a strategy of organic growth and bolt-on acquisitions, he has opted to keep that company intact, albeit with a narrower focus.

Since January 1 last year, the market capitalisation of BTR has fallen 21 per cent to £9.75bn. The combined value of Hanson and its demerged entities, meanwhile has diminished by just 10.3 per cent to £8.94bn.

The principal reason for the falls is that during the course of the demerger process at Hanson, and normal trading statements from BTR, it became clear that each was pursuing an unsustainable dividend strategy.

Shareholders in the demerged Hanson businesses will receive in total about 6p of dividends – half the level paid out by the business in the past. Last year, investors in BTR saw their dividends fall by approaching 40 per cent.

'The Hanson break-up was an acknowledgement that they had to change because the old formula was no longer working,' says one leading analyst.

'Could they have bitten the bullet and slashed the dividend without the demerger?' asks another.

Listing particulars for the demerged Hanson businesses have released far more detailed information about their activities than ever before. Each now makes its own presentations to fund managers and City analysts.

The difficult markets faced by Millennium Chemicals are clear to all. Its share price has reacted accordingly. But shares in US Industries have performed well, and Imperial Tobacco, despite its unfashionable trade, has been well received.

But the most telling test of the demerger decision will be the performance of the most promising parts, the Energy Group and the rump Hanson – essentially a building materials business.

Mr Bonham is visibly more relaxed now that the demerger process is all but complete. He has chosen to stay with the Energy business as executive chairman. But there is one consequence of the restructuring he rues. 'I regret that I am not taking over as chairman when Lord Hanson retires,' he says. 'That is what I thought I was going to do.'

*Source*: *Financial Times*, 21 February 1997. Reprinted with permission.

**Questions for exit case study 4.3**

1. In the 30 years prior to the article date, how had Hanson achieved growth? In what ways was Derek Bonham's approach different from that of Lord Hanson?
2. What were the advantages or disadvantages of splitting Hanson into four separate companies?
3. What organisational structures might be appropriate for the new companies?
4. In what ways might the internal culture of each of the four former parts of Hanson change along with the new structures and operations? Is it important that all of the four new parts operate in the same way?

## Short-answer questions

1. Explain the term 'a centralised organisation'.
2. Explain the term 'a decentralised organisation'.
3. What are the differences between organic and external growth?
4. List the five generic organisational structures.
5. Draw the five generic organisational structures.
6. Summarise the key features, strengths and weaknesses of one of the five generic organisational structures.
7. Define the term 'organisational resource conversion'.
8. What are the three stages of a resource conversion process chart?

9 What are the three main types of resource inputs? Give one example of each.

10 Name Handy's four organisational culture types.

11 What is the main danger for the organisation of the power culture?

12 Which two structures are likely to develop a role culture?

13 What sort of work are task culture organisations likely to be involved in?

14 In your opinion, what are the potential difficulties of managing people in the person culture?

## Assignment questions

1 Illustrate the structure of each of the following organisations in diagrammatic form. Name the type of generic organisational structure on which you have based your structure and explain your choice for each organisation:

- A newsagent's shop in London run by the owner, and employing a morning paper boy, an evening paper girl and two shop assistants.
- A garden centre on the outskirts of Glasgow selling a range of plants and gardening equipment to the general public. The garden centre is supplied with plants by a commercial nursery attached to it, which also supplies other shops and garden centres in the West of Scotland. The other gardening equipment is bought in wholesale from a number of suppliers. The garden centre employs 30 people, including a nursery manager, a marketing manager, an accountant and a human resource manager.
- A market research company operating across the UK, with offices in Aberdeen, Leeds, Southampton, Edinburgh, Manchester, Cardiff, Swansea, Nottingham, Glasgow and two offices in London.

2 Apply two of the five generic organisational structures to each of the following organisations. Compare and contrast the two structures you have chosen and comment on which is the more appropriate for each organisation:

- A science department in a university with undergraduate BSc students doing full-time three-year degrees and some MSc and PhD students undertaking postgraduate research qualifications.
- A management consultancy operating throughout the UK, Hong Kong, Singapore and Malaysia and which has recently opened a large office in Johannesburg, South Africa. The company offers consultancy services and expertise in the areas of auditing, taxation and business planning.

3 Compare and contrast the power culture with the role culture. What advantages and disadvantages do you see in each of these? Comment on the link between culture and structure, and also on the effects of culture on the organisation's operational abilities.

4 Examine the transactional dependencies between the three stages of the organisational resource conversion chart. Use examples from the private or public sector to illustrate your answer. In what ways is it possible to ensure that outputs are of the highest possible quality?

## References


1 Gore, C, Murray, K and Richardson, B (1992) *Strategic Decision Making*, London: Cassell.
2 Worthington, I and Britton, C (1997) *The Business Environment*, 2nd edn, Harlow: Financial Times Pitman Publishing.
3 Gore et al., op. cit.
4 Worthington and Britton, op. cit.
5 Gore et al., op. cit.
6 Worthington and Britton, op. cit.
7 Stevenson, T (1995) 'Hanson to break up into four companies', *Independent*.
8 Tieman, R (1997) 'Evolution plays its part in Hanson's big bang', *Financial Times*, 21 February.
9 Cope, N (1995) 'Retailers add value for the nineties', *Independent*, 12 April.
10 Dobie, C (1992) 'Breaking up is hard to do', *Independent*, 31 July.
11 Connon, H (1992) 'ICI shake-up plan sends shares soaring', *Independent*, 31 July.
12 Tyerman, R (1997) 'ICI in the cracker', *Sunday Telegraph*, 9 February.
13 Lynch, R (2000) *Corporate Strategy*, 2nd edn, Harlow: Financial Times Prentice Hall.
14 Thompson, J L (1997) *Strategic Management Awareness and Change*, 3rd edn, London: International Thomson Business Press.
15 Lynch, op. cit.
16 Thompson, op. cit.
17 Handy, C B (1993) *Understanding Organisations*, 4th edn, London: Penguin.


## Further reading

The following articles all look at organisations, their resources, processes and outputs:


Brooks, I and Reast, J (1996) Re-designing the value chain at Scania Trucks', *Long Range Planning*, 29 (4), August.

Carmichael, J (1992) 'Managing inputs', *Long Range Planning*, 25 (1), February.

Darwant, C (1996) 'Bangers and cash', *Management Today*, June.

Darwant, C (1996) 'Pier pressure', *Management Today*, August.

Dowling, G R (1993) 'Developing your company image into a corporate asset', *Long Range Planning*, 26 (2), April.

Oliver, J (1996) 'Which numbers count', *Management Today*, November.

van de Vliet, A (1997) 'Are they being served', *Management Today*, February.

Vyakarnam, S (1992) 'Social responsibility: what leading companies do', *Long Range Planning*, 25, October.


The following books and articles contain comprehensive coverage of organisational structures:


Brocklesby, J and Cummings, S (1996) 'Designing a viable organization structure', *Long Range Planning*, 29 (1), January.

Daniell, M (1990) 'Webs we weave', *Management Today*, February.

Johnson, G and Scholes, K (1999) *Exploring Corporate Strategy*, 5th edn, Chapter 9, Harlow: Prentice Hall.

Karabadse, A, Ludlow, R and Vinnicombe, S (1988) *Working in Organizations*, Chapter 11, London: Penguin.

Lynch, R (2000) *Corporate Strategy*, 2nd edn, Chapter 20, Harlow: Financial Times Prentice Hall.

Thompson, J L (1997) *Strategic Management Awareness and Change*, 3rd edn, Chapter 19, London: International Thomson Business Press.

The following articles all look at organisations and changing structure:

Foster, G (1996) 'Three over thirty', *Management Today*, May.
Fry, J M (1990) 'Abbey National becomes a company', *Long Range Planning*, 23 (3), June.
Greiner, L (1998) 'Evolution and revolution as organizations grow', *Harvard Business Review*, July/August 1972, reprinted May/June.
Kennedy, C (1993) 'The ICI demerger: unlocking shareholder value', *Long Range Planning*, 26 (2), April.
Lorenz, A (1996) 'Time to unpick GEC', *Management Today*, October.
Lorenz, A (1997) 'ICI's new chemistry set', *Management Today*, January.
Lynn, M (1996) 'Fortress Zeneca', *Management Today*, May.
Lynn, M (1996) 'The Hanson inheritance', *Management Today*, June.
Ogden, S and Glaister, W K (1996) 'The cautious monopolists – strategies of Britain's privatised water companies', *Long Range Planning*, 29 (5), October.
Pekar, P Jr and Allio, R (1994) 'Making alliances work – guidelines for success', *Long Range Planning*, 27 (4), August.

The following books and articles all look at organisational culture and changing culture:

Burack, E H (1990) 'Changing the company culture – the role of human resource development', *Long Range Planning*, 24 (1), February.
Curteis, H (1997) 'Entrepreneurship in a growth culture', *Long Range Planning*, 30 (2), April.
Deal, T and Kennedy, A (1988) *Corporate Cultures*, London: Penguin.
Humble, J, Jackson, D and Thomson, A (1994) 'The strategic power of corporate values', *Long Range Planning*, 27 (6), December.
Ishizuna, Y (1990) 'The transformation of Nissan – the reform of corporate culture', *Long Range Planning*, 23 (3), June.
Johnson, G (1992) 'Managing strategic change – strategy, culture and action', *Long Range Planning*, 25 (1), February.
Kono, T (1990) 'Corporate culture and *Long Range Planning*, 23 (4), August.
Morgan, M J (1993) 'How corporate culture drives strategy', *Long Range Planning*, 26 (2), April.
Nakajo, T and Kono, T (1989) 'Success through culture change in a Japanese brewery', *Long Range Planning*, 22 (6), December.

# 5 Culture and organisations

Claire Capon and Andrew Disbury

## Organisational context model

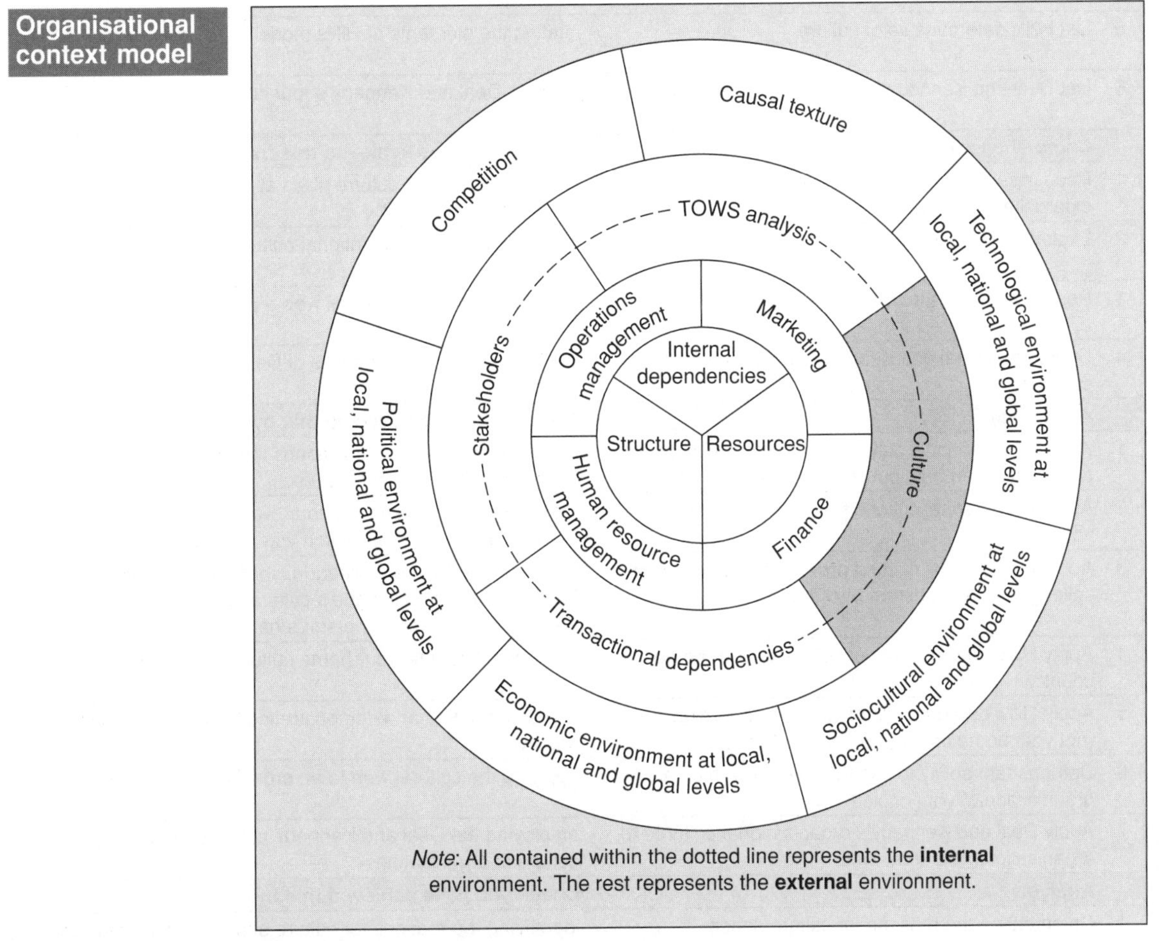


*Note*: All contained within the dotted line represents the **internal** environment. The rest represents the **external** environment.

**Exhibit 5.1**

## Learning outcomes

While reading this chapter and engaging in the activities, bear in mind that there are certain specific learning outcomes you should be aiming to achieve as set out in Exhibit 5.2 (overleaf).

**Exhibit 5.2 Learning outcomes**

| | Knowledge | Check you have achieved this by |
|---|---|---|
| 1 | Define organisational culture | stating in one sentence what organisational culture is |
| 2 | Define a 'weak' culture organisation | listing the traits of an organisation with a weak culture |
| 3 | Define a 'strong' culture organisation | listing the traits of an organisation with a strong culture |
| 4 | Define personal cultural provenance | summarising the influences on an individual's culture |
| 5 | List Hill's determinants of culture | stating the elements of Hill's model |
| 6 | List Deal and Kennedy's organisational cultures | naming Deal and Kennedy's four cultural types |
| | **Comprehension** | **Check you have achieved this by** |
| 1 | Recognise that culture exits at different levels of the external environment | discussing the role culture plays at different levels of the external environment |
| 2 | Explain the determinants of organisational culture | discussing how organisational culture is formed |
| 3 | Recognise what 'the cultural web' is | explaining what the cultural web represents |
| 4 | Understand Deal and Kennedy's organisational cultures | explaining the characteristics of Deal and Kennedy's four organisational cultures |
| | **Application** | **Check you have achieved this by** |
| 1 | Apply organisational cultural models to a variety of organisations in both public and private sectors | examining different organisations in the light of theory and discussing their internal culture |
| 2 | Apply the personal cultural provenance factors to yourself | discussing the influence of your own cultural background on determining the sort of person you are today |
| 3 | Apply the personal cultural provenance factors to a colleague with a different background to yours | discussing, with your colleague, your understanding of the influence of your colleague's cultural background on determining the sort of person s/he is today |
| 4 | Apply Hill's determinants of culture to your home country | demonstrating how your home culture has evolved |
| 5 | Apply Hill's determinants of culture to a foreign country (not your home country, Britain or China) | researching data and demonstrating how a foreign culture has evolved |
| 6 | Demonstrate ability to use the cultural web on an organisation of your choice | applying the cultural web to an organisation you know well |
| 7 | Apply Deal and Kennedy's organisational cultures to organisations you know well | employing the Deal and Kennedy cultures to examine the culture of organisations |
| | **Analysis** | **Check you have achieved this by** |
| 1 | Comment on the reasons for culture shock | identifying the reasons for culture shock in specific situations and how it can be managed |
| 2 | Differentiate between managing culture in the international arena and home arena | comparing and contrasting the management of culture in the home and international arenas |
| 3 | Analyse the key issues for managing organisational culture | identifying and discussing the key issues for managing organisational culture |
| | **Evaluation** | **Check you have achieved this by** |
| 1 | Evaluate the usefulness of organisational cultural models | determining the relevance of models of culture to working life for both the individual and the organisation |
| 2 | Comment on the impact of organisational culture on efficiency and effectiveness | appraising the efficiency and effectiveness of organisations with different organisational cultures |

ENTRY CASE STUDY 5.1

FT

# Remove the iron rice bowl: management investing in China

**By James Kynge**

What would you get for your money if, as a foreign investor, you decided to buy a former state enterprise in Shenyang, the centre of China's north-eastern industrial rustbelt?

The question is more than academic. Such businesses have, in the past, been largely off-limits to foreigners. But China's faltering economy has forced the authorities to search for investors wherever they can find them.

This month Mu Suixin, Shenyang's mayor, toured Europe to find foreign investors for 18 large state companies, with a total workforce of 309 436. Other cities in the north-east are also planning mass sales; Harbin, near the Russian border, is to put about 1000 medium and small enterprises under the auctioneer's hammer in June.

As to the question of what a foreign investor could expect, the answer is likely to be an appallingly managed business. But this is the main attraction, according to Jiang Enhong, a leading corporate turnaround expert in the region, who has himself taken on such companies.

Rebellious workers, deception and years of inertia bred from central planning define the opportunities for any new investor – local or foreign – in China's crumbling state-owned sector, he believes. 'I have straightened out the management of companies within 44 days, and set them on the way to profitability,' he says. 'This just shows how bad their management was.'

The work is arduous, but foreigners may never get the chance to buy into state assets as cheaply again.

Foreign investors can also benefit from preferential terms, such as municipal tax concessions, which are not on offer to locals. Joint venture and foreign-invested companies also have advantages in raising local bank finance – which private companies in China find difficult to secure. 'If only I could get banks to lend to me, I would not just be a tiger, but a tiger with wings,' says Mr Jiang.

His main problem has been in rationalising the many managers who clog most state-owned enterprises. When he bought the Shenyang Antibiotics company, which had not made a profit nor paid any tax since 1979, there was one manager to every four workers.

Some managers had been awarded offices out of favouritism but their roles had not been clearly defined.

Mr Jiang has removed two-thirds of the managers in all three state companies he has bought, and defined clearly the task of those remaining. Anyone who failed to perform their duties was fired, he adds.

A measure of shock therapy was necessary to teach workers that when their factory passed from state to private hands, their 'iron rice bowl' of socialist-era benefits was taken away. When he took over the Xincheng Pharmaceutical factory, workers were supposed to arrive at 8am but most came at 10am and many would return home for the day before lunch.

At the opening ceremony for the factory, one worker sat down on a seat reserved for dignitaries. When he was told to move, he smashed the chair. Mr Jiang told him he would be sacked if he did not donate a new chair by the following Monday. 'He brought the chair, but we fired him anyway,' says Mr Jiang.

Laying off workers in China has been fraught with ideological and social impediments. It is not as simple as Mr Jiang makes it sound, but a Communist Party congress in September ushered in a phase of faster free market reform. Redundancies are increasingly seen as unavoidable.

The September congress, and the subsequent National People's Congress in March, also helped to overturn an ambiguous official stance on private ownership by permitting 'diverse forms of public ownership'. In Shenyang, this has been taken as a cue for the rapid and comprehensive sale of state assets. Thousands of state enterprises are to be sold this year and next.

Gai Ruyin, the deputy mayor, says that a severance allowance of, say, 10 000rmb (£738) could be paid per worker by the new owners of state companies, and that the cost of providing for those who are made redundant may be set against the purchase price of the factory.

Morale among the remaining workers is a complicated issue. But Mr Jiang believes that some of the control mechanisms found in state enterprises, such as the Communist party cell and the trade unions, should be retained. The Communist party has been invaluable in resolving industrial disputes because it carries the authority of China's most powerful body.

One worker had lost his legs when he was run over by a train. He was causing trouble outside the factory, inciting others to militancy. With the party cell's intervention, a settlement was found. Mr Jiang's company bought him a mobile telephone and a stall: he now rents out the phone for calls and sells wine by the factory gate.

But Mr Jiang is conscious that the heavy hand with which he has put his corporate empire in order should at some point give way to a lighter touch. Eventually his three factories are to be 'democratised', with each worker owning shares.

*Source*: *Financial Times*, 26 May 1998. Reprinted with permission.

## Introduction

The aim of this chapter is to introduce the complex issue of culture and its effects on organisational as well as personal efficiency and effectiveness in the workplace. This chapter necessarily combines aspects of a variety of specialisms, each with its own vast literature: sociology, psychology, management development and international business. Here some of the main strands are brought together and interwoven to form a useful framework for the business studies student wanting to examine the impact of culture on the way organisations work.

## Culture at different levels of the external environment

**Culture at the global level of the external environment**

At the global level of the external environment, issues emerge with the interaction of more than one national culture. When one leaves one's home country to work, live or even visit abroad, one is faced with different ways of doing things. In Chapter 2, it was mentioned that changes in lifestyle and behaviour can lead to feelings of culture shock, where individuals have to recognise and cope with experiences that are different from those to which they are accustomed. Transferring one's life from Milton Keynes to Tokyo when one is sent off to work in Japan, having grown up in Milton Keynes and never left, is likely to include dealing with culture shock. However, with such a dramatic move we expect things to be different in Tokyo but anticipate we will be able to cope. This may not be the case. The thorough pre-departure orientation of executives posted overseas is one of the key issues in the management of international business operations that can contribute greatly to the success of overseas postings. Nevertheless, it is often overlooked.

**Culture at the national level of the external environment**

The chapter begins with an illustration of culture in society that aims to explain the concept of culture predominantly at the national level of the external environment. Once an individual's cultural influences from the national level of culture have been identified and analysed, the individual is then furnished with a vocabulary and consciousness that allows them to make coherent comparisons with other nations' cultures, giving that global-level view of cross-cultural issues. National cultural characteristics are adopted by one national grouping and attributed by one nation to others, leading to stereotype and prejudice. National culture, however, belies the fact that within it are local differences between communities, providing a more local perspective to cultural influences on organisations.

**Culture at the local level of the external environment**

Local cultural issues in the community are complex. The national and local levels of the sociocultural external environment can often intertwine, due to the fact that some people's national characteristics make them the people they are in their local communities. Although we recognise cross-cultural issues between nations, we sometimes do not recognise that equivalent culture shock can occur within the same country. This might have a particular resonance for a full-time student who, at a relatively young age, has left the family and is attending a higher education course in another part of the country far from home for the first time.

As communications, transport and technology have developed in western industrialised societies in the last 200 years, the provenance of this personal and social identity has changed. A sense of identity used to come from home, family and village, all elements of the local external environment. As society developed, these local-level sociocultural influences emerge less importantly from influences located at the local level, since personal movement and individual horizons are no longer as limited as they once were. Now, because of the access to travel and communications media, an individual's cultural identity could be said to be much more strongly located at the national level, or possibly emerged from more global influences, with the strong lifestyle messages emanating from global brands such as Coca-Cola and McDonald's.

## Culture inside organisations

It can be seen from the organisational context model at the start of this chapter that as well as forming part of the external environment, culture can also be identified in the internal environment of the organisation. This chapter therefore also presents models for identifying and analysing internal organisational or corporate culture. Organisations can and do develop their own internal operational culture. Sometimes internal culture emerges as a result of the internal structure of the organisation (*see* Chapter 4) and sometimes it is as a result of the type of people employed by the organisation.

Where organisational managers identify internal cultural problems, they may make structural or personnel changes in an attempt to alter the internal culture and achieve through these alterations the desired organisational objectives. Internal culture can be as explicit or implicit within the organisation as it is in society. Whatever the case, internal culture is a vital factor for the newly recruited employee to understand if he or she is to learn 'the way we do things around here'. Culture shock can occur within organisations when two whole organisations or sections of organisations are merged and two completely different sets of working practices and behaviours are expected to operate together.

The issues of culture presented in this chapter are extremely complex. Simply examining the external or internal environments alone is already a difficult enough topic to grasp. Marrying the two into the study of organisations in context in this way presents intellectual and practical challenges to the student in order to achieve the learning objectives for this chapter.

## Personal cultural provenance

To help with understanding the topic of culture, the study begins at the individual level. Personal cultural provenance is the origins of individuals' culture. It is another way of saying, 'Where do I get my culture from?' (*see* Exhibit 5.3). External cultural influences can be identified at a basic level as being our system for understanding and expressing what we think is our identity.

Identity is a key concept from a sociological perspective, but here we will examine its influence on our sense of who we are, where we come from and how we do things. A first-year university student might identify him- or herself in all or some of the following ways. These could all be said to be expressions of culture that have their location in sources outside the individual. They are all facets of individual identity that can provide a useful insight to personal cultural provenance.

### Name

People respond differently to the question 'Who are you?' according to context. With young people one's own age in an informal setting, it is most probable that the answer would be one of the names given by one's parents, and that the surname or family name would not be given unless specifically asked for later. If people do not use the full name their parents gave them, they use a version of it that they have chosen themselves, e.g. Chris instead of the full Christine or Christopher. Some people prefer to use a new and individual name that they have chosen for themselves. Others have a new name chosen for them by their peers and contemporaries, like a nickname that is a shortened version of their original name, or one based on a particular skill or habit they have, their appearance or their preferences.

**Exhibit 5.3 Determinants of personal cultural provenance**

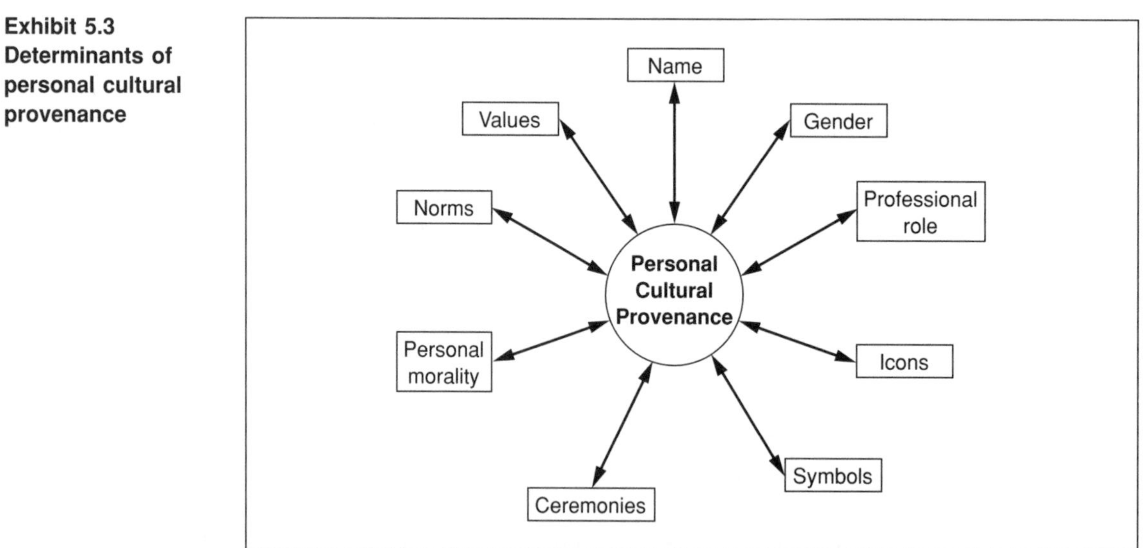

**Gender**

Our gender is a fundamental element of our nature that also determines the cultural experience we have throughout our lives. In whatever ethnic culture we are born, the raising of children conveys clear gender identities and roles, and instils stereotypes and role models from an early age. The simple attribution of qualities or skills to either gender dictates the way a child interacts in society: boys may be discouraged from crying if they are hurt; girls may be taught to cook; boys may learn that fighting in self-defence is 'manly'; girls may be bought dolls to play with. This leads to differences in attitudes towards education as well as variations of treatment of the different genders at school. Ultimately this leads to different expectations of performance and ability in the workplace, and in society in general. Women, for example, bear the brunt of the responsibility for childcare in UK society far more often than do the fathers of their children. This is a social norm to which most people adhere.

**Professional role**

When asked 'What do you do?', people identify with their occupation. Despite unemployment statistics most people have a job and, again, that is a social norm. For the three or four years of full-time undergraduate study, the words 'I'm a student' give many signals as to your lifestyle, income bracket now and in the future, level of intelligence, most likely habits and leisure activities and even political affiliations. Replying to the question that one does not have a job can again have many automatic connotations, depending on context: a woman with children who does not work may be praised for having devoted her time to raising the family; a man who has children and no job may be criticised for not providing for them and expecting the state to do it for him.

**Icons**

At the first-year stage in your undergraduate career, when you are just beginning to realise how much effort is involved in gaining your qualification, someone you respect could be someone who has already gained the qualification for which you are working, or at least someone who is in the final year. At the same time, many national icons will have resonance. These come in the form of pop and rock stars, actors, television personalities, sports stars and supermodels. All of them have influence over culture, as they have a bearing on what we wear, where and how we spend our leisure time and our attitudes to religion, drugs, politics and a whole host of issues.

**Symbols**

For many people, an obvious and visible symbol of culture is the clothes they choose to wear. These often make a statement concerning their cultural identity. For example, many students choose to wear jeans and trainers, which are casual and acceptable attire for their relatively informal lifestyle. Other people, including students, may choose to identify more closely with their ethnic background and wear clothing that identifies them as belonging to a particular ethnic culture. In the UK this is often done by people whose families originate from India or Pakistan. Alternatively, people may view their values and professional role as an important part of their cultural identity and choose to symbolise this by wearing smart clothes to work. Some employers require their employees to identify with the organisation they work for by virtue of the job they do and/or the organisation. For example, hospitals require nursing staff to wear uniforms. This is for two reasons: first, so nursing staff can be easily identified as such; secondly, because the work nursing staff undertake is sometimes messy and dirty, so a clean, easily washable uniform helps prevent the spread of infection and is preferable to getting one's own clothes dirty. The requirement on nurses to wear uniforms is a longstanding historical example of staff being required to wear clothes that identify them with a particular profession and hospital, since different hospitals often have different coloured uniforms for different types of nurse. In contrast, since the 1980s, the idea of a corporate uniform in the UK has caught on with banks, building societies, shops and some restaurants requiring staff to wear corporate uniforms, which identify the individual as working for an organisation with its own organisational culture.

**Ceremonies**

Most students will attend their graduation ceremony on being awarded their qualification, and will enjoy the formal occasion with its mortar board, academic gown and procession before the Chancellor of the university to collect the piece of paper bearing the university seal and stamp as proof of the studies they have undertaken. The graduation ceremony marks, in a formal manner, the successful completion of higher education studies, which will in turn influence the professional role one adopts throughout life and hence also an individual's personal cultural provenance.

**Personal morality**

Cultural influences also give us our innate sense of right and wrong. Thus in our culture it may be wrong to murder others, to steal from them or for adults to have sex with children. We all agree with this and those who do not adhere to the code contradict our moral code and our laws, which are the political expression of our moral code. Culture may also express itself in those whom we respect. Therefore in a free-market or mixed economy, where self-reliance and providing for one's family's material well-being are considered admirable, people who are prudent and work hard all their life are admired. Those who do well for themselves and become 'self-made' by enriching themselves and their families through their own efforts are considered worthy of our respect. Entrepreneurs are admired and ennobled via political honours to the House of Lords as people who created wealth for themselves and for the country, and thus have done a public service. Linked to religion, this is embodied in the 'work ethic', where it is seen as a good thing in northern European society to work hard for a living and provide for one's dependants.

Culture also emerges in our system of faith or beliefs, usually through the expression of formal religious belief. Once we have a common system for what we think is right or normal, we then have a language for expressing what we consider to be abnormal. Once we can identify people the same as us, who share our norms and values, we can then identify those who differ and so do not fit in with our norms and values. This gives rise to another key aspect that culture dictates, which is our moral standards. As we grow we learn society's norms and values, and with this embedded sense of normal and abnormal, right and wrong, we begin to judge the correctness of our own and others' behaviour. Thus the group becomes judgmental, dictating that certain modes of personal action are acceptable while others are not.

An example of this in many societies is the emphasis placed on the family. It is considered the norm, and thus deemed valued and morally correct by the group, for people to form lifelong heterosexual partnerships and to have children within that context. Thus the religious and politico-legal systems are designed to support this notion, and marriage, with its accompanying wedding ceremony, marriage certificate and wedding reception, has become the traditional as well as the normal way of demonstrating that this desired legal and social state have been entered into. Social welfare systems, including pensions, sickness and unemployment benefit, and taxation systems are also designed around the notion of a single breadwinner, a dependent adult and dependent children. Backed up by religion, this notion of the nuclear family becomes the moral, as well as social and legal norm. Refer to Chapter 2 for the changing position of the nuclear family in society.

**Norms**

It is from all the elements in society that affect us as we develop that we obtain our cultural norms and values. Norms are literally ways of behaving or attitudes that are considered to be normal. Norms can be defined as the social rules and guidelines that prescribe behaviour. Furthermore, norms not only affect how an individual behaves, but are the shared sense of what a group of individuals thinks is the normal way to behave. They begin to be taught by parents to babies, and are learned by children as soon as they come into contact with other children. Thus to behave against the group's norms is abnormal and shunned, leading to ostracisation.

For many students a key element taken into account when choosing a university is its location and night life. Socialising thus becomes the first-year norm, with the vast majority of first years going out frequently to pubs, clubs and the students' union with new-found friends and peers. Many town or city-centre pubs and clubs aim to maximise income by cutting entrance and drinks prices on otherwise unpopular nights of the week. They introduce 'student nights', normalising socialising during the week instead of staying in and studying. Late nights and excessive intake of alcohol lead to the inability to rise early the following morning, and so missing lectures or seminars becomes a social student norm.

**Values**

Values are things, people or attitudes that groups of individuals think are important or to be revered or respected. Thus many people might consider that loyalty and honesty between friends, not letting a friend down, not cheating on a boy- or girlfriend or trying to conform to the norms of the peer group are all values that the group would share. Values indicate what society sees as important collectively. While norms may be ways of behaving, values could be qualities that society looks up to.

Thus cultural norms and values contain a variety of elements and attitudes that can be said to come from external influences in society on the individuals who form the group of people constituting that society. These external influences can be formal and explicit, such as the teacher in the classroom rewarding good standards with public praise and a gold star, or punishing bad behaviour with public criticism and humiliation. Or external influences can be informal and implicit, such as the ways in which children observe adults' behaviour and then imitate that, taking their cues for the right and wrong way to behave either in the family context or in public from what they perceive around them.

## Identifying British culture

The first thing to declare about defining British culture via Hill's[1] determinants of culture (*see* Exhibit 5.4) is that Britain is in no way a homogeneous society. It is recognised that there are many cultural groupings populating the British Isles. The United Kingdom of Great Britain and Northern Ireland comprises three countries and one province: England, Wales, Scotland and Northern Ireland. Arguably the Celts in Wales, Scotland and Ireland form the indigenous population of these islands, who suffered Anglo-Saxon, Roman and Norman invasion and conquering as history unfolded. In the twentieth century, there has been large-scale immigration from territories within the former Empire and current Commonwealth, making the UK a multiracial

**Exhibit 5.4 The determinants of culture**

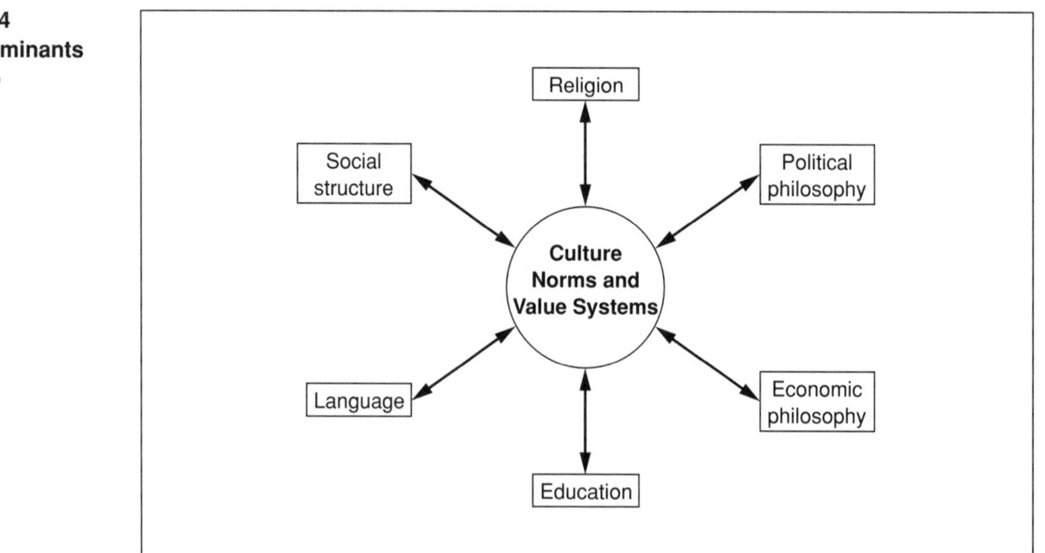


*Source*: Charles W L Hill, *International Business: Competing in the Global Marketplace*, 2nd edition, 1994. Irwin, reproduced with the kind permission of The McGraw-Hill Companies.

society, with large populations whose families originated in the Caribbean, the Indian subcontinent and Hong Kong. Thus any attempt to apply a model such as Hill's to 'British society' is liable to be relevant to some British people and irrelevant to others. This supports the notion that the study of culture in society and organisations is extremely complex.[2] Where certain elements included are irrelevant to the reader, it is expected that they will consider each determinant in relation to their own personal cultural provenance and will arrive at their own conclusions about how they have been influenced during the acquisition of culture from external influences.

## Religion

Many people have some kind of religious faith or have been brought up following the edicts of one of the world's major religions, be it Christianity, Islam, Judaism, Hinduism or Buddhism. The UK has a formal, state religion built into its constitutional monarchy, referred to as an established religion. It is a Protestant Christian religion, divided into Churches representing the countries of the Union: England, Scotland, Wales and Northern Ireland. There are many other religions, and a long history of political struggle between the dominant Protestant faith and the previous historically dominant Catholic faith. In addition there are Methodists, Baptists and all the religions of the UK's immigrant communities. However, neither multiculturalism nor ecumenicalism is reflected in the country's political institutions, as the current Queen, who is head of state as well as head of the Anglican churches, has the formal title 'Defender of the Faith', meaning the established church, and not of all the faiths within the country. Prince Charles, heir to the throne, has stated it is his intention to be enthroned 'Defender of Faith' in an effort to be more inclusive of all Britain's many religions.

Because of the establishment of the church, it is the norm for religion to play a part in all major state occasions. At the coronation of a new monarch oaths are sworn before God to undertake the duties of sovereign. At the State Opening of Parliament the Queen gives the Queen's Speech, where she reads aloud the intended legislative programme of the government elected to govern in her name, and prayers are said. Royal weddings and state funerals take place in religious locations, such as Westminster Abbey or St Paul's Cathedral. Indeed, these major religious buildings are part of the country's cultural heritage and tourist landscape, linking state and church ever more tightly. There is at least one church in most towns or villages in the country.

At the local and more personal level, the norm for many, but not all, families is to mark the significant events of life, birth, marriage and death with ceremonies that are religious in nature. For the majority of these people the religion in question will be a form of Christianity. In this tradition, a baby's birth is celebrated with a christening, whereby the new infant is formally inducted into the church through a ceremony representative of the Biblical ritual washing known as baptism. Godparents are appointed to offer spiritual and moral guidance, as well as having the functional role of taking over the children should the parents die. Other religions have their own similar ceremonies to welcome newly born family members, including, in many cultures, ritual circumcision of male babies. Even those who have grown up in a religious tradition but have rejected it still feel the need to celebrate the birth of a child with family and friends through some kind of secular naming ceremony following a recognised pattern, even though the mention of any God is omitted.

Similarly, the pair bonding that marks human relationships is institutionalised in celebrating the joining together of couples through legal marriage and wedding ceremonies. The Christian image of this religious ceremony has become the international symbol for marriage: the white dress of the bride and the church building as a backdrop. In the UK until recently there were only two places licensed to hold legal marriage ceremonies: registry offices and churches. This changed in the early 1990s to allow marriages to take place in any suitable venue holding a licence to hold weddings, which includes venues like hotels and stately homes.

Finally, it is unusual in the UK to mark the passing of a relative or friend without the involvement of Christianity. Funerals, even if they do not consist of ritual burial of the corpse within consecrated ground such as a churchyard, are 'normally' presided over by a priest who invokes the care of God for the individual who has entered Heaven. This is the case even at municipal crematoria. It should be noted that in the UK at the end of the twentieth century, participation in organised religion on a regular basis is low. Religion also has a significant influence on language, as outlined below.

**Politics**

The link between culture and nationalism, and by implication politics, has been mentioned above. However, the political system under which we grow up has explicit and implicit effects on our personal cultural norms and values. The UK is a liberal democracy with a largely conservative tradition. While the monarchy's power has been limited over the centuries, and universal suffrage and the rule of Parliament have emerged, this has been achieved with the agreement and deference of the people and not as a result of large-scale popular revolt, apart from the brief Civil War of 1642–52. Following Oliver Cromwell's short-lived republic, the monarchy was restored in 1660 but severely restricted and subjugated to Parliament from 1688 onwards.

There are certain traditions that instil in the UK's citizens certain standards of civic behaviour and conduct. We consider democracy to be the norm, and that political life should be conducted with fairness and justice, decisions being made on a majority basis. We expect fair play, equity of access to power and that everyone's views will be taken into account.[3]

Nevertheless, there are some differences in approach to the state's institutions, with particular activist or terrorist factions taking more direct and undemocratic action in an attempt to achieve their goals, for example the terrorist groups in Northern Ireland. On the surface, it seems that the UK's political culture is largely homogeneous and respectful of power and authority. The national consensus could be said to be breaking down at the end of the twentieth century, with votes in favour of devolution from central government in Westminster to regional Parliaments for Scotland, Wales and Northern Ireland. Kingdom writes, 'The geographical, class, gender and racial cracks in the social fascia are only smoothed over with political Polyfilla and concealed beneath unwritten constitutional wallpaper'.[4]

**Economics**

The UK is a regulated free market, as discussed in Chapter 2. This also gives us a set of economic norms and values that affect the way we behave and what we consider to be right or wrong behaviour. This was particularly evident in the western media reporting of the changes in Eastern Europe when the Communist governments lost power and former planned economies introduced free-market reforms. This was reported in many media as the 'normalisation' of their economies, rather than changing from one economic system to another, because now these countries were beginning to do things the way 'we' do and so were now deemed to be normal.

**Education**

The way in which we are educated (or, in some societies, whether we are educated) affects our comprehension and synthesis of the world around us. The word 'educate' originates from the Latin meaning 'to lead or draw out', and in western societies the focus for education is to draw out intelligence and understanding that are deemed to be inherent in all. The western tradition, founded in that of ancient Greece, is to teach people how to think, to question, debate and argue their point with philosophical underpinning and supporting individual freedom to differ.

**Language**

When considering the affects of language on the development of culture,[5] there is a fundamental philosophical aspect to consider that cannot be resolved in this text: does language control thought or does thought control language? If language controls thought, then the way we use our language has some sort of control over the attitudes we hold and the norms and values that underpin our culture. If thought controls language, then the language we use is a symbol of our cultural norms and values.

In fact, when examining British culture it is possible to identify both. Aspects of its historical, political and religious development are evident everywhere in its language. First, the fact that until recent decades all British languages other than English were suppressed is evidence of English colonisation (Wales is now bi-lingual). Nevertheless, English is itself an impure language, having origins as it does in the Scandinavian languages spoken by the Angles, German spoken by the Saxons, Latin spoken by the Romans and French spoken by the Norman conquerors. So we have a formal, high-register language full of polysyllabic words of Latin or French origin, reflecting the language of our rulers, while we have more Anglo-Saxon[6] monosyllabic slang and swear words, reflecting the social position of the indigenous peasants. Even the language of food contains class distinctions: beef, mutton and pork for the meat eaten by Norman aristocrats, but ox, sheep and swine for the animals tended by the

Anglo-Saxon serfs. Due to this heritage, when inventing words for new technologies English reverts to scholarly words of Latin or Greek origins, while German or Chinese simply use words to describe the function of the new invention. Thus television (which caused a scandal at its coinage for being a hybrid of Greek and Latin) comes from tele ('far') and vision ('sight'), while in German the original word was simply *Fernsehen* ('far seer') and in Chinese it is *dianshi* ('electric sight'). Either of these would seem ridiculous in English.

In UK society, however, it is the way in which an individual uses the English language that says more about him or her than the mere words being used. As Britain remains a class-oriented society, dialect and accent can be a social advantage or disadvantage, depending on the context. There are different types of English that are taken as the standard or benchmark language: the Queen's English, Oxford English or BBC English. These are in fact not the standard language, but are particular class or regional accents that are considered to reflect 'received pronunciation' (RP), or the way we ought to speak. The Queen's English is an aristocratic accent that evolved from her German ancestors, the Hanoverian Georges (I, II, III and IV), whose German accents the English courtiers imitated in order that their sovereign did not feel alien when in England. Oxford English refers to the English exhibited in the dictionaries and grammars written by scholars at one of the UK's oldest universities, a seat of learning that by definition has been invested with the authority to set the national standard. The national broadcaster, the British Broadcasting Corporation (BBC), has changed in its attitude towards accent over its 75 years of broadcasting. In the early days, only the King's English was broadcast across the nation and Empire. In fact, the King was one of the early broadcasters, using the new technology to send messages to his subjects around the world. During the Second World War, broadcaster Wilfred Pickles read the news bulletins and was deemed to have a strong Yorkshire accent for the time, although to the modern listener he sounds as 'posh' as all of his contemporaries. However, regional accents are now commonplace among presenters.

Nevertheless, accent is a passport giving access to different milieux in UK society. If you speak in an accent that sounds upper class, you are immediately accepted by 'posh' people into their society. If you speak colloquially and with a strong regional accent, you are accepted in that part of the country as 'one of us', but may be shunned elsewhere. Some accents have national reputations: Scottish or Yorkshire people sound trustworthy and reassuring; people from Birmingham are widely deemed to be amusing and less intelligent. These language and culture relationships are being used by large companies when selecting locations for national call centres for their direct telephone services. Thus as soon as an individual opens his or her mouth to speak, those around make immediate value judgements about social class, profession, education, status, ability and personality.

**Social structure**

British social structure has changed vastly over the last century. It is now a more fluid, dynamic and meritocratic society, with possibilities for social mobility through the classes depending on effort and ability. Up to the end of the Second World War, the British population knew its place in society and did not expect to undergo social change: once an aristocrat, always an aristocrat; once a manual labourer, always a manual labourer. As post-war social attitudes changed, so did traditional attitudes to authority, the family, our elders, and the more disadvantaged members of society such as the poor or the ill.

An aspect of social structure that has changed considerably is the family. Grown-up children move away from home and settle down in other parts of the country or even the world, following economic trends and the necessity to work. This means that they have no help from their parents with childcare, and the parents have no family members to care for them in their old age. This puts pressure on society in terms of providing healthcare for the elderly and childcare for pre-school toddlers. As divorce statistics grow and people marry for the first time later in life, the number of single-person households is growing (29 per cent of all households in England in 1996 compared to 18 per cent in 1971;[7] for further detail *see* Chapter 2), which affects both the way social structures operate and also government reaction to taxation, healthcare provision and education. In addition, the UK at the end of the twentieth century is a far more informal place than ever before. Attitudes on the part of young people towards their peers and their elders are more egalitarian and tolerant than at any time in the past.

## Applying Hill's model in a Chinese context

The value of Hill's model (*see* Exhibit 5.4) is as a framework to help us consider what our culture is and how it affects who we are and how we behave as individuals in society. To aid in this understanding, it is now re-applied to a generic Chinese cultural context.

As part of this it is necessary to examine the issue of what it means to be Chinese and what being foreign means to a Chinese. In the standard Chinese language the most common name for China is *zhong guo*, meaning 'middle country'. Thus China is the country at the centre of its universe, which equates with many cultures' own view of the world. Logically, everything that is not inside the middle country is outside it, hence the Chinese term for foreigner is *wai guo ren*, literally 'outside country person'. This can be compared to traditional Chinese life, especially in the countryside, where people are immobile and generations live and die in one locale. Anyone not from the same village or town is known as *wai ren* or 'outsider'. It can further be observed that *wa guo ren* is habitually used to mean 'white people'. The traditional image of foreigners for Chinese, often seen in the media, is of a *wai guo ren* with white skin, blond hair and blue eyes. When referring to other races, Chinese will usually specify for example 'Japanese', 'black people' or 'Arabs'. In the wider Chinese diaspora the term *wai guo ren* is used to refer to non-Chinese (and specifically white people), irrespective of whether the latter are in China or not.

Chinese remain Chinese and *wa guo ren* (white people) are *wai guo ren* (foreigners) even when nationality is shared or the white people in question are the indigenous population. Chinese never refer to themselves as *wai guo ren*. In contrast, English native speakers use nationality as a determinant, and are comfortable with referring to themselves as 'foreigners' when they are in another country.

After 'Liberation' (the Communist Party takeover in 1949), most foreigners left China, apart from some committed to the revolution's aims. From 1949 to the mid-1980s any foreigners visiting for business or pleasure were closely supervised by cadres of the Foreign Affairs Office to maintain an official filter between the bourgeois, capitalist outsiders and any Chinese people with whom they had contact. Interactions between foreigners and Chinese became carefully crafted and scripted events designed to put both parties as little at ease as possible. During the Cultural Revolution (1966–76), Chinese were persecuted to death for having had contact with foreigners or even for having relatives abroad, thus being culpable in the eyes of the Red Guards of bourgeois rightism and counter-revolution. While recent years have seen the normalisation and humanisation of Sino-foreign relationships, with many Sino-foreign marriages resulting, there is still to some extent a psychological hang-over from the fervently anti-foreign dogma of the post-Liberation years that can affect the operational effectiveness of foreign workers in China.

#### Religion

The main religions affecting social and economic behaviour are Confucianism, Buddhism, Islam and Christianity. Attitudes towards issues of crime and punishment, sex and the family, the position and respect for the aged and how society cares for them can be set by ideological dictate. Buddhism, with its belief in reincarnation, leads the Chinese to be quite fatalistic in their view of the individual's importance in the grand scheme of things. Religious influences are stronger in Chinese societies outside China than within it, where religion has been banned and punished heavily in certain periods since 1949.

#### Politics

Irrespective of political colour, political systems in most Chinese societies are largely authoritarian, non-democratic or oligopolistic. In mainland China there is no aspect of life on which politics has not had a huge impact at some time in the last 50 years. It is hard, looking from the perspective of the West, to imagine a society where even the morning delivery of milk could depend on your having demonstrated the correct ideological stance at the workplace's weekly political study meetings. Indeed, the idea of a political study meeting at all in the workplace would seem inappropriate.

#### Economics

Early entrepreneurs tackled the problem of business with the People's Republic of China (PRC) or 'New China' unaided by source materials, secondary data or effective diplomatic relations. Doing business with the Chinese was a very unfamiliar process and experience, for two main reasons. First the system, with its interminable bureaucracy, vertical integration and complete lack of flexibility, interspersed with periods of complete economic breakdown due to the supremacy of political dogma, makes coping with officialdom difficult for a western businessperson. Secondly the culture, with the uniformly and impossibly inscrutable Chinese people and their unwritten, impregnable, yet unbreakable rules of engagement, led to paralysis through analysis on the part of foreigners interacting with them. Old China hands learnt their way around through trial, error, good luck and good judgement.

The Deng era, with burgeoning official foreign trade organs at central and local levels, saw the rapid growth of the number of organisations empowered to deal directly with overseas organisations. Hence new entrants to the marketplace in China have been able to use the services of consultants, 'how to' guides, agencies, information bureaux and government departments to seek the appropriate Chinese opposite number. While the system grows ever more complex as the volume of trade increases and hurdles are undoubtedly still numerous, there are now mechanisms in place to help foreigners understand and manipulate the system in order to meet corporate and individual needs on both sides. Even though there is increasing economic liberalisation, the economic systems in China and elsewhere in Confucian societies have featured interventionist government macro-economic policies to promote stable and rapid economic growth. It was only after the economic crises of late 1997/early 1998 that this interventionist stance was beginning to be questioned.

**Education**

Education is very highly valued and greatly prized. Parents spend a large amount on ensuring the education of their offspring and their advancement to professional success and social security. This is a legacy from imperial days, where the Mandarins ruling China on the Emperor's behalf were all scholars who had passed rigorous entrance examinations. Confucian teaching also respects and values education and intelligence. The teachings of the ancient Chinese philosopher Confucius left a huge legacy in China, Japan and Korea as well as in Chinese societies in other countries. The tradition of master and pupil is much more didactic, in that the teacher is always right and should be copied and emulated by the student at all times. The role of and respect for formal education, training and qualifications remains strong in most countries of the Far East. There are clear expectations that families will provide the funding and support for education and the student will put in the hard work necessary to succeed.

**Language**

Chinese is one language where the written form is not only a tool for communication but an art form as well. Chinese is an ideographic not an alphabetic language, meaning that each word is a picture. Having a good handwriting, or calligraphic style, is the sign of a good scholar and thus the sign of a good ruler. Otherwise, Chinese is a language of simple structure and ambiguous meaning. Unlike English, the fewer the number of words, the more formal the language being expressed. Thus each word has many meanings and is open to the interpretation of the individual and many nuances.

**Social structure**

The regional socialisation process in the Far East is strong and dominant, perhaps much more so than in Europe and North America. Social structures are still rigid and well defined, giving everyone a clear picture of their position and role in society. This fixed view of outsiders, described earlier in this chapter, comes from a cultural source as well as from the fact that the Chinese state classifies the nationality and ethnicity of its citizens automatically at birth. China's 1.2bn citizens are categorised officially into 56 ethnic groups,[8] 96 per cent of the population being Han Chinese and 4 per cent divided into 55 official ethnic minorities. This is of key importance to the people themselves,[9] as Han Chinese are bound by the 'one child only' population control policy, while ethnic minorities may have two children.

## The determinants of organisational culture

We now move from personal to organisational culture. The Johnson and Scholes cultural web[10] identifies and draws together many aspects of organisational culture: 'The cultural web is a representation of the taken-for-granted assumptions or paradigm, of an organisation and the physical manifestations of organisational culture.'[11] In other words, organisational culture is determined by the entities taken for granted by the people in an organisation. If an organisation is to function effectively it must develop a coherent culture. This is supported by Deal and Kennedy,[12] who identify two types of culture, strong and weak. The strong culture is highly cohesive and coherent and has a system of informal rules, which indicates to people exactly what is expected of them, so that employees will know how to react and what to do in given situations. In contrast, people operating in a weak culture, one lacking in cohesiveness and coherence, will waste time working out what to do and how to do it.

The entry case study for this chapter showed how changing the determinants of organisational culture in Shenyang Antibiotics Company resulted in a stronger organisational culture. The changes in culture are illustrated by changes in the organisational and power structures, for example the new owner Mr Jiang sacked two-thirds of managers and clearly defined the roles of those remaining. The routines and rituals were also changed by Mr Jiang, with consequent alterations to the behaviour of workers. Workers were now required to arrive at work at 8am and do a full day's work, in contrast to the past when they turned up at 10am and went home at lunchtime. Those failing to work proper hours and perform their duties correctly were sacked. There was also the symbolic sacking of an employee who smashed a chair reserved for a dignitary at an opening ceremony, which sent a very clear message to remaining employees as to what behaviour was and was not going to be tolerated by the new order.

**Routines and rituals**

Routines (*see* Exhibit 5.5) are the scheduled and deliberate practices carried out as a matter of course and forming the habits of day-to-day life in an

**Exhibit 5.5**
**The cultural web**

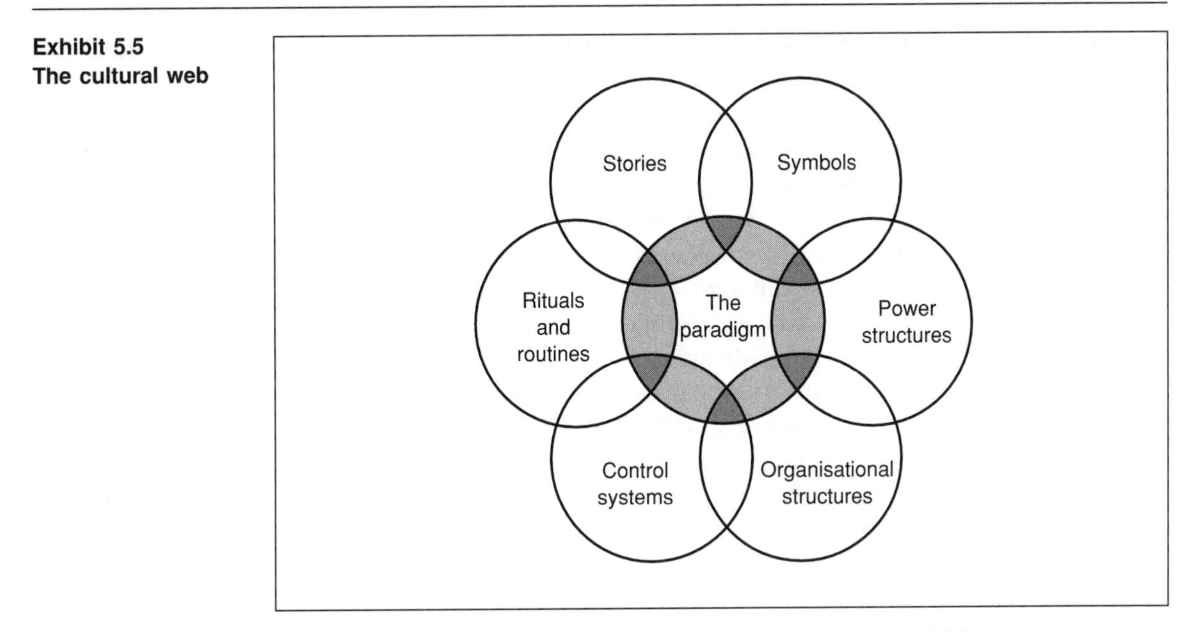


*Source*: Johnson, G and Scholes, K (1999) *Exploring Corporate Strategy*, 5th edn, Prentice Hall Europe.

organisation. In normal circumstances routines ensure the smooth running and operation of the organisation. In organisations with a strong culture, routine behaviour is very clearly spelt out and allows employees, particularly new employees, to know and understand 'the way we do things around here'. A good example of an organisation where routine is important is a fast-food restaurant such as Burger King and/or a chain restaurant such as Pizza Hut. Here the actions employees have to take in preparing the food and taking orders are very clearly and explicitly laid down, e.g. frying the french fries for exactly seven minutes or always asking the customer if they would like side orders of garlic bread or salad with their pizza.

Rituals in organisational life are used to reinforce the routines and 'the way things are done around here'. Rituals can be formal events that employees are subjected to such as induction courses, training courses or periodic assessments to ensure an employee's performance is up to scratch and conforming to the routine way of doing things. Rituals may also be more informal in nature, for example the office Christmas party, drinks in the pub on Friday at the end of the working week, or gossiping around the office coffee machine. However, they still promote the common routine of the 'way we do things around here'.

**Stories**

In any organisation, stories will be told by employees to each other, to new recruits who join them and to others outside the organisation. The stories represent the organisation's history and typically highlight significant events and characters in its past. They characteristically focus on the achievements and failures of the organisation and the individuals involved, be they heros or villains. Stories summarise the meaningful and key aspects of an organisation's past and tell people what counts as acceptable conduct today.

**Symbols**

The symbols present in an organisation can be many and varied and often symbolise someone's position in the organisation or how much that individual is valued by the organisation. Symbols can include titles, office size, company car and salary scales, with all indicating the power and value that an individual possesses with respect to an organisation. In long-established organisations like the Civil Service, many symbols will exist and indicate the power and importance of employees. In such organisations there will be a rigid structure, comprising different jobs at different grades with different salaries, with office accommodation directly dependent on job, grade and salary. Individuals with better offices, higher salaries and further up the hierarchy will have greater power and may be perceived to be of greater value to the organisation. In contrast, a newer organisation, such as an architects' practice which has all staff on performance-related pay, all working in an open-plan office, displays a different culture by virtue of the symbols that do or don't exist. The message in such an organisation is that all employees are equally valued and succeed on merit.

**Power structures**

Power structures evolve in organisations over time and consist of individuals with power, who all share a common set of beliefs and values that underpins the way they work together. Membership of the power structure is often determined by seniority and/or length of service in an organisation. Alternatively, power may be based on expertise, with a common source of power being technical expertise that is in short supply and highly valued in the organisation. This may occur particularly in firms where innovation is a key success factor; *see* Chapter 9. This type of power base will be strengthened if there are many valued experts who group together to promote or resist particular issues in the workplace.

**Organisational structure**

The structure that an organisation adopts will determine where the power exists within it; *see* Chapter 4. The location of power in an organisation will define the power relationships and designate the fundamental linkages between the seats of power and control. Nevertheless, in any formal organisational structure there will exist smaller, more informal structures and networks, which are equally important to the culture of the organisation.

**Control systems**

The term control systems denotes systems for control, measurement and reward within the organisation. The systems that an organisation puts in place and monitors indicate what is important to it. Control systems include financial control and accounting systems such as cashflow and budgeting, which are systems for regulating expenditure. Measurement systems examine the output of organisations and their efficiency and effectiveness. The output of an organisation can be the amount of product manufactured or the throughput of customers. Efficiency and effectiveness relate to the aspects of time and resources used to produce the final output/throughput. In some organisations the control of expenditure will be more important than the measurement of output, in others both will be equally important.

The reward system in an organisation will determine how employees behave with respect to their work and jobs. A reward system that pays for a large volume of work will elicit very different behaviour to a reward system that pays for high-quality work.

## Deal and Kennedy's organisational cultures

Deal and Kennedy[13] examined hundreds of companies and claim to have identified four generic cultures: tough-guy macho, work hard/play hard, bet your company and process. These cultures are defined by two factors in the marketplace, the degree of risk associated with the organisation's activities and the speed at which the organisation and its employees receive feedback on their performance. These cultures are summarised in Exhibit 5.6. Deal and Kennedy also acknowledge that no organisation will exactly fit one of their four generic cultures and some may not fit any at all.[14] However, they maintain that such a framework is a useful initial step in assisting managers to recognise the culture of their own organisation.

**Tough-guy macho culture**

In organisations exhibiting the tough-guy macho culture, it is customary for staff to take high risks and receive rapid feedback on the effectiveness of their actions. Deal and Kennedy indicate that police departments, surgeons, management consultants and the entertainment industry may all exhibit a tough-guy macho culture.[15] The key characteristics of this culture are rapid speed and the short-term nature of actions. This results in great pressure being placed on the individual culture to achieve success in the short term. This is often illustrated by such organisations having young staff achieving financial rewards early in life if they are successful.

The consequences of this type of culture are that burnout is common and failure is harshly condemned, often by dismissal. Those who do succeed and avoid burnout often do so by taking a tough stance with regard to their work

**Exhibit 5.6 Deal and Kennedy's organisational cultures**

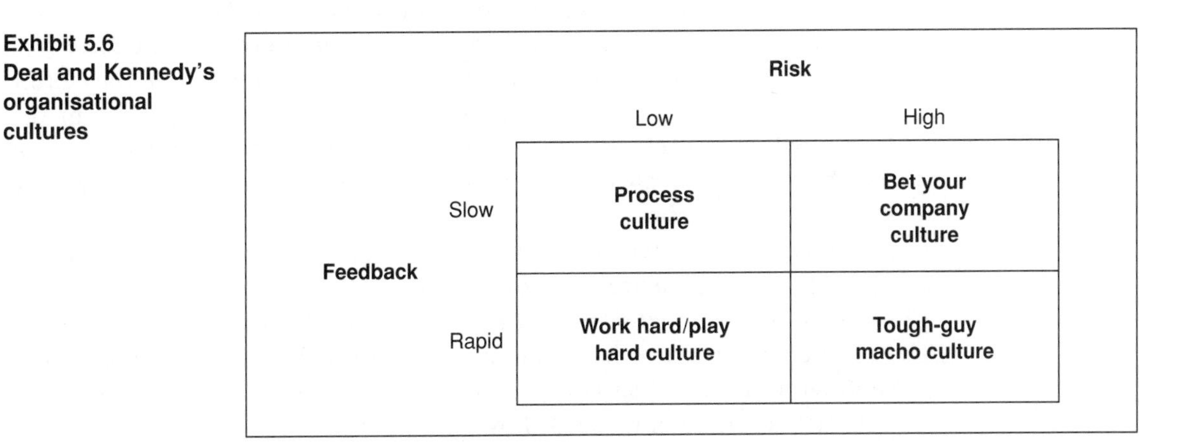

and their colleagues, and pace themselves accordingly. This results in internal rivalries occurring, which in turn produce tension and conflict between staff, which is normally expected in these organisations. An organisation with this type of culture achieves quick returns from its high-risk environment, but it finds it difficult to achieve success via long-term investment. This is due to rapid staff turnover, along with very limited co-operation and tolerance between staff. Hence the creation of an organisation with a strong and cohesive culture is almost impossible if the predominating culture is tough-guy macho.

**The work hard/ play hard culture**

Organisations for whom there is low risk and quick feedback on performance are those with a work hard/play hard culture. Sales along with a sense of fun and action are key characteristics of this culture. Typical work hard/play hard organisations include both manufacturing and service companies, such as fast food and computer companies. In both types of organisation the risks are small; in a service provision organisation failing to sell a single item will not severely damage the salesperson, and in a manufacturing organisation examinations and inspections will ensure that departures from the normal standard of product are minimised. Quick feedback on performance is easily obtainable in such an organisation, e.g. whether staff have achieved sales or production targets. Hence organisations with a focus on sales and meeting targets are often customer oriented. This may be reinforced by the use of contests, games and rallies that focus on the achievement of individuals and teams of employees and are meant to motivate staff to succeed. There may be an inclination to focus on the sales volume achieved by individuals or teams, at the expense of service/product quality, and thus there is a focus on the short-term rather than the long-term future.

**Bet your company culture**

The bet your company culture organisation takes high risks and waits a long time for the response to actions and decisions. This is because the investment is huge and long term and the outcome is seen in the long-term future. Examples include the manufacture of aircraft as undertaken by Boeing, or the finding and refining of oil as in British Petroleum. Hence there is an enormous amount of detailed planning that has to take place, evidenced in the ritualistic business meetings that occur. Decision making in this culture focuses on the future and is top down, reinforcing the hierarchical nature of the culture. The person who fares best and survives in this type of culture is the mature worker with a respect for authority and technical ability. They will also possess the skills to operate effectively in a team with similar-minded people and to cope under pressure.

The result of this type of working environment is that many high-quality innovations and scientific discoveries are made. However, innovation and scientific breakthroughs are long-term goals and this makes such companies vulnerable in an economy and to a stock market that are more interested in short-term profit and success. Nevertheless, it could be that companies that innovate are those most needed by western economies.

**The process culture**

The process culture is a low-risk and slow-feedback culture. The response to actions and decisions seems to take forever. This type of culture is typically seen in the Civil Service, public-sector organisations, banks and insurance companies. The lengthy feedback time means that employees focus on how something is done, i.e. the process, rather than the reason for doing it or the outcome. The employees who survive best in such a culture are methodical and punctual. In the process culture there will be significant emphasis on job titles and roles and this will be symbolised by the size of someone's office and the style of office furniture that the rigid, strict and hierarchical organisation will allow that individual. This illustrates one further point concerning the process culture: the position that a person occupies determines the amount of power that individual wields. This is the type of culture in which remaining with the organisation and enduring will be rewarded by long service awards.

The process culture is most successful if the organisation operates in a predictable and stable environment and is perhaps likely to struggle if asked to react quickly to rapidly changing circumstances, as the organisation lacks the creativity and vision to do so. An alternative view is that such a culture offers a balance to the other three cultures, which all have either high risk and/or rapid feedback.

## Understanding and managing culture

An understanding of personal cultural provenance and national culture is important, as it allows managers to develop the cultural awareness that is needed in the world of work at the beginning of the twenty-first century. This understanding of personal and national culture allows today's employee to appreciate the differing personal and national cultures of employees in a firm in a foreign country (competitor or customer) or individual consumers in a foreign country. They may have different expectations and different needs and wants (*see* Chapter 6) to those in the home market. In contrast, the entry case study for this chapter clearly demonstrates that overseas managers in China are faced with different organisational cultures resulting from a different national culture with regard to work – the 'iron rice bowl'.

Differences in national culture will be reflected in the way that organisations collaborate with one another in the international arena and evaluate the outcome of their activities. The compatibility of national cultures may influence an organisation as to the nationality of a collaborating partner company. For example, British companies least like to partner Japanese companies, primarily because of language difficulties.[16] The management of corporate or organisational culture is as important as that of individual and national culture and hence has a role to play in the overall management of culture. This is looked at in the section prior to the conclusion of this chapter.

## Managing culture in the international arena

### Evaluating success

The way in which organisations evaluate their success or failure can reflect their home country and culture; *see* Exhibit 5.7. Companies for whom the USA is their home country are most likely to measure performance on the basis of key quantitative measures such as profit, market share and other key financial benefits. In contrast, Japanese companies evaluate success or failure via skills improvement and how that has strengthened the organisation's strategic position. In European companies a balance between profit and the meeting of social objectives is more often sought.[17] The existence of such variations in the way performance is evaluated can produce difficulties if two collaborating companies both have very different expectations as to how success or failure is measured. The most extreme situation would be if one partner viewed the collaboration as a success and the other partner a complete failure.

Problems may also arise from differences in the corporate cultures of collaborating organisations. A collaboration between two companies is unlikely to be successful if the organisations involved have very different corporate cultures and neither is prepared to change. Referring back to Deal and Kennedy's organisational culture types discussed earlier in this chapter, merging an organisation with a tough-guy macho culture (high risk and rapid feedback) with a process culture organisation (low risk and slow feedback) is clearly unlikely to be successful. Hence organisations may only agree to collaborate on large long-term projects with each other once the water has been thoroughly tested by working together successfully on smaller projects, over a significant period of time. Therefore cultural compatibility is critical to ensuring the consolidation of business relationships.

### Communicating with consumers

Consumer buying behaviour is complex and in the international arena there are many potential constraints that the marketer has to overcome to be successful in a foreign country. These include differences in language, taste and

**Exhibit 5.7**
**Issues for managing culture in the international arena**

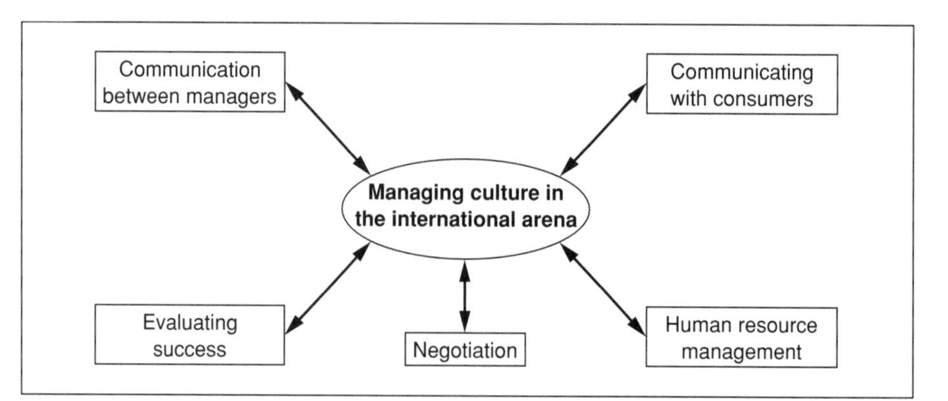

attitudes of the target market, as well as variations in government control, media availability and local distribution networks. Hence it is difficult to determine in advance whether new or different products will be accepted by an international or overseas market.

At the start of the twenty-first century there exists a large number of global brands that are familiar to people in many different cultures, such as McDonald's, Kodak and Levi's. However, even successful global brands have experienced difficulties in being accepted. For example on entering China, Coca-Cola provided shopkeepers with signs in English to advertise the soft drink. This was a mistake, as the Chinese shopkeepers translated the English signs into written Chinese, with the literal Chinese translation being 'Bite the wax tadpole'. This, not surprisingly, held limited appeal for the Chinese and was revised to read 'happiness in the mouth', which is more acceptable and appealing to the target market.[18]

**Human resource management**

If a manager is posted abroad to manage a subsidiary of a parent company, they are likely to find that they have wide-ranging responsibility for all functions of the business and relations with external stakeholders such as government, the local community, suppliers and customers. Selecting a manager (*see* Chapter 9) to fulfil such a role needs to be done with care. This is because managers with similar profit or cost responsibility at home in the larger parent company are only middle-level management, and lack the skills and abilities to perform as a top manager in a foreign environment.[19]

The other type of foreign experience that a manager may have to undertake is as an international manager who finds him/herself frequently interacting with very high-level authorities in foreign countries. For example, this may occur when a construction company negotiates with a foreign government for a contract to build major infrastructure projects such as new roads or bridges, or if a company is negotiating to expand current facilities in a foreign country or selling a new, innovative technology. The tasks of an international manager are even more complex than those of subsidiary managers based in one foreign country, as an international manager has to gain trust and build relationships with officials in many foreign countries. Therefore the international manager will have to deal with the cultures of many countries. Appreciation and understanding of one's own personal cultural provenance and national culture are good starting points from which to build a comprehension of the different cultures in which one may work.

**Communication between managers**

International managers or those interacting on a regular basis with cultures different from that of their home country must ensure that messages between headquarters and subsidiary operations are clearly understood. The advent of technology such as e-mail and faxes makes written communication with people almost anywhere in the world possible in an instant. However, there may be instances when direct contact and verbal communication are preferable to ensure a complex message or idea is correctly understood by its overseas recipients. This is achieved either by international travel, careful use of the telephone (take account of time difference) or video conferencing.

It should also be noted that the language of communication may influence how it is received and understood. A manager receiving a message in a non-native language is likely to take longer to read and comprehend it. Equally, a manager working abroad and having to carry out at least some of their work in a second language will take longer for the same reasons and have to work harder than when at home to produce the same quality of work.[20] Therefore in recruiting overseas managers and international managers, the language part of one's personal cultural provenance is important.

**Negotiation**

A country's national culture is likely to influence the way managers from that country behave in negotiating contracts with managers from a different national culture. For example, negotiations between the Saudi government and a British company wishing to secure defence contracts will be very different in nature to negotiations between an American and British firm wanting to merge. Hence the type of issue under negotiation and the national culture of negotiators are both likely to influence the nature of the negotiations. In some national cultures it will be normal for individual negotiators to have the power to make decisions, in contrast to other national cultures where referring back to those behind the scenes and head office will all be seen as part of the negotiating process. In some cultures negotiators are required to go through every line of a contract and every possible contingency, in comparison to other negotiators who will be satisfied with a holistic view and understanding of the contract.

The behaviour of the individual negotiators is based on their national culture and this can influence social behaviour in the negotiating process. For example, in some cultures eating and drinking will form part of the negotiating process, in others it will not, or will only occur once a contract has been settled. Equally, some cultures place great importance on punctuality and others do not. Therefore understanding the national cultures of the different parties involved in negotiations will help those involved discern if the negotiations are based on their own culture, another party's culture or some hybrid of the different cultures involved.

## Managing organisational culture

If the relevance of personal and national culture to business today has been understood, then the same level of perception regarding organisational or corporate culture is required if an individual manager is to work and manage within the context of an organisational culture. The human resource management function is the most powerful of all the four key business functions in influencing the management of organisational culture. For example, the human resource management function will help determine organisational

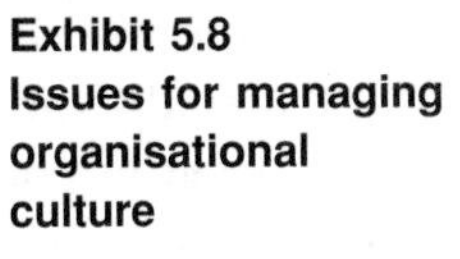
**Exhibit 5.8**
**Issues for managing organisational culture**

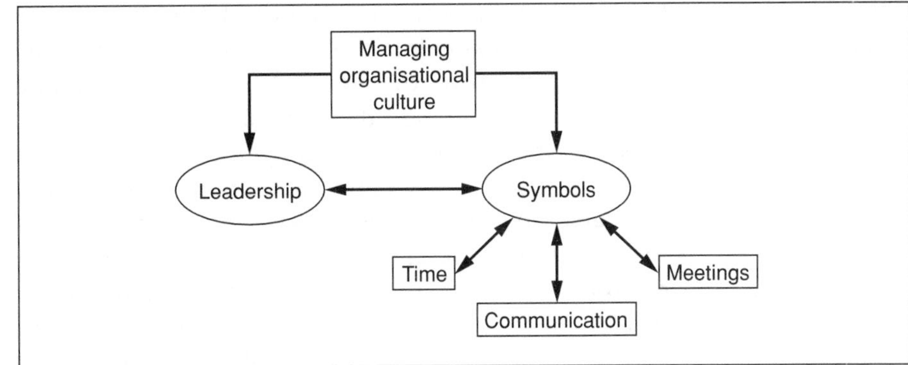


rituals, such as induction courses, training courses and appraisal (*see* earlier in this chapter), as well as cultural symbols such as the allocation of offices, furniture, company cars, job grades, salaries, promotions and dismissals. Hence human resource policies and procedures can have a great influence on an organisation's culture. The outcomes of an attempt to manage culture using the complex influence of human resource policies and procedures are difficult to predict.

If a culture emphasises the importance of teamwork and innovation (*see* the bet your company culture earlier in this chapter) as crucial for success, the managers responsible for culture will want to create one that rewards imaginative and inventive technical behaviour and co-operation and collaboration with others in the workplace. This will mean that rewards, salaries, promotions and bonuses will have to reflect this focus. This may appear to be simple and straightforward, but it is not necessarily so. The lack of simplicity is due to the difficulty in foreseeing the full implications of a specific reward system or promotions policy. There are two main reasons for this: first, the full workings of any policy or procedure are often not determined in enough detail prior to implementation; secondly, those implementing the system and procedures do not always follow directions to the letter, putting a different interpretation on the policy and procedures and hence introducing unexpected outcomes. This variation in interpretation of the policy and procedures compounds the complexities of managing culture in this way.[21]

**The role of leadership**

The successful management of organisational culture requires the support of top managers, as they play a key role in setting the vision (ideal culture) that the organisation is aiming to adopt. Top management also has responsibility for allocating tasks, activities and resources, and determining the organisational structure, which affects the power structures and control systems in the organisation. Therefore if human resource policies and procedures, as discussed in the previous section, are to be used in the management of culture, it is essential that top management is involved in the design of human resource policies and procedures, as part of its leadership responsibilities.[22]

## The role of symbols

Peters[23] recognises a variety of different characteristic practices that can enable an executive to influence the culture of their organisation. These include how top executives spend their time, their use of communication and their use of meetings. The dominant theme in all these is personal enactment. Managing directors who seek to model an organisation's culture should individually personify the beliefs, values and assumptions that they seek to inspire in others. The same applies to all managers endeavouring to influence employees in the departments or divisions they manage. People generally comprehend a remarkable amount from modelling the conduct of those they respect, particularly if other benefits are derived from doing so. Leaders should seek to maximise the impact of such symbolic actions continually and regularly. Symbolic actions need to be positively reinforced in the shape of praise, money, status and other rewards to champion behaviours in line with the desired culture.

### The use of time

Senior and middle managers are generally perceptive and sensitive to the activities of their leader and will spend time determining the consequences of what is seen and heard for their current and future careers. Hence a chief executive is able to communicate influential messages to employees through his/her actions.

### The use of communication

A good managing director seeks to understand employees and their views on all elements of organisational life, including work activities, colleagues and the marketplace. If a managing director makes a public announcement that quality is the organisation's most pressing problem, then employees will listen. If the managing director raises the issue on an ongoing basis and in a memorable fashion by the use of anecdotes and stories, then in time people may begin to alter their view of the organisation and the key issues affecting it.

### The use of meetings

Organisational leaders enjoy significant authority in determining the key issues, quality, innovation and marketing, which are vital to an organisation and its success. For example, a top manager can communicate the relative importance of an event or meeting by simply turning up or by turning down an invitation to attend. If a managing director always attends quality meetings, but virtually never attends a meeting of the marketing team, then the relative importance of quality over marketing is clearly indicated. Accordingly, leaders have the power to convene, postpone and cancel meetings, fashion agendas and the manner in which minutes are written up. These devices have a part to play in moulding employees' understanding of what is required of them; what beliefs it is deemed acceptable to hold; and how they are expected to perform their work activities.

## Conclusion

There is a close link between culture internally and other elements of the internal environment, as well as elements of the external context. The size and structure of an organisation dictate the ways in which the people in the organisation are able to operate. Culture in society is generated by us and it also controls us. Whatever its provenance, external culture has an impact on the human resource inputs to the resource conversion process and in turn the organisation's operational abilities. Organisations strive for harmony between the work they have to do, the structure they adopt to make the work possible and the culture of the human resources they employ in order that goals can be achieved. Organisations can be multicultural internally, and may need to be both global in their outlook on business, operating in many markets simultaneously, as well as local in their sensitivity to the needs and wants of employees and customers in a particular environment. These cultural issues are revisited in Chapter 12, where issues of changing culture are discussed.

### ETHICAL ISSUES CASE STUDY 5.2

FT

# Home in on office culture

**By Alison Maitland**

Ted Smith realised his life was out of kilter when he came home from a US trip and saw a picture his son Tom had drawn of the family. When he asked why the Daddy figure was so small, the six-year-old replied: 'You're not here very much.'

The irony hit him hard, since he had been devoting 70 hours a week to introducing flexible working practices in his company, Glaxo Wellcome. He decided his own working pattern had to change. For the past two years Mr Smith has been leaving home at 6am to beat the M25 traffic jams, aiming to be back by 5pm to spend supper and bedtime with his two children.

His working day is still long – he spends about an hour e-mailing and phoning colleagues in the US once the children are in bed. But he has three hours a day with them instead of seeing them only at weekends.

Leaving work at 4pm was alien to the company culture. 'A lot of people said it was suicidal,' says Mr Smith. But last autumn he was promoted to human resources director with responsibility for the 4000 information systems staff in the UK and US.

Mr Smith will be talking about his own experience and the changes made at Glaxo at a seminar tomorrow at Roffey Park Management Institute. The meeting will focus on the importance of a sympathetic culture within organisations to make flexible working compatible with continued career success.

The institute will be publishing research showing that managers' desire to have more time for their lives outside work does not necessarily imply lower interest in their careers.

The institute had responses from 405 employees, most of them middle and senior managers, in private and public sector organisations.

Nearly a third worked at least 56 hours a week. Workloads had generally increased in the past two years and commitment was still equated with long hours in the office.

The prevailing corporate culture sees work and family as two separate domains, with employees either completely work-focused or home-centred, says Caroline Glynn, author of the report. 'There is not an accepted intermediate position.'

One hard-pressed respondent defined the desired balance as 'being recognised as a parent rather than it coming as a shock to people'.

Among male respondents, a surprising 39 per cent said they had chosen their personal lives over their careers, compared with 54 per cent of women.

Nearly 60 per cent of all respondents agreed or strongly agreed they would turn down promotion if they felt it would harm their personal lives.

But respondents disliked having to make that choice. Those least happy with their work-life balance were the ones who felt they had been forced to choose between work and home.

**Ethical issues case study 5.2 *continued***

Ms Glynn says people want both a fulfilling personal life and a successful career. 'They do not perceive their lives to be balanced if they feel unable to achieve their full potential in one of these areas . . . An ideal balance does not appear to involve compromise and sacrifice.'

Her research suggests organisations may be sending mixed messages about what is acceptable. Sixty-four per cent of respondents said their manager respected them as an individual with a life outside. Yet 43 per cent said they had to be ready to make big personal sacrifices to succeed in their organisation.

'It may be that at the one-to-one level people feel able to communicate their needs but at the organisation-wide level they feel acting on this will be punished,' she says.

Are people simply expecting too much of their employers? Or will employers have to change because growing numbers of staff feel there is a limit to the sacrifices they are prepared to make?

Ms Glynn argues that the business case for organisational change is all the greater given that demands for balance are coming from committed, senior managers.

BAA, the airports group, was surprised to find that the need for a balanced life was one of the two main motives mentioned by 209 senior managers in a survey examining their career choices. The other was the desire to learn specialist skills. Motives such as security, challenge and autonomy trailed behind.

Tom Leigh-Wilson, BAA's management development adviser, says there appears to be no link between age and the desire for balance. He believes it reflects the trend, accelerated by the last recession, for employees to be less committed to a single employer while still wedded to career success.

So how should organisations that want to retain gifted staff go about making the culture more conducive to flexible working?

Glaxo's experience suggests that crucial ingredients are leadership by example and trust in employees. The company asked focus groups of research scientists what arrangements would make their lives more manageable and then encouraged them to decide how best to implement the changes.

About half the UK workforce of 13 000 now has access to flexible working. More than 100 people are job-sharing and a further 200 are working part-time. Similar options are available in the US.

But Mr Smith warns that culture change takes time. He recommends pilot schemes as a way of winning over sceptical line managers – they can be wound up if they fail, though that rarely happens.

Symbolic gestures can help to change a culture of long hours. 'One of the things we were trying to break down was the culture of looking out of the window waiting for the boss's car to leave to see if it was safe to go,' he says. The directors' parking spaces have gone – and they use the main car park which cannot be observed from inside the building.

*Source*: *Financial Times*, 25 February 1999. Reprinted with permission.

**Questions for ethical issues case study 5.2**

1 Identify the advantages and disadvantages of changing your work pattern in the way Ted Smith did in the case study. Indicate which advantages apply to the individual, the organisation and both.

2 Examine your lists of advantages and disadvantages from your answer to Question 1. Do they apply equally to men and women? Explain your answer.

## EXIT CASE STUDY 5.3

FT

# Facing the pain in hope of gain

**By Peggy Hollinger and Alison Maitland**

Veteran staff facing the chop. Sweeping office closures. Sales reps ordered to work from home. Could corporate managers get away with the kind of brutal shake-up planned for the Conservative party machine?

What Archie Norman is doing at Central Office echoes his work in turning round Asda, a supermarket business which, like the Tories, had lost touch with its core customers and relied too heavily on its past success. 'Central Office is a 1950s office with a 1950s culture,' says Mr Norman, party vice-chairman and Asda chairman. 'That has got to change. Modern organisation has moved on and no one noticed.'

Patrick Moylan, partner at Kurt Salmon Associates, a management consultancy that has worked with leading retailers, says there is no question that drastic measures are sometimes justified. 'It took something as radical as

**Exit case study 5.3** ***continued***

Archie did to make Asda what it is today,' he says. 'That's what it's going to take to make Conservative Central Office function properly.'

Suffering seems to be an integral part of the process. Already Tony Garrett, the Tories' long-serving chief agent and director of campaigns, has resigned and dissenting voices are being raised among local activists.

Mr Norman, writing two years ago about the Asda experience, said change was inevitably 'radical and painful, not minor'.

There are other striking corporate parallels with the Tory party experience. For example, Ricardo Semler, the Brazilian who now has cult status on the international management circuit, inherited his machine manufacturing company Semco when it was on the verge of bankruptcy in 1980. Mr Semler created havoc by dismantling the company's highly conservative structure – a third of the management left over a 14-month period. The surgery worked and Semco became one of Latin America's fastest growing companies.

Mr Semler's maverick style allows staff to elect managers and boot them out. Employees use green flags for 'good mood' days and red flags to say 'not today thank you'. At Asda head office, staff wear red hats when they do not want to be disturbed. The mind boggles at how true-blue Tory officials might react to the idea.

Mr Norman, a former McKinsey consultant, dismisses as nonsense reports that most party staff over 50 will be sacked. 'We want to get rid of stick-in-the-muds but not necessarily older ones,' he says. But he also wants an office that is 'buzzing and crackling with excitement' and 'a fun place'. Tory insiders say older people are more likely to be in the firing line when voluntary redundancies are decided.

Helen Garner, campaigns manager at the Employers Forum on Age, warns that organisations make costly mistakes and risk losing their 'corporate memory' when they get rid of older, more experienced staff. Often companies have ended up trying to recruit back the knowledge they have lost, she says. Some might argue that after last year's stultifying election defeat, the Conservative party needs to wipe its collective memory clean and reinvent itself.

Mr Moylan of Kurt Salmon says: 'It could be very liberating for the people left behind to think some dead wood has gone and that avenues for career opportunities and progress have been opened up.' He says the workforce at Asda is far more motivated than it was eight years ago.

But Stefan Stern of the Industrial Society warns that disaffected employees can undermine change. At first sight, Mr Norman's Central Office reforms look like early 1990s re-engineering, from the top down, he says. 'We're in favour of consulting people and not imposing change.'

Philip Cox-Hynd, senior partner at Harley Young, consultants in change management, has doubts about the methods of the party leadership. He cites last week's launch of the 'Listening to Britain' exercise and says there appears to be a 'huge disconnect' between this and news of job cuts in the party. 'Organisations can't get away with saying one thing and acting another way.'

Everyone agrees timing is of the essence for the Tories to have any chance of winning the next election. 'We have 18 months to two years to really reconstruct the Conservative party. It might take longer but that is as long as we have got,' says Mr Norman.

'You have got to face up to the pain early and give the organisation a chance to heal itself before you go on to create new values. If you try to do the second phase too early the values become tainted with the pain.'

Mr Norman has transformed the atmosphere of Asda's Leeds head office. 'It is almost like a cult,' says one analyst. 'Archie Norman can do it because he does it with a messianic look in his eyes. He turned 70 000 people from thinking they were absolute losers to believing they could beat anyone.'

Can William Hague's Conservative party be 'saved' in the same way? Mr Cox-Hynd points to Tony Blair's transformation of the Labour party, and Mr Semler's revolution at Semco. 'In most cases, culture change rests or falls on one person,' he says. 'In both these cases it was done by individuals who had both vision and power.'

*Source*: *Financial Times*, 22 July 1998. Reprinted with permission.

**Questions for exit case study 5.3**

1 Identify the type of organisational culture present at Tory Central Office prior to the arrival of Archie Norman. Specify the type of organisational culture that Archie Norman wanted to see at Tory Central Office.

2 Mr Norman has transformed the atmosphere of Asda's Leeds head office. 'It is almost like a cult,' says one analyst. 'Archie Norman can do it because he does it with a messianic look in his eyes. He turned 70 000 people from thinking they were absolute losers to believing they could beat anyone.' Can William Hague's Conservative party be 'saved' in the same way?

In your opinion, 'Can William Hague's Conservative party be "saved"'? Summarise the changes you think Archie Norman will have to put in place if the Conservative party is to win the next general election, which will probably be held in 2002.

## Short-answer questions

1 Define culture at the global level of the external environment.

2 Define culture at the national level of the external environment.

3 Define culture at the local level of the external environment.

4 Define organisational culture.

5 Name five of the nine determinants of personal cultural provenance.

6 Name four of Hill's six determinants of culture.

7 Draw a detailed diagram showing Deal and Kennedy's four organisational cultures.

8 Indicate four areas in the international arena where culture can influence achievement for an organisation.

9 Indicate five issues for the management of organisational culture.

## Assignment questions

1 Write a 2000-word essay that compares and contrasts the relative importance of personal cultural provenance and national culture to the successful performance of the individual in the workplace.

2 Compare and contrast Handy's cultures (end of Chapter 4) and Deal and Kennedy's organisational cultures. In your opinion, which is a more realistic representation of organisational culture? Justify and explain your answer. Present your response in a 2000-word essay.

3 The university or college you attend is to be privatised and students charged the full fees of about £5000 per annum. Apply Johnson and Scholes' cultural web before and after privatisation. Summarise how the culture of the university or college would have to change if it were to be successful and maintain student quality and numbers in the face of competition. Present your findings in a 2000-word report.

## References

1 Hill, C W L (1994) *International Business*, Burr Ridge, IL: Irwin.
2 Hofstede, G (1984) *Culture's Consequences: International Differences in Work Related Values*, London: Sage.
3 Hofstede, G (1980) 'Motivation, leadership and organisation: do American theories apply abroad', *Organisational Dynamics*, Summer.
4 Kingdom, J (1991) *Government and Politics in Britain*, Cambridge: Polity Press.
5 Hill, op. cit.
6 Ibid.
7 Mintel Special Report (1996) *Single Person Households*.
8 Hill, op. cit.
9 Ibid.
10 Johnson, G and Scholes, K (1999) *Exploring Corporate Strategy*, 5th edn, Harlow: Prentice Hall.

11 Ibid.
12 Deal, T and Kennedy, A (1988) *Corporate Cultures*, London: Penguin Business.
13 Ibid.
14 Ibid.
15 Ibid.
16 Cartwright, S and Cooper, C (1998) 'Why suitors should consider culture', *Financial Times*, 1 September, quoted in Daniels, J D and Radebaugh, L (1998) *International Business*, 8th edn, Reading, MA: Addison Wesley Longman.
17 Bleeke, J and Ernst, D (1991) 'The way to win in cross-border alliances', *Harvard Business Review*, quoted in Daniels and Radebaugh, op. cit.
18 Kotler, P, Armstrong, G, Saunders, J and Wong, V (1996) *Principles of Marketing: the European Edition*, Upper Saddle River, NJ: Prentice Hall.
19 Daniels and Radebaugh, op. cit.
20 Ibid.
21 Brown, A (1998) *Organisational Culture*, 2nd edn, London: Financial Times Pitman Publishing.
22 Ibid.
23 Peters, T J (1978) 'Symbols, patterns and settings: an optimistic case for getting things done', *Organizational Dynamics*, 3 (23), Autumn, quoted in Brown, op. cit.

## Further reading

Black, S J and Gregersen, H B (1999) 'The right way to manage expats', *Harvard Business Review*, March/April.
Brown, A (1998) *Organisational Culture*, 2nd edn, London: Financial Times Pitman Publishing.
Burnes, B (2000) *Managing Change*, 3rd edn, Financial Times Prentice Hall.
Deal, T and Kennedy, A (1988) *Corporate Cultures*, London: Penguin Business.
Gracie, S (1998) 'In the company of women', *Management Today*, June.
Johnson, G and Scholes, K (1999) *Exploring Corporate Strategy*, 5th edn, Chapter 5, Harlow: Prentice Hall.
Lynch, R (2000) *Corporate Strategy* (2nd edn) Chapter 8, Harlow: Financial Times Prentice Hall.
Mitchell, A (1998) 'The dawn of a cultural revolution', *Management Today*, March.
Wickens, P (1999) 'Values added', *People Management*, 20 May.
Senior, B (1997) *Organisational Change*, London: Financial Times Pitman Publishing.
Thompson, J L (1997) *Strategic Management*, Chapter 4, London: International Thomson Business Press.
Sparrow, P (1999) 'Abroad minded', *People Management*, 20 May.
Trompenaars, F and Woolliams, P (1999) 'First-class accommodation', *People Management*, 29 July.
Warner, M (1996) 'Managing China's enterprise reforms', *Journal of General Management*, 21 (3), Spring.
Wright, N J (1996) 'Creating a quality culture', *Journal of General Management*, 21 (3), Spring.
Yan, R (1998) 'Short term results: the litmus test for success in China', *Harvard Business Review*, September/October.

# 6 Marketing

Claire Capon with Andrew Disbury

## Organisational context model

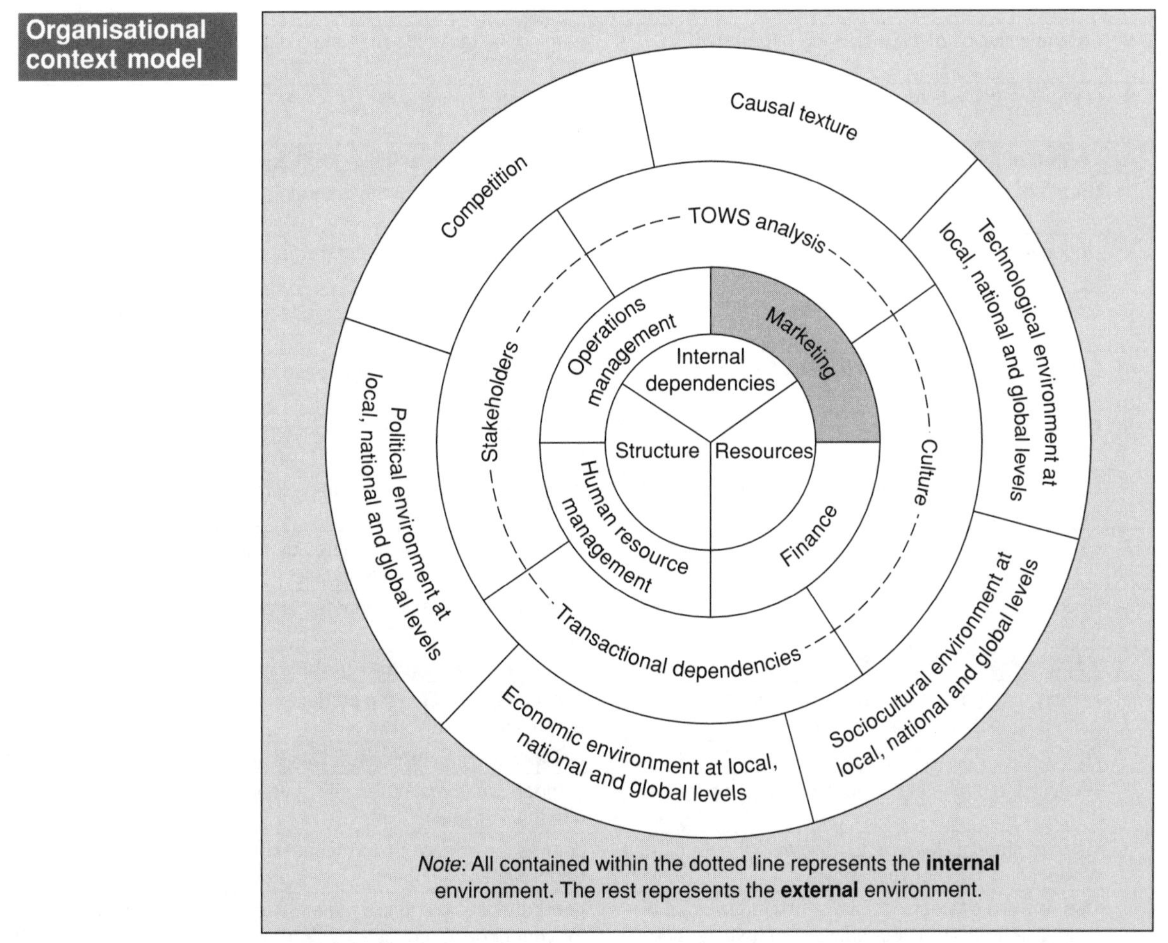


*Note*: All contained within the dotted line represents the **internal** environment. The rest represents the **external** environment.

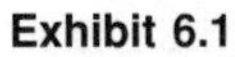
**Exhibit 6.1**

## Learning outcomes

While reading this chapter and engaging in the activities, bear in mind there are certain specific outcomes you should be aiming to achieve as set out in Exhibit 6.2 (overleaf).

**Exhibit 6.2 Learning outcomes**

| | Knowledge | Check you have achieved this by |
|---|---|---|
| 1 | Define marketing | stating in one sentence what marketing is |
| 2 | Define the production era | specifying the key characteristics of the production era |
| 3 | Define the sales era | specifying the key characteristics of the sales era |
| 4 | Define the marketing concept | listing the key characteristics of the marketing concept |
| 5 | List the activities of the marketing department | naming the tasks the marketing department performs |
| 6 | Name Ansoff's marketing strategies | list Ansoff's strategies |
| | **Comprehension** | **Check you have achieved this by** |
| 1 | Understand the organisational context of marketing | identifying the role of the marketing department in the organisation |
| 2 | Understand needs and wants | indicating specific needs and wants that customers may have |
| 3 | Explain Maslow's hierarchy of needs | naming and illustrating the needs in Maslow's hierarchy |
| 4 | Explain what each of the activities of the marketing department involves | summarising the key points of each activity carried out by the marketing department |
| 5 | Explain the term 'marketing mix' | naming and illustrating the components of the marketing mix |
| 6 | Understand what is meant by the term 'the extended marketing mix' | naming and illustrating the fifth element of the marketing mix |
| 7 | Explain the term 'product life-cycle' | summarising the key attributes of products in the different stages of the product life-cycle |
| 8 | Explain the term 'BCG' | summarising the key attributes of products in the different quadrants of the BCG |
| | **Application** | **Check you have achieved this by** |
| 1 | Identify a range of products and categorise them according to their position on a BCG | applying the BCG to the products and services of an organisation that you know |
| | **Analysis** | **Check you have achieved this by** |
| 1 | Analyse the marketing mix of a product of your choice | comparing the marketing mix of your chosen product with one that is similar |
| 2 | Use the product life-cycle to anticipate the future of a product or service of your choice | contrasting current performance with predicted future performance |
| 3 | Use Ansoff's strategies to show how the product life-cycle of a product of your choice can be extended | generate possible future marketplace behaviour and position for your chosen product |
| | **Integration** | **Check you have achieved this by** |
| 1 | Explain the similarities between the product life-cycle and BCG | summarising and discussing the similarities of the product life-cycle and BCG |
| | **Evaluation** | **Check you have achieved this by** |
| 1 | Evaluate the accuracy and usefulness of a variety of marketing tools in anticipating their future behaviour in the marketplace | critically judging the relevance of marketing tools for the marketing department in organisations today |

ENTRY CASE STUDY 6.1 FT

# China's urban cool develop a thirst for a cappuccino

**By James Harding**

'Coffee brings its own special atmosphere, its own sense of romance,' says Yang Qingqing, Shanghai's best-known fashion columnist and self-appointed style guru. 'The young generation drink coffee not because they prefer it to tea. The real reason is that coffee represents modern, western culture.'

In China's largest city, thousands of cafés have opened in the last couple of years and a taste for a cappuccino has become a badge of cool. Much more than a beverage, coffee has become a mark of cosmopolitan culture and a milestone on the path to prosperity.

Rural China, by contrast, remains a devoutly tea-drinking society. Instant coffee is practically non-existent in the poorer households of the Chinese countryside, but has become a regular feature of urban homes. According to a Gallup market survey, just 1 per cent of rural homes have a jar of coffee granules, but in Shanghai 51 per cent of houses do.

The fashion for coffee in more prosperous urban areas, though, has been enough to multiply the coffee business in China several times over.

Theo Klauser, managing director of Nestlé China, says sales of Nescafé, by far the best-selling brand of instant coffee, have 'increased five times in the last five years . . . although from a small base'. Sales from the Nescafé plant in Dongguan were worth RMB450m ($54.4m) last year, he says.

The International Coffee Organisation recently sent a delegation to Beijing and Shanghai to promote coffee consumption and provide support to China's coffee producers in Yunnan, in the country's south-west.

As with so many other consumer goods in China, the massive potential of the market can have a dizzying effect on the industry.

Coffee consumption in mainland China is one cup a person, a year, according to Nestlé, but wealth and greater integration with the west has tended to develop a thirst for coffee in Asia.

In Taiwan, people drink 38 cups each year, in Hong Kong 44 cups and in Japan 330 cups. This compares with 440 in the UK, 463 in the US and 1100 in Sweden.

Jonathan Eisenberg, who set up a gourmet coffee shop in Shanghai and a sister shop in Beijing called The Daily Grind, says: 'The Chinese will go for coffee. Look at what has happened in Japan, Korea and Hong Kong. But they will probably go for a US-style coffee, something that is diluted with milk.' Making a good cup of coffee in Shanghai, though, has its problems. 'To do Asian-style coffee you need foam,' which means buying a lot of good quality milk, notes Mr Eisenberg. For a take-away coffee shop, one of the chief headaches is cups and lids, he says: 'They are often more expensive than the beans.' Tariffs on coffee beans are also very high, prompting Mr Eisenberg to buy beans roasted in China.

The country's largest processing operation, the Yunnan Coffee Processing Plant, aims to increase annual production of 1500 tonnes last year to 5000 tonnes in 2000.

But the company expects the business environment to get tougher as more foreign brands and coffee importers target the market. 'We are trying to set up a sales network throughout China's major coastal cities, but we have a long way to go. Cheaper and more famous foreign coffee companies promise to be a big challenge.'

Mr Eisenberg cautions against excessive enthusiasm about the coffee market: 'Coffee for coffee's sake is still a novelty. The coffee market is still premature for serious growth.'

Likewise, Mr Klauser at Nestlé warns that while the potential of the Chinese market may be beguiling, it is a long way from being realised. 'Coffee drinking is still foreign to the Chinese, who are essentially tea drinkers and their habits are not so easily changed,' he says.

In large part, coffee producers and café owners say, this is because Chinese people are not yet convinced by the taste. After a lifetime of tea, coffee seems bitter and acrid, which helps explain the popularity of the sweeter and milkier cups of coffee such as latte, cappuccino or instant coffee ready-mixed with milk and sugar.

But, then, for the time being, the modish Ms Yang suggests that people drink coffee often for something more than the taste: 'Coffee has its own mood . . . for example, if a young man invites a young lady for a cup of coffee, this is a good way for him to express his feelings.'

## Introduction

This chapter examines the relationship between the marketing department and the external and internal environments of organisations. Conclusions will be drawn about the interactive nature of this relationship. Organisations and their context will be examined with the aim of assessing how the external environment dictates the activities of the marketing department, or whether marketers inside organisations are able to exercise their marketing talents to influence the external environment. In addition, how the marketing department relates to other departments in the effort to achieve organisational goals will be examined. Basic marketing tools will be presented and these will be applied to practical examples in order to facilitate an understanding of marketing and of how marketing tools play their part in this process.

## Marketing

Marketing is the first of the departments to be considered for two reasons. First, the nature of its activities is such that marketing could be said to be the activity best placed to interact with the external environment. Through its market research activities, the marketing department is the internal area of the organisation most explicitly required to scan, analyse, monitor and contend with what is going on in the outside world; *see* Chapters 1 and 2. It is therefore the role of marketers to understand the size and nature of the marketplace, to know the organisation's customers and to undertake market research into customer requirements. While scanning the environment in this way, the marketing department will be continually monitoring the activities of competitor organisations and the latest developments in the same or similar fields. These research and monitoring activities are the way in which the marketing department's efforts ensure that the organisation is first into the market with the products or services to meet its customers' identified needs or wants. It could therefore be said that the marketing department is most aware of the external environment, and of the environmental linkages between the external and internal environments; *see* Chapter 3. It also follows that marketing is the department most aware of any changes in this relationship and of the effects that these changes may have on the strategic or operational management of the organisation.

Secondly, it is clear that the marketing department has several roles to play in the relationship between the external and internal environments. It has to identify changes in the external environment and predict the kinds of products or services the outside world requires. In addition, the causal effects of these changes mean that marketing also has to manage closely the links between the marketing department and the other departments of the organisation in order to meet the organisational objectives of satisfying customer wants or needs. Without close collaboration and communication within the organisation, referred to here as functional convergence, the marketing department or function will not be able to marshal the tangible, intangible and human resources necessary to achieve organisational objectives. Without functional convergence, the organisation risks wasting organisational energy on internal tension and conflict and not being able to meet the goals it has set itself. Thus this chapter will look closely at the relationship between marketing and the other elements of the internal environment and will draw some conclusions about how the organisation's internal efficiency affects its ability to achieve its business goals.

## The development of the marketing concept

**Production era**

From the industrial revolution onwards, the urbanisation of an ever larger part of previously rural populations in Britain resulted in a workforce able to produce large quantities of goods to supply mass markets. At the same time, lifestyle patterns altered, with the long hours of factory work instead of agricultural activity. Therefore urban workers had to buy products that they previously would have grown or made themselves, thus creating a demand for the goods that other workers produced. As technology advanced, labour-intensive activity gave way to mechanised mass production. The new range of products included mechanised products and labour-saving devices that had previously been unimaginable and unavailable to a wider public at prices they could afford. This meant that new markets for these new products were easily created and satisfied. This is referred to as the production era, when organisations treated their customers as a plentiful and captive supply of people who were easily pleased with the new-fangled devices offered to them by manufacturers. In the production era, if organisations did not make it, customers could not buy it. Further, if organisations did not make it, customers could not conceive of it.

**Sales era**

As industrialisation progressed through the eighteenth, nineteenth and twentieth centuries, competition between manufacturers increased, and organisations were obliged to work harder to woo their customers away from competitors' products. The British Empire expanded all the while, driven and fuelled by trade imperatives. The Empire brought to British companies ever-expanding sources of raw materials and labour and endless captive markets for their goods. Like the production era, this sales orientation still did not consider the actual needs or wants of customers before deciding what products or services to provide. Rather, organisations concentrated on making the public and thereby potential customers aware of what was available from Company A so that they would be enticed to purchase Company A's products, rather than Company B's or C's. Advertising focused on reliability and brand loyalty, in its efforts to retain customers once hooked. Little was done to survey customers and find out if there were unfulfilled needs and wants that the company itself could work towards meeting. Services such as transport, health and education, particularly on the election of a Labour government after the Second World War, became more and more centralised and less responsive to need or want when nationalised by the government into monopolistic public providers. In the private sector, however, it became the prime focus of organisations to sell as much of what they could make as possible.

**Marketing era**

Given the ever-increasing rate and scale of the competitive environment in the twentieth century, the sales orientation has given way to the marketing orientation or marketing concept, which is the subject of this chapter.

## Defining the marketing concept

In order to understand the activities of the marketing department, it is first important to define the marketing orientation or marketing concept that underpins its activities. At its most basic, a marketing orientation leads organisations to consider the needs and wants of their actual or potential customers before considering what services and/or products to offer. More than this, the marketing concept could be said to be a holistic approach to managing organisations. Marketing-oriented organisations focus on gearing all their internal activities towards achieving their goals and objectives by satisfying customers' needs and wants. Marketing can be likened to a philosophy or a firm set of beliefs. There is a huge literature on and by marketers defining their role in the environment and within the organisation.

Peter Drucker[1] defines marketing as:

> not only much broader than selling, it is not a specialised activity at all. It encompasses the entire business. It is the whole business seen from the point of view of its final result, that is from the customer's point of view. Concern and responsibility for marketing must therefore permeate all areas of the enterprise.

This gives an idea of the holistic nature of the marketing concept, dealing as it does with the organisation's activities from the very beginning of product research and planning through to after-sales service and customer comments.

Kotler et al.[2] defined marketing as:

> a social activity and managerial process by which individuals and groups obtain what they need and want through creating and exchanging products and value with others.

This introduces the idea of the role of the marketing department as contributing to adding value to the organisation's inputs, an essential element of the resource conversion process; *see* Chapter 4. Kotler goes on to suggest that marketing is the area of a business that identifies and quantifies current unfilled needs and wants, before determining the markets that the organisation can best serve with its current and future products and services. Thus marketing links a society's needs and the organisation's response.

The professional body for marketers in the UK, the Institute of Marketing,[3] identifies the marketing department as 'the management process which identifies, anticipates and supplies customer requirements efficiently and profitably'. This introduces to the marketing concept the notion of prediction and anticipation of customer needs and wants. In the realms of innovation and technology, marketers have a role in providing for needs and wants before potential customers know that they could need or want the product or service on offer. Who, for example, in the heyday of the long-playing record, could have known that they would have been updating their collection to compact disc before the end of the twentieth century, until the invention had been made?

According to the marketing concept, it is the *raison d'être* of the marketing-oriented organisation to allow the marketing department the power and resources to dictate what should be done, for whom and when, to the organisation's other departments. This focus is achieved through marketing's research activities, which are the mechanism for identifying customer needs and wants and competitor provision. Marketing can provide further evidence of how effective it has been in doing this through its customer service and feedback capabilities, closing a circular loop of information and data gathered from the outside world. This is clearly illustrated by the entry case study for this chapter, showing that companies such as Nestlé have identified China as a growth market for coffee.

## The organisational context of marketing

The marketing concept demands not only that marketing should focus on meeting the wants and needs of customers, but that nothing less than the full dedication of people and resources from all sections of the organisation to the marketing concept will ensure that the organisation can achieve business success. In order to achieve the customer satisfaction it heralds as the key to business success, the marketing concept dictates that any effort in the organisation is wasted if it is not directed towards customer satisfaction and providing the products and services that it says the customer wants or needs. In this, all the organisation's departments need to work together in harmony to achieve organisational objectives. We refer to this harmony as functional convergence, as the internal departments or functions of the organisation must converge their efforts to succeed.

Because the marketing concept deals with the satisfaction of customer needs and wants through organisational activities, it can be seen in organisations' in-house programmes that are focused on achieving 'customer orientation' 'customer satisfaction' or 'customer care'. Hence marketers hold as the tenets of their philosophy the consideration of three basic points:

1 What do customers need or want?
2 How can the organisation meet these needs or wants?
3 How does the organisation make money doing so?

This last is true not only of private-sector, profit-making enterprises, but also of contemporary public-sector or voluntary-sector organisations. In the latter cases, the aim would be to make enough money from products or service to break even or increase activity level, rather than making profit for redistribution to owners or shareholders.

This marketing concept demands that the only point of the organisation's existence is to try to make or do whatever market research has proved that customers need or want. The organisation must have whatever resources or technology are necessary in order to achieve this. Any organisational resources, whether tangible, intangible or human, that are not focused on meeting the needs and wants of customers are therefore redundant. Although to the business studies student this approach may seem obvious, the opposite attitude may still be found in some organisations researched. Their approach might be summed up in the more production- or sales-oriented phrase: 'We make this product because we know how to. Now, who can we find to sell it to?'

This latter approach is familiar in the case of the early mass-production pioneer, Henry Ford, who famously offered his customers any colour Model 'T' Ford car, as long as it was black, as that was the only colour of car his company made. In the early days of mass production this approach was successful due to the simpler nature of the competitive environment; *see* Chapter 3. This meant that customers and markets were more easily satisfied by the new products that technological advances brought, as the vast majority had never had access to such products before.

## Marketing assumptions

Underpinning the marketing concept are three basic assumptions about the external environment, marketplace and customers. First, marketers must assume that their actual or potential customers have some element of free choice. In politico-economic systems containing free markets or a mixed economy with commercial competition supported and regulated by the politico-legal system, customers or consumers will have the opportunity to exercise free choice between products or services when spending money; *see* Chapter 2. Secondly, actual or potential customers have disposable income that can be spent on whatever product or service is being offered. This crucial financial aspect empowers customers to follow their needs or wants and make decisions about the deployment of spare income. Finally, if customers have both disposable income and the freedom to spend it as they wish, it is the job of marketers to attempt to persuade them to alter their choice and move from something to which they are loyal or with which they are familiar to something new in the same line. The entry case study to this chapter highlights the efforts that have to be made by companies like Nestlé to tempt Chinese people to drink coffee instead of the tea with which they are familiar.

True marketing orientation could be renamed customer orientation. If organisations are truly marketing or customer oriented, then they are achieving the necessary focus of all organisational activities on customer satisfaction. In achieving marketing orientation, the organisation has recognised that the only point of organisational effort is to try to satisfy customer needs and wants through the combination of marketing, finance, human resource and operations management activities. In order to do this, the organisation needs to aim to achieve a level of integration of the efforts of all departments. This is the concept of functional convergence, the co-operation and collaboration of all internal elements of the organisation towards its corporate mission of customer satisfaction. To achieve this, the organisation must understand the needs and wants of the customer.

### Needs and wants

It is useful at this stage to consider the difference between customer needs and customer wants, as occasionally marketers will be appealing to a perceived need, but more often they will be focusing marketing activities on appealing to customers' wants or desires. A need is something that cannot be done without, like basic food and water. A want could be said to be something to which someone aspires or which they desire. People need grain and water to survive, but might want to have a variety of grains and vegetables and fruit juice or alcohol to drink.

This difference can be illustrated by using Maslow's hierarchy of needs.[4] This American psychologist defined need at various levels. People all have in common certain basic requirements for food, drink and sleep, referred to by Maslow as 'physiological or basic needs'. In less well-developed countries it is these needs that are the prime concerns of citizens. Keeping enough food in their stomachs is how all of the family's income is spent, whether that income be earned through economic activity or handed out in subsidy from the state or from private charity. Once basic physiological needs are met and satisfied, the next priority is to keep a roof over the family's head. Maslow refers to this as 'security or safety needs', which also include self-defence and saving for future eventualities.

Once these needs are satisfied, essential life is not threatened by homelessness or hunger and the future is planned for in terms of the upbringing of children and social security for the elderly and sick in the family, then less fundamental needs can emerge. Maslow's next level refers to the human proclivity for socialisation activities, e.g. going to the pub, cinema or theme park. Maslow refers to this as 'belonging or affection needs'. From this, natural ambition and aspiration lead to spending surplus income on more luxurious accommodation or better food, or improving social status through acquisition of possessions or land. Maslow calls this 'esteem or ego needs'.

Finally, the pinnacle of Maslow's hierarchy of needs is the stage of self-fulfilment, not only having achieved that fundamentally needed for basic food and shelter, but also being able to obtain things that are desirable for complete fulfilment. Thus the ultimate dream home with all modern conveniences, a top-of-the range sports car, designer-label clothes, regular five-star holidays and eating only in the best restaurants might be the height of aspiration for the successful entrepreneur or lottery winner. Maslow calls this 'self-actualisation'.

Therefore while organisations can never make customers actually need anything, the job of the marketing department is to tempt potential customers to deploy their financial resources to meet their needs by purchasing Company A's products or services instead of Company B's. For example, in the entry case study to this chapter, the International Coffee Organisation, in promoting coffee drinking in Beijing and Shanghai, is tempting urban people in China to forgo their traditional tea and on occasions drink coffee instead. Hence while marketers cannot create need, they can sometimes make customers want something through their marketing activities, activating the higher echelons of Maslow's hierarchical model. The marketing department therefore needs to focus its efforts on appealing to the various needs and wants of the target market. This is why market research (*see* later in this chapter) is important.

## The customer

'Customer care', 'customer orientation' and 'customer service' are very much part of the rhetoric of organisations striving for business success in Britain in the twenty-first century. To what extent this orientation is sincere or successful will be examined later in this chapter. In the private sector many truly customer-oriented examples can be cited. Car manufacturers have come a long way since the days of Henry Ford's Model 'T', largely under the influence of the Japanese car manufacturers whose post-war domination of global markets is legendary. Nissan UK Ltd is an example of a marketing-driven car manufacturer. No car going through the factory on the production line in Sunderland has not been ordered by a sales showroom for a customer. Therefore each car being made is bespoke, i.e. constructed with a particular customer's needs in mind.

In the UK public sector, service delivery has changed radically since the election of the first Thatcher government in 1979. With the introduction of markets in the NHS, education and local government services, public-sector organisations have had to begin considering customer needs and wants in the provision of their services. The university sector also faced radical change in its markets with the expansion of higher education, and the introduction of student loans and tuition fees directly payable by students instead of by their local education authority. However, despite having introduced marketing departments, managers and officers, universities have considerable work to do before they could be considered to have a marketing orientation. They are still more fixed in the production or sales era, expecting students to want to take the courses they already teach, rather than truly designing courses that meet identified customer needs or wants.

## Marketing activities

If, put simply, marketing can be said to be the carrying out of business operations to steer the flow of goods and services from manufacturers to consumers, it is then useful to define exactly what it is that the marketing department does in order to achieve this goal. There is a temptation to associate marketing with selling, but sales is in fact just one of the activities that may be located within the marketing department; *see* Exhibit 6.3. Housed within the marketing department are likely to be many individual subdepartments, including market research, product planning and development, advertising and sales promotion, sales, distribution and after-sales customer service.

This section will examine the marketing department's relationship to the external environment, its activities and its relationship to other elements of the internal environment of the organisation. In its relationship with the organisation's external environment, the marketing department must seek to answer the following questions:

1 Who is in the marketplace and where is the market?
2 What are the changing needs and/or wants of the marketplace?
3 What resource inputs are necessary to meet those needs/wants?
4 How will the organisation make money doing this (even voluntary, public sector or other not-for-profit organisations have to make their running costs)?

### Market research

Having stated at length the importance that the marketing concept places on establishing customer needs or wants, a company will only be as successful as its market research activities. Through market research, organisations can identify who is in the marketplace, where they are located, what they need or want and how products and services can be developed to meet these identified needs and wants.

Organisations can carry out market research in two basic ways: through generation of primary data, or through use of available secondary data. The generation of primary data requires the organisation to invest heavily in terms of time, money, people and in the execution of extensive and detailed surveys of current and potential customers through direct contact with them. This may take the form of postal or telephone surveys of a database of current customers. Alternatively, the organisation may purchase commercially produced marketing

**Exhibit 6.3 Marketing activities**

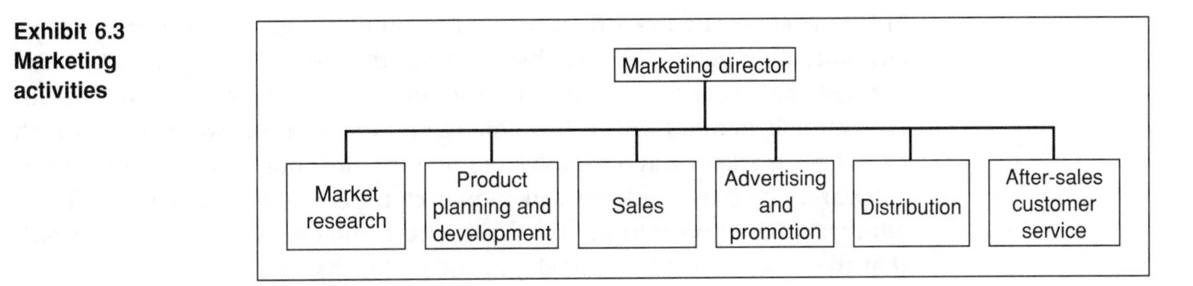

databases of potential customers who fulfil certain criteria in terms of location, income, profession or family composition and then survey them directly.

Examples of primary market research include Cable & Wireless's 1998 survey of households regarding the type of telephone and television services they wish to see developed in the future. After the privatisation of British Telecom, Cable & Wireless, a renowned international telecommunications player, had established the first private competition to BT, called Mercury Communications. Later, it decided to relaunch Mercury under its own Cable & Wireless brand name. Therefore the customer survey, along with wide media advertising, was a way of conducting useful market research and introducing the new brand name to a wide audience. Water industry regulator OFWAT also conducted a survey of water company customers in 1998 to ascertain their attitudes to the ways in which they wished the regulator to direct water companies to reinvest their profits.

The advance of technology has assisted organisations in knowing their customers better and contacting them more frequently, as well as increasing the amount and quality of primary data available to help organisations with their market research. The use of itemised bills and customer loyalty cards in supermarkets and shops enables those organisations to build exact customer profiles, and to target the marketing of various products and services at particular customer groups. This technique is called data warehousing, and is a growing market research trend in many sectors. A customer identified from the checkout terminal as having begun to buy baby products might be sent details of the supermarket's discount scheme on baby products. Unlike Sainsbury, Tesco does not offer its loyalty discounts directly at the checkout, but rather mails discount vouchers up to the amount 'saved' on the loyalty card directly to the customer's home address, thus enabling it to include details of other relevant products, services or offers.

Sainsbury's, Safeway and Tesco have all invested large sums of money in data warehousing to track customer preferences and behaviour, thus providing a rich seam of market research. The speed and sophistication of the technology allow millions of customer records to be searched and processed relatively quickly. This method enables supermarkets to break down their customer base by category of shopper, e.g. young families, thirty-something singles, elderly couples, elderly single people. Thus the supermarket can compare actual shopping habits with a perceived theoretical shopping basket that each average consumer type 'should' be purchasing. This then enables the retailer to tell which products customers do and do not buy from that particular store. This market research can then be used to promote certain products that the customer may be buying elsewhere outside the store. The supermarkets can also tell the frequency of visits and the impact this has on shopping habits. For example, families with young children may shop at the weekend, at which time supermarkets may promote economy brands, bulk purchase discounts, baby and children's products, including clothes as well as foodstuffs. These efforts are based on information arising from the data and customer records that the supermarket has stored and collated in its data warehouse.

Organisations may also choose to use specialist market research firms to conduct market research for them, particularly when they may not have the skills or resources in house specifically to devote to carrying out the research themselves. The market research firm may carry out bespoke market research expressly and confidentially for its client, or may carry out generic market research that is of use to and available to more than one customer. Alternatively, secondary data may be collected and compiled by private market research companies such as Mintel and Euromonitor. This will be topic specific, e.g. retailing of children's clothing, but not specific to any one company. The secondary data will be collected through mailshots to business customers or private households, through cold calling by telephone or in person to business or private addresses, or by in-person interviews on the street. It will then be sold to any interested companies or libraries. There are also publicly available forms of secondary data, available to all, including census data and information on social and demographic trends.

## Product planning and development

Having gathered the required market research data on customer needs and wants, the next stage in the marketing process is to try to ensure that the organisation is in a position to introduce new products or services designed to meet the needs and wants it has identified. This may seem an obvious and simple stage to achieve, but there are in fact many complex factors to take into account. The organisation may first want to consider how it might meet those needs or wants with its existing product or service range, through utilising the marketing tools explained later in this chapter. However, if existing products or services are no longer suitable, then the marketing department must clearly work closely with the other departments to achieve a balance between what is required and what can be provided. These internal relationships are dealt with more explicitly in the next section.

In terms of relationships with the external environment, product planning and development are responsible for a number of factors:

- identifying new technology that improves design or production capabilities;
- identifying suitable suppliers of necessary raw materials or components;
- liaising with customer or consumer focus groups at various stages to take into account changes in taste or design capability.

Product planning and development are included here as part of the marketing department, but considered under the banner of research and development (R&D) it could equally comfortably be located in the operations management department of an organisation. Without close liaison with the operations managers, the marketers' dreams remain dreams. The operations management department must realise the ambitions of the marketing department. Therefore in different organisations the responsibilities for R&D might be found in marketing or operations management. Whichever is the case, there must be close co-operation between marketing and operations management. This is the functional convergence mentioned before. Functional convergence could be defined as the situation where two or more departments or functions of an organisation work in harmony towards meeting organisational goals and objectives without rivalry or internal competition.

## Sales

Once the new product is up and running, there then comes the task of selling it to the people who were identified as needing or wanting it by the market research. However, this is by no means a foregone conclusion. During the market research and product development stages, customers have had time to develop other wants or needs that may supersede those originally identified. Additionally, competitors have had time to enter the market with their own new products or services. At this stage it is vital to decide on appropriate pricing and marketing strategies to get the finished product or service delivered to the target market at a price it can afford and that also makes money for the organisation. There are different and conflicting approaches to costing and pricing that are covered in Chapter 8.

The marketing-oriented approach to pricing examines what prices the market might support through its market research. If potential customers find a certain price acceptable for certain products, marketers will want their new product's prices to fit in with the public's perception of 'normal' prices. For example, if the range of prices for a 1.5kg box of washing powder in the supermarket is between £4.89 and £5.29, then any new powder introduced has to be priced within or close to this range in order to be successful. For a higher than normal price to be acceptable to the marketplace, the product has to be differentiated; *see* Chapter 3. This is where the manufacturer relies on additional product features ('Now destroys grease'), quality reputation ('Still the market leader') or brand loyalty ('Your favourite washing powder') to be able to sell the new washing powder.

Therefore the organisation's challenge is to be able to produce the product and make a profit within this price range, ensuring that all inputs, costs and overheads are covered. This is one aspect where the idea of functional convergence might falter. While marketing is dictating to the organisation not only what products must be made but also what price the marketplace will stand, the operations management department's reaction can be to state what it is possible to produce and what the basic cost of this will be.

A production-oriented approach to pricing examines the costs of all inputs, including overheads and labour, and then adds profit to this to achieve a sales price. This appears to be a logical approach, even if the resulting price is higher than the market norm for a similar product or service. It is not logical, however, where the ultimate price of the good is higher than the customer is prepared to pay for such a product.

## Salesforce

From the manufacturer's point of view, the typical method for achieving sales is through a salesforce of representatives who are able to travel to existing or potential customers with a view to achieving new or continued sales. In addition to staff costs, there is the investment in a fleet of company cars and mobile phones to be considered. Managing the salesforce, with its quasi-independent status, cut off geographically from the internal culture of the organisation, is a complex issue; *see* Chapter 4. There are important concerns concerning the quality of information passed between the salesforce in the field and the organisation, as orders must be processed correctly and quickly and organisational developments communicated to sales personnel. Salesforces work on salary-plus-commission contracts, adding a motivational incentive to encourage them to meet and/or exceed their targets.

**Wholesalers**

A large amount of goods are sold not directly to the consumer but through an intermediary such as a wholesaler. This is convenient for the manufacturer as, although reduced profits are made since wholesalers command discounts from manufacturers, they have a simpler task in only selling to centralised wholesalers rather than having to identify and target a variety of customers to whom to sell. Wholesalers need discounts from the manufacturers because they too wish to make a profit from the transactions of buying and selling on to retailers, and they have their own operational costs to consider. The wholesaler requires efficient distribution networks, including appropriate transportation. The wholesaler also removes the need for the manufacturer to keep a large amount of finished goods in stock, as it is the wholesaler who needs the large warehousing capacity, centrally positioned to service a network of retailers.

**Retailers**

The retail trade sells goods to the end user or consumer of the product. In order to be successful in retailing, it is crucial to make correct decisions about a number of factors. Location is a key issue, as the retailer must position itself in the place the consumer would expect or would like to find the goods on offer. Thus city-centre shops are faced with tough decisions when a new out-of-town shopping mall opens, taking trade away from the city-centre location: do they remain loyal to the city centre; do they move sites into the new environment of the shopping mall; or do they attempt to make both locations successful?

Department stores in Sheffield are an interesting case in point. Meadowhall opened outside the city in 1992. This had a further significant impact on city-centre shops, given that the economic situation of Sheffield as a whole has never fully recovered from the economic downturn in the 1970s. The department chain House of Fraser initially opened a second store in the Meadowhall complex and kept its city-centre store open. However, after five years of operating two stores, it finally closed its town-centre store in favour of the Meadowhall location in 1997. Shoppers had demonstrated a preference for visiting the shopping centre-based store over the town-based one. John Lewis, however, remained loyal to the city-centre location with its Cole Brothers store. It resisted the challenge of Meadowhall and retained its band of loyal customers.

**Direct selling**

Many organisations are avoiding intermediaries such as wholesalers or agents in order to reduce costs and thus prices, hence passing on cost savings to the consumer. In this way, manufacturers or service providers can also increase profit margins by charging the end price directly to the customer rather than selling at a discount to a wholesaler or agent. Examples of organisations now dealing directly with their customers or consumers are home and car insurance services, who sell their services by telephone instead of in a shop or office, hence reducing overheads and providing a more customer-oriented service. Customer service is also improved by giving customers access to services at a time convenient to them, e.g. from home by telephone in the evenings or at weekends. These services are often notable by the inclusion of the word 'direct' in their company or service title, e.g. the insurance company Direct Line.

**Mail order**

There is a 'chicken and egg' link between customer orientation and customer demands. As organisations improve their products and services, customers become used to a high level of service and good-quality products. As they become accustomed to better services and products, they demand more and more from the organisations they patronise. One of the manifestations of this is longer opening hours in shops (late-night shopping in city centres, supermarkets opening 24 hours a day at busy periods, such as just before Christmas, and even year-round in some locations).

Another method of selling that is becoming increasingly popular is mail-order purchasing. Clothes stores such as Next and department stores such as Debenhams have directories for home shopping and Marks & Spencer has a range of home furnishings and household goods available through its mail-order service. Marks & Spencer is also introducing its full range of in-store clothes to mail-order customers. This is beneficial not only to those who cannot get out regularly to shop, but also to the busy professional customer who cannot find time physically to go shopping, even during extended opening times. It is likely that Internet shopping will become more popular and increase the amount of remote shopping that people do.

**Advertising and promotion**

In order to bring new products or services to the attention of the potential market, there has to be advertising and promotion. These, along with sales activities, are the subdepartments most easily associated in the public conscience with the term 'marketing'. Advertising and promotion are not necessarily the same activities. They are the part of the marketing department responsible for planning strategies to retain existing customers and attract potential customers. Advertising and promotion are carried out in a variety of media appropriate to the product or service and to the target audience.

Advertising takes two forms: institutional or product. That is to say, advertisements either promote the company as a whole and do not focus on specific products; or focus on particular products, irrespective of where they are sold. The advertising media most familiar to the general public are print media (newspapers and magazines), broadcast media (television, radio and cinema) and public spaces, such as roadside billboards and street furniture. The cost of advertising space varies according to the size of audience likely to be reached and the date or time of the advertisements placed. Many advertisements take on a life of their own quite beyond or apart from the popularity of the product. An example of this is the Renault Clio advertisements featuring 'Nicole' and 'Papa', whose names have become quite familiar to the British public. This was extended in such a manner that the 1998 advertisement for the Renault Clio, and thus the latest remodelling of the car, was announced well in advance of its first television broadcast, in the commercial break of *Coronation Street*, a popular soap opera, on a Friday evening. In a sense, this went beyond advertising and into the realm of promotion.

Promotion is the way in which the organisation attempts to manipulate the external environment by combining advertising with special offers and particular benefits or service packages intended to attract customers. There are many famous examples of organisations that have failed to match the demand created by special promotions and have come unstuck as a result. The Hoover promotion of free trips to America seriously underestimated the number of people prepared to purchase a new vacuum cleaner in order to benefit from the free offer, and the company ended up in the courts having to justify why customers had not benefited from the holidays promised them. Advertising and promotion constitute one part of the marketing department that is strictly regulated by law and the Advertising Standards Agency (ASA).

**Distribution**

Having whetted the customer's appetites with advertising and promotion, and potentially having made sales to them by remote methods, ensuring that products or services are physically available to customers who wish to purchase them is the next stage in the successful marketing process. The 1998 World Cup was dogged by problems of ticket distribution. First of all there were the restrictions placed by the French organising authorities on ticket allocations to national football associations to be sold in participating countries. Then they announced that tickets could only be bought over the telephone by customers with an address in France, effectively debarring all non-residents. In order to remedy this situation, they offered an extra allocation of tickets by international telephone over a certain period, only to find the lines so over-subscribed that few callers could get through. This was obviously a clear case of demand outstripping supply, but also one of extremely poor distribution management.

Efficient distribution networks are particularly important for mail-order companies or companies moving towards direct selling to the customer at home, rather than in the shop. Consumer protection organisations and radio and television programmes are kept particularly busy in the early part of each year investigating complaints from customers who have ordered and paid for goods as Christmas presents that were not delivered in time.

**After-sales customer service**

The final stage in the holistic customer-oriented package is how customers are dealt with once their purchase has been made. Dealing effectively with follow-up enquiries or complaints is a key indicator of the truly marketing-oriented organisation. This enables organisations to target customers for current or future alternative products or services. Organisations often include aspects of their after-sales service as part of the promotion package, offering money-back guarantees, 24-hour helplines, or free installation or home visits. The standard of this type of service is often the factor enabling customers to judge whether repeat business will be placed with the company or not. The cable communications company that keeps the complaining customer on hold for half an hour before putting them through to 'one of our service operatives' does not demonstrate excellent customer service or good telecommunications equipment.

Privatised utilities have invested heavily in this end of their customer service, which may be partly due to the fact that they have many dissatisfied customers who find service levels below par and prices rising. Partly due to customer service initiatives, and partly due to external regulation, utilities offer *per diem* refunds for any disruption of service in water, electricity or gas, as well as compensation schemes for complaints received.

## Relationship to the internal environment

As well as operating on a set of assumptions about potential or actual customers, the marketing concept also makes some basic assumptions about the organisation for which the marketing effort is being made. The first of these assumptions is that the organisation is able to be flexible and alter its products and processes in order to meet newly identified customer needs and wants. Having identified customer needs and wants, in theory it should be a simple process for the organisation to plan the production or service delivery required to provide customers with the products or services that they have indicated, via market research, are a need or a want. However, as will emerge later in this chapter, the internal culture or structure of the organisation could make this more difficult than could be anticipated.

The second assumption about the organisation is that it is able to react quickly enough to the consumer's perceived needs or wants and redirect the necessary resources away from their current focus towards meeting newly perceived needs and wants in the marketplace. Without the necessary financial resources to purchase new technology or recruit new people, and without the production flexibility to rearrange current production lines to produce the new product lines, the marketing effort is wasted. Again, the marketing department's effort is dependent on the other departments for support and implementation.

The marketers' internal challenge, therefore, is to ensure that their organisations only produce what they have identified as the customers' needs or wants, and are flexible enough to react to perceived changes in customer requirements and other external environmental elements. The functional convergence mentioned before is the ultimate goal, i.e. all departments of the organisation work rationally towards meeting the ultimate goal of customer satisfaction. However, this is not always so easy to achieve. With the marketing department's specialist knowledge gained through its market research of the external environment, and with its philosophical conviction of the needs of marketers to take precedence within the organisation, departmental tension can replace functional convergence. Therefore, the relationships that marketing has with the other departments are crucial not only to customer satisfaction but also to the cultural and structural survival of the organisation.

**Marketing and operations management**

As mentioned previously, the R&D stage is the point at which marketing clearly requires a close and supportive working relationship with operations management, to decide on product or service design and to agree the resource inputs needed. Once resource levels are agreed, operations management can decide costs and marketing the sales price, thus working towards deciding together how profits can be made within cost and price constraints. Operations management obviously also requires an input into the design of new products and services, to indicate from a practical point of view what can technologically and physically be manufactured.

**Marketing and finance**

The link between finance and marketing is also crucial, but is also dependent on the link between marketing and operations management. The finance department is responsible for ensuring that the organisation has enough money to perform all the organisation's activities. Operations management has the responsibility of manufacturing within cost constraints. Marketing must bring the products to market and sell enough of them at a sustainable price to make profits for the organisation.

**Marketing and human resource management**

If new products and services are being introduced, marketing must keep the human resource management (HRM) department fully informed of developments. The message concerning new products or services must be communicated, as the HRM department needs to know the type of skills and experience that current or new workers will have to possess to be able to deliver the new products or services. HRM will develop its own strategies in order to plan for recruitment and selection of any new staff or formulate training programmes for existing staff who lack the necessary skills or expertise.

## Marketing tools

So far this chapter has examined the relationships that the marketing department has with the external and internal environments of the organisation. Marketing has been examined in terms of its responsibility for monitoring changes in customer needs or wants. The marketing department must convince senior management and the other departments that the organisation should be producing new products or delivering new services to meet these perceived changes. Therefore the marketing department uses an array of marketing tools to monitor elements of the external environment.

Marketing is not only about purely reacting to perceived changes in the external environment. It also has a responsibility to aim to be proactive in relation to the external environment. Thus marketing may aim not only to provide the products or services the customers know they want, but also attempt to influence customer choice by anticipating what customers are likely to need or want in the future and providing it ahead of competitors. This proactive stance is to a large degree reliant on assumptions about customers and the organisation. Marketing tools can thus be used both to react to and to influence the marketplace. Some of these basic tools are presented and applied here to a variety of products and services.

## The marketing mix

Borden[5] developed the idea of a marketing mix to describe the marketing elements that could affect the way a product performed in the marketplace. McCarthy[6] summarised the marketing mix as the four Ps of marketing. Today the marketing mix or 4Ps is one of the traditional tools used to manipulate the organisation's relationship with the external environment. The 4Ps are product, price, promotion and place. That is to say, there are four basic ways in which organisations can affect the relationship they have with their customers to increase sales and profitability.

### Product

Once the initial investment in research and development has been paid back and resulted in a successfully launched product, it is in the manufacturer's financial interest to recoup as much profit on that investment as possible. Thus the manufacturer will want to make as much of the product for as along as possible with little new investment or alteration in order to keep sales high and reap profits. If sales do begin to decline after a certain time, it is not necessarily inevitable that a brand new product needs to be found straight away. In order to keep sales of the existing product buoyant, it is possible to manipulate certain aspects of that product at little cost in order to offer a newer, fresher face to it. The clear aim is to continue to attract new customers or tempt existing customers not to begin to buy a competitor's version.

Product aspects that may be manipulated include style, performance, quality, branding, packaging and after-sales service. Examples of such product manipulation abound in the washing powder industry. Famous and familiar brands are often relaunched as 'new', 'improved' or as version 2 or 3. While alterations to the basic washing powder have undoubtedly been made in order that trade description legislation is not infringed by the use of these words, the basic product is still washing powder, with additional features. Altering the product slightly offers the manufacturer the opportunity to make a statement to the marketplace about continual innovation, improvement and customer

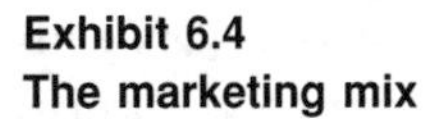

**Exhibit 6.4**
**The marketing mix**

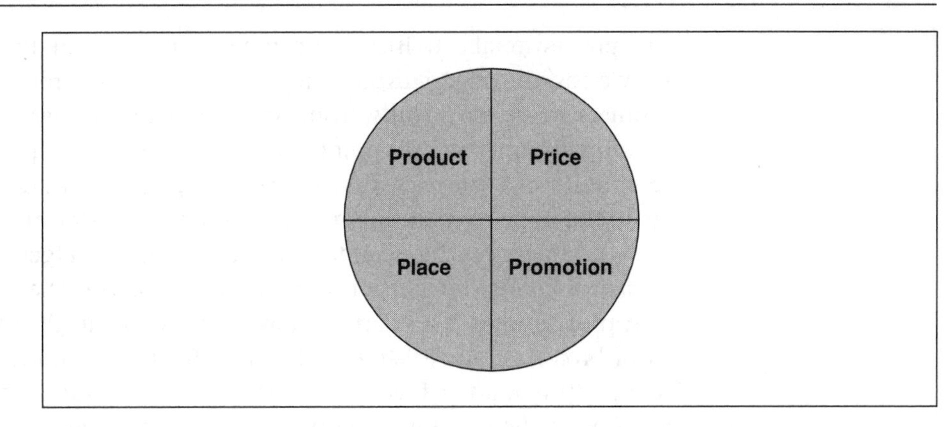


orientation. Hence this type of manipulation of the product is a reasonably simple yet effective marketing tool. Product manipulation requires co-operation between the marketing and operations management department.

**Price**

The second of the four Ps is price, and there are various ways in which organisations can use the price they charge to influence sales. The music industry has used pricing strategies aggressively to influence the positioning of new releases in the pop charts. Most singles on release are now sold at a discount price in the first few weeks to entice customers to purchase them. The national music charts are compiled out of sales figures at monitored outlets, and these subsidised sales cause new releases to have a high entry position in the pop charts in the first week, which encourages more radio play. This produces continued sales as radio listeners go out to buy the music they have heard.

Price can be used not only when introducing new products, but also when products are perceived to be near the end of their natural life (*see* the discussion of product life-cycle in the next section). Special offers and finance deals can be used to affect customers' buying habits. The mass television advertising by DFS/Northern Upholstery of its dining-room and lounge furniture is an excellent example of this. The frequent, prime-time advertisements offer cheaper prices, 'buy now, pay later' deals, 0 per cent finance on hire purchase and long payment terms of four to five years. All these are offered in special sales promotions that 'must end Sunday 5pm', and yet seem very similar from week to week. Price manipulation depends on co-operation between the marketing and finance departments.

**Promotion**

As well as manipulating price to influence customers, a product can be advertised and promoted with the aim of encouraging sales. Decisions have to be made concerning the advertising media to be used: press, magazines, television, radio or Internet. Alongside the advertising decisions, the promotional activity for a product has to be decided. A combination of advertising and promotional activities is required to create and support a successful product. The combination of activities needs to create an awareness and interest in the product and acceptance of it by the marketplace.

The promotional activities that can be used are varied. In supermarkets with loyalty cards, the offer of extra bonus points on certain products or goods entices customers to switch loyalty from one brand to another or to buy products not normally on their shopping list. Magazines are frequently used to offer free samples of cosmetics, cassette recordings or even books to their readership, in the hope of capturing new and potentially loyal customers. The launch of new products can be heralded by the delivery to target households of free samples, discount vouchers or promotional literature. The sponsorship of television programmes is a relatively new activity in the UK following deregulation of broadcast advertising, and is an effective promotional method. Thus Cadbury, 'the nation's favourite' chocolate manufacturer, sponsors *Coronation Street*, one of the leading soap operas, and the holiday company Going Places sponsors *Blind Date*, where winning contestants are sent off on holiday dates around the world. Neither of these television programmes explicitly advertises the companies' products, but implicitly links their products with the programme in the viewers' minds.

The importance of an organisation's advertising and promotion activities is clear if a product is a leading brand and the organisation seeks to maintain that position. The importance of good advertising and promotion is heightened if a competing brand advertises heavily and is easily substituted; *see* Chapter 3. Also the advertising and promotion surrounding a product will remind consumers making frequent repeat purchases to purchase the same brand of product as before, which is crucial if customer loyalty to a brand is low. A product's added value or low cost, which will be important to particular groups of customers as explained in Chapter 3, can be emphasised by the use of advertising and promotion.

**Place**

The fourth element of the traditional marketing mix is the location of the interface between customer or consumer and product or service. This links with the issue of distribution, and getting the right amount of product to the right place at the right time is essential. This is fundamental to each stage of the distribution process, whether from manufacturer to wholesaler, wholesaler to retailer, retailer to customer or manufacturer direct to the customer. This is particularly so for those organisations that are reacting to customer demand for more products that can be delivered direct to the home without the need to go out shopping, with for example mail-order or Internet shopping.

The manipulation of the place element of the marketing mix can also take place outside the home in shops. For example, the simple positioning of sweets at a supermarket checkout constitutes the use of placement to attract customers or their children who might be tempted to buy that bar of chocolate for the journey home. Supermarkets also use place on a grander scale by frequently changing the position of everyday basket goods by locating them next to more unusual or aspirational goods that might not be purchased on a regular basis. Placing products not normally associated with a particular location is also included. Thus being able to buy Haagen-Dazs ice cream at a Warner Brothers cinema links the two brands in the customer's mind. Similarly, supermarkets have begun to offer petrol stations as part of their overall package and petrol stations now sell groceries on a 24-hour basis.

## The extended marketing mix

The extended marketing mix or 5 Ps (*see* Exhibit 6.5) is a development of the traditional marketing mix with the addition of a fifth element: people.

**People**

It is initially difficult to see how marketers could manipulate people in the same way as they could manipulate the other 4 Ps, as the latter are internal elements within the direct control of the organisation, whereas people in terms of customers are not part of the organisation's internal environment. The people element inside an organisation is constituted by the employees. However, the consideration of people in the marketing mix reflects the importance to successful marketing of both the person who is the customer and the person doing the selling. Interpersonal skills play a large part in achieving a successful relationship between customer and organisation. Hence there is a crucial relationship between the marketing and human resource management departments in ensuring that appropriate staff are recruited to do the selling and marketing. In addition, recognising who customers are and what they want is an explicit part of the people element of the marketing mix.

**Exhibit 6.5**
**The extended marketing mix**

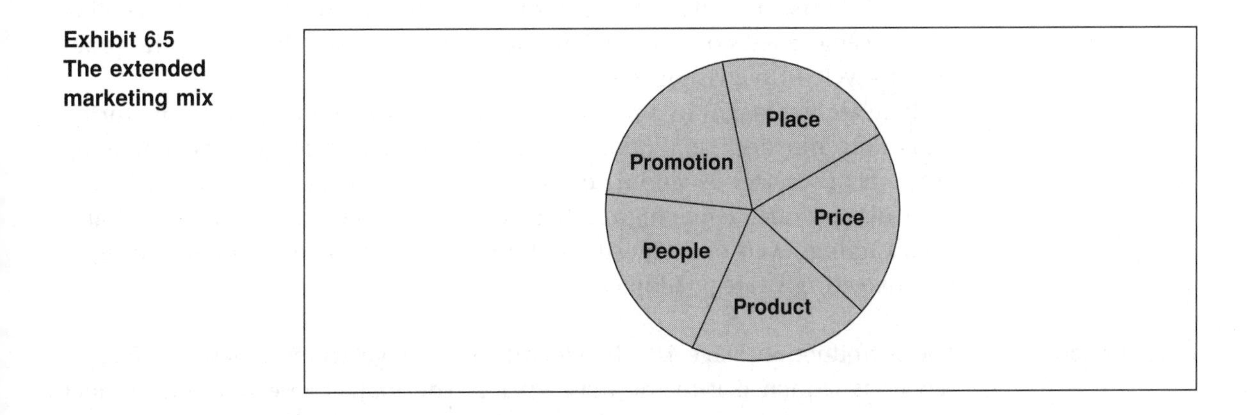


## Product life-cycle

The product life-cycle is the 'natural' lifespan of a particular product or service and may last for a few short months or many years. For example, a Walt Disney *Lion King* tee-shirt has a life-cycle of, at most, a few months or until the next Walt Disney film is released and associated merchandise is marketed. In contrast, a product like a television set has a product life-cycle many years

**Exhibit 6.6 The product life-cycle**

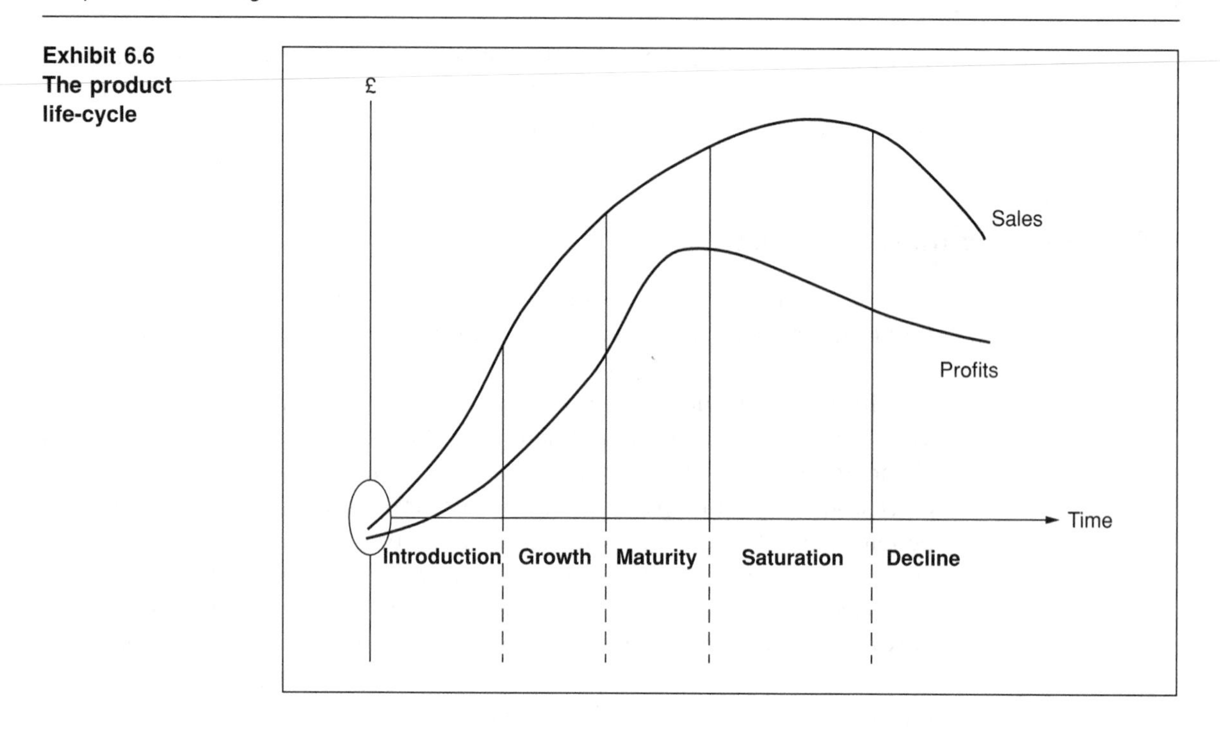


long due to continued product development in the field of television and broadcasting. These developments have included the very first black-and-white television sets that became available in the 1950s, through colour, stereo and digital television sets. The development of associated products such as satellite dishes, cable television and digital television has also help extend the product life-cycle of television sets.

There are five stages in a product life-cycle: introduction, growth, maturity, saturation and decline. Companies adopt different strategies for marketing products or services depending on the stage in the life-cycle they identify that particular product as having reached. Various products and services will also need varying levels of financial and human resource investment according to the marketing strategy adopted.

## Introduction

The introduction stage is self-evidently that stage when a new product or service is launched into the marketplace. The market research and product development have been completed and the product has been designed with its target market in mind. This has necessitated considerable investment on the part of the organisation with no attributable profit, since, until the product is sold, there is no income from it. Thus the sales and profit curves run together at the bottom of the profitability axis in Exhibit 6.6. It is clear that for new organisations this is the 'make or break' period, as the investment for new products or services is likely to have come from bank loans or other borrowings. For existing organisations the investment for research and development may come from profits earned by other products. This issue is examined under the Boston Consulting Group matrix later in this chapter. Research and development will have considered the product and price elements of the marketing mix for the new product or service.

A product in the introduction phase of the product life-cycle will have relatively few buyers and those that do purchase the product will buy the product to try it. Thus at the introduction stage promotion of the product is crucial if it is to move on to the growth phase. The introduction of Dr Pepper's to the UK market was a case in point. The drink itself was a long-term successful soft drink in the USA, as familiar as either Pepsi or Coca-Cola, but had never been sold in Britain. Therefore its introduction to the UK market relied on successful promotion, since the product, a soft drink, with a price comparable with other soft fizzy drinks, meant that product and price aspects of the marketing mix were to a large extent unalterable. In television advertisements, Dr Pepper's product was promoted as something that had arrived literally from another planet and was proven safe to a suspicious public by a risk taker wearing a silver suit in a sealed environment testing the drink.

**Growth**

If the research and development work has been done correctly, a newly launched product will sell well in the early period of its life-cycle. Thus sales will rise quite dramatically. However, the profit curve remains relatively low due to the cost of the continued promotion needed to allow sales to grow. Nevertheless, profits are made and market share starts to be accumulated in the growth phase of the product life-cycle. The marketing department will use a variety of promotional tools to achieve growth in sales, some of which will include manipulation of other elements of the marketing mix, such as place. For example, a new fragrance could be promoted through a number of locations: the distribution of free samples in fashion magazines; high-profile sales stands in up-market department stores; large advertisements on roadside billboards as well as television and cinema advertisements. The unisex fragrance CK One achieved growth in sales through such a variety of promotional methods.

**Maturity**

The dramatic growth in sales of a product or service slows down when the maturity section of the product life-cycle is reached. Depending on the type of product and the type of market, the maturity stage may last a long or a short time. A product in the maturity stage still has limited growth, but rapid and significant progress such as that achieved in the growth phase of the product life-cycle is unlikely. The maturity stage is where maximum profits are achieved and the outlay required to maintain sales is minimal compared with that in the growth stage; *see* Exhibit 6.6. Advertising and promotion of a product in the maturity phase of the product life-cycle are aimed at retaining existing customers and persuading others to switch from competitor products. The overall aim is to keep the product at the peak of the maturity stage of the product life-cycle for as long as possible as this is when profits peak; *see* Exhibit 6.6.

**Saturation**

The saturation stage of the product life-cycle is reached when growth tails off and the market for a product is no longer growing. Nevertheless, sales volume may be kept buoyant and loyal customers retained by price competition and special offers, although this will mean reduced profits; *see* Exhibit 6.6. These are competitive options that it is easy for rivals to replicate. An alternative would be for the marketing department to choose to implement an extension strategy. These are discussed briefly in the next section of this chapter.

**Decline**

Following the saturation of the marketplace, products eventually lose sales volume through being replaced, in the customers' eyes, by new products introduced by the same organisation or by competitors; *see* Exhibit 6.6. A product or service in decline may be withdrawn from the market altogether if it is losing money. Alternatively, it may find a small, loyal, niche market that either breaks even or makes a limited profit for the organisation. An example here is the vinyl LP record. Most people today prefer their music to be on tape or compact disc, therefore most music produced today is in those formats. However, there exists a small group of consumers who still buy vinyl LP records, either vinyl buffs or DJs in clubs, therefore record companies continue to release a certain amount of material on vinyl. The decline of vinyl LPs will therefore be long and slow.

## Extending the product life-cycle

The initial research and development programme, the costliest part of the process, should be recouped by the organisation during the growth and maturity phases of the product life-cycle. A product in the saturation phase will continue to generate profit, but profits will decline towards the end of the saturation phase. Therefore the majority of the profit that a product can expect to make in its life-cycle will have been made by the end of the saturation phase; *see* Exhibit 6.6. Hence extension of the product life-cycle should be considered before the product reaches the end of the saturation phase; *see* Exhibit 6.7.

There are various methods for accomplishing this and Ansoff (1987)[7] summarises four different marketing strategies that organisations may follow and presents them in the Ansoff matrix. The first three strategies, market penetration, market development and product development, can all be used to extend a product's life-cycle; *see* Exhibit 6.7. The fourth, diversification, involves changing to a significantly different product.

**The customer growth matrix**

In following Ansoff's marketing strategies, organisations are seeking to increase the number of sales and/or the number of customers. Jenkins (1997)[8] presents four different types of customer growth options that organisations may follow to achieve sales and customer growth; *see* Exhibit 6.8.

**Exhibit 6.7 Product life-cycle and extension strategies**

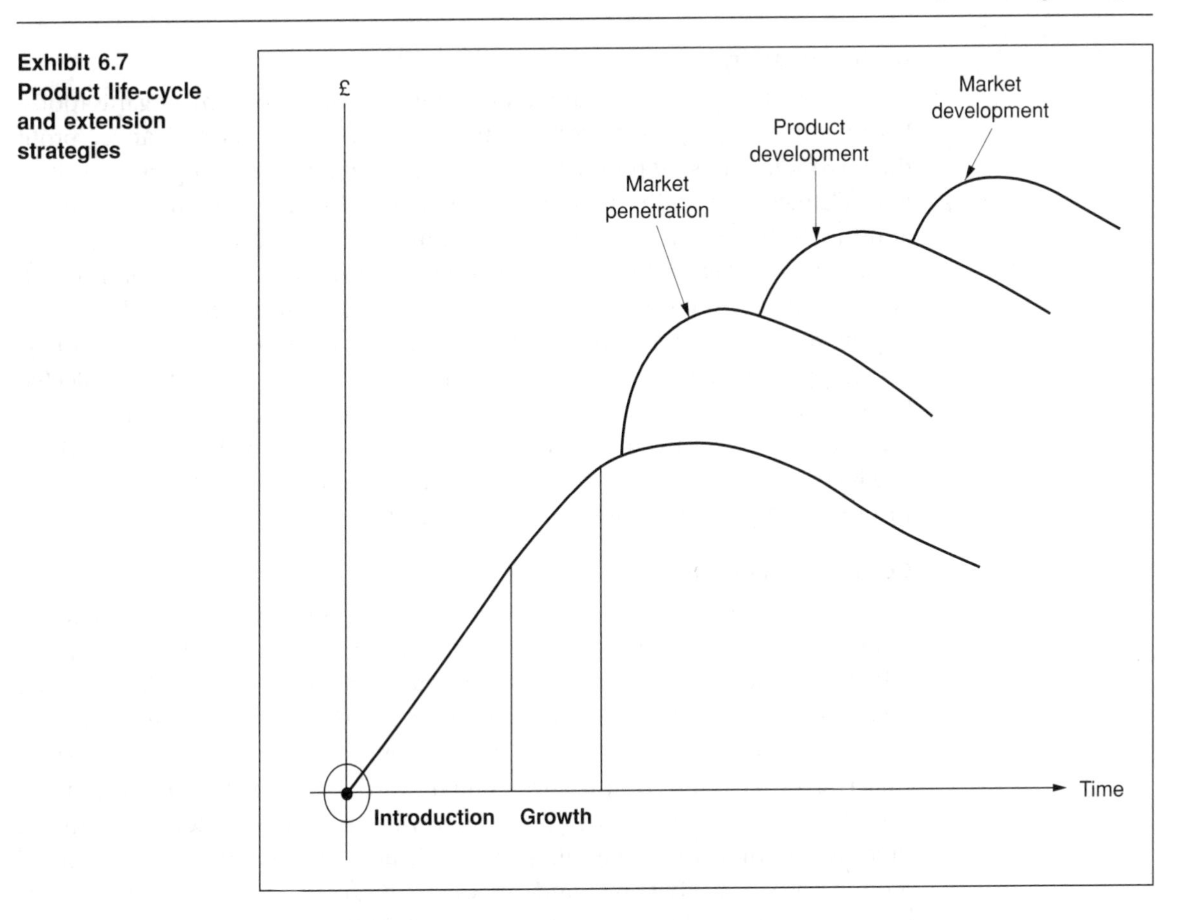


**Exhibit 6.8 The customer growth matrix**

| Customer \ Product or service | Existing | New |
|---|---|---|
| Existing | **Customer loyalty** | **Customer extension** |
| New | **Customer acquisition** | **Customer diversification** |

*Source*: Jenkins, M (1997), *The Customer Centred Strategy*, Prentice Hall. Reproduced with permission.

### Customer loyalty

Customer loyalty is important and should be developed by organisations, because it brings greater profitability.[9] Faithful customers will affect profit directly by purchasing products and services over an extended period of time, and indirectly by telling friends, neighbours and relatives about the benefits and satisfaction they derive from the company's products and services.

The supermarkets in the United Kingdom are examples of organisations which attempt to create customer loyalty via their use of so-called 'loyalty cards'. Loyalty to the service provided by the supermarket is created by offering money-off shopping when a certain number of 'loyalty' points have been collected by a customer. The development of loyalty to products involves manipulation of the marketing mix. Alterations may be made to the price at which products are being sold, or to promotional activities or distribution locations in order to try to increase sales to existing customers.

### Customer extension

Customer extension is concerned with extending the range of products or services available for a customer to purchase from the organisation. Increasing sales by extending the range of products and services available to the customer involves following strategies of product development and diversification.

Product development is likely to be the preferred choice of the organisation that is good at research and development and strong in the area of innovation. Other organisational features which will make product development the preferred choice for expanding sales include an organisation structured around product divisions and products with short product life-cycles. Consumer electronics companies fit this profile very well. The original Sony Walkman, a portable cassette player, was the result of product development by Sony and extended the range of entertainment products which customers traditionally purchased from Sony. In 2000 the launch of the DreamCast2 computer games console once again extended the range of Sony products available to customers.

Diversification is an alternative option for extending the organisation's range of products and services. Diversification is risky, as both organisational effort and capacity are stretched. There are two basic types of diversification, related and unrelated. Related diversification occurs when development is beyond current products and markets but still in the same broad industry. For example, if Sony were to diversify into producing other electronic goods for the home such as fridges, freezers and washing machines, this would be related diversification, extending the range of products available to include white goods but remaining in the broad industry of providing electrical consumer goods. In contrast, unrelated diversification for a company like Sony would be a move into running a rail franchise, which is completely unrelated to electrical consumer goods but still increases the range of products and services available to customers. Richard Branson's empire is a good example of an organisation which has expanded by unrelated diversification and currently offers customers a wide range of products and services including air travel, train travel, cola, cinemas and financial services.

### Customer acquisition

Customer acquisition is expanding the number of customers for existing products. This could involve expanding customer numbers in home markets, which will be easiest if home markets are growing in size. If home markets are mature, then expansion into growing overseas markets may provide the best opportunity for increasing the number of customers. This was one of the reasons for expansion into the overseas market of China by companies like Nescafé, which was seeking to increase the number of customers drinking and buying instant coffee; *see* the entry case study for this chapter.

Customer acquisition in overseas markets requires the organisation to engage in international business activities such as exporting or internationalising its operations.

- Exporting involves selling existing product ranges, which incur no further development costs, to new customers abroad.
- Internationalising operations involves locating activities overseas, such as manufacturing, distribution and promotion. The benefits of doing this can include overcoming import controls, lower labour costs and lower distribution costs. An organisation may choose to internationalise its operations if the number of customers in a particular geographic market is large and home markets are mature or in decline.

Attempting to increase the number of customers in a static or mature market will be difficult, because there will be few or no new customers available. In a static or mature market increasing the number of customers can only be realistically achieved if customers can be persuaded to switch products or brands. This requires customer loyalty to a competing product or brand to be broken. The only other opportunity for attracting customers in a static or mature market will arise if the market leader becomes complacent and allows performance to slip. Increasing customers while operating in a declining market is only possible if competitors leave the marketplace early and their customers transfer their business to those remaining in the industry. However, it should be noted that seeking to increase customers while operating in a declining market is a short-term option with a limited lifespan.

Organisations may choose to follow a combination or hybrid of customer growth options. An organisation may expand its sales and customer base by following both customer loyalty and customer acquisition options.

### Customer diversification

Expanding customer numbers by customer diversification is the riskiest of all of Jenkins' options. Customer diversification is achieved if sales are increased by selling a new product or service to new customers. The availability of a new technology or process is usually required if customer diversification is to occur.[10]

The best recent examples of customer diversification to occur are those being achieved through e-commerce by many of the so-called dotcom companies offering products and services over the Internet. It is the use of new technology (computers and the Internet) which allows the provision of services to be offered in an entirely new way. For example, the provision of financial services by many of the high street banks via the Internet and bookshop services by companies such as Amazon.

March 2000 saw the launch of the company lastminute.com, selling last-minute travel, trips and gifts by the process of booking via the Internet. These dotcom companies are pursuing a risky strategy in hoping to sell to customers via the Internet. The risk arises from the way such companies provide their service, via the Internet, which is largely untried and untested in the marketplace, so it is not known if customer numbers will grow. The risk is also heightened as many of these dotcom companies currently have small sales revenues and have not yet made a profit.

## The Boston Consulting Group matrix

Henderson[11] of the Boston Consulting Group developed seminal work on categorising products in a useful way that then enables the marketing department to decide appropriate strategies for products in different stages of the life-cycle. The categories are based on the rate of market growth identified compared to the volume of market share the product has achieved in the marketplace; *see* Exhibit 6.9. It should be noted that large market share alone

**Exhibit 6.9 The Boston Consulting Group matrix**

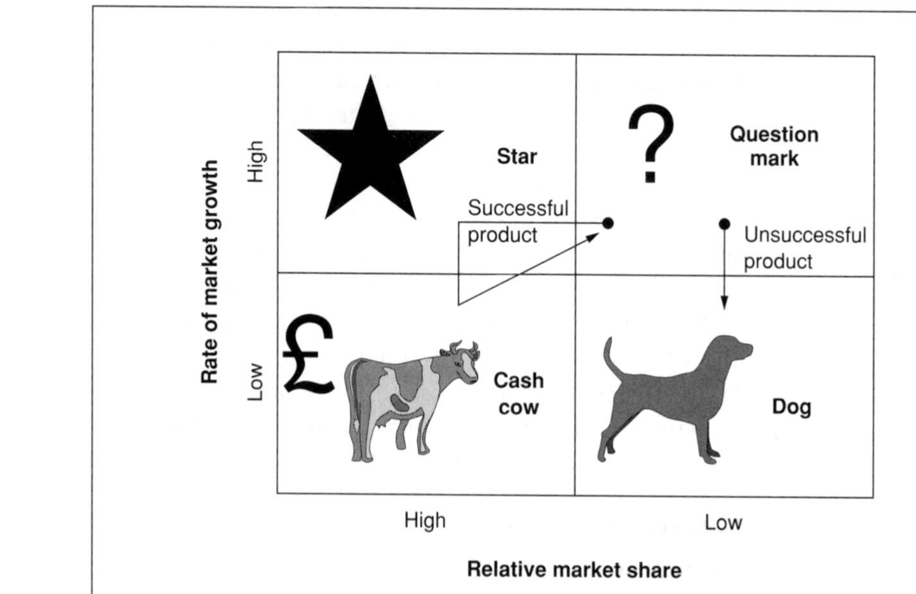


*Source*: Henderson, B (1970) *The Product Portfolio*, Boston Consulting Group. Used with permission of The Boston Consulting Group.

will not generate low costs and high margins. The relative comparison of costs and margins to those of competitors is also important; *see* Chapter 3.

**Question marks**

The question mark is a product located in the introduction stage of its product life-cycle that is likely to achieve high market growth, but currently holds low market share. If a question mark is to become a star, then high expenditure will be required to promote the product or service such that higher market share is achieved. The funding of a question mark will come from other successful products in the organisation's product range; *see* the later section on cash cows. A product that is a question mark is not necessarily certain to be successful in the future.

**Stars**

The successful question mark will gain market share and become a star. The star product is located in the growth stage of its product life-cycle. The market at this stage may still be small in terms of overall sales or market size. However, the star has a sizeable portion of overall market share. It is likely that star products will be successful, based on predicted growth in sales and continued domination of the growing market. The successful star product will be on its way to being a cash cow of the future.

**Cash cows**

The cash cow will be in the maturity or saturation stage of its product life-cycle. The rate of market growth will be low, but the volume of market share will be high, as shown in Exhibit 6.9. Hence sales of the product are at maximum levels, as are profits. Therefore the cash cow is the bread-and-butter product of the organisation, and its profits provide the finance for research and new product development. The organisation is thus reliant on cash cows and needs to maintain them in its product portfolio.

**Dogs**

Dogs are products that are definitely in the decline stage of the product life-cycle and may have previously been successful cash cows. Dogs have no growth in terms of sales and do not have significant market share. Therefore the organisation has to decide what to do with its dogs. If the dog product is profitable and has a small but loyal band of consumers still willing to continue purchasing it, then keeping it in the product portfolio is a viable option. Alternatively, the organisation may decide to kill off the dog. This would be the more viable option if the dog product had become loss making or if the organisation had a substitute product that was a star.

**Successful and unsuccessful products**

A successful product will move around the Boston Consulting Group matrix in the following order: question mark, star, cash cow, question mark. Hence a successful product will at times cost the organisation money and at other times make it money. A product that ceases to be successful may return to the right-hand side of the matrix, moving from being a star or cash cow to being a dog. A product that is never successful will stay on the right-hand side of the matrix (low cash generation) and go from being a question mark directly to being a dog and will never make significant profits for the organisation.

## Conclusion

This chapter examined the concept of the marketing-oriented firm, the marketing activities that organisations carry out today and finally some of the best-known marketing tools.

ETHICAL ISSUES CASE STUDY 6.2 FT

# The giant who lurks behind a smokescreen

**By Richard Tomkins**

Some weird-looking cigarettes have hit the streets in the US. With names like Politix, Planet and Icebox, they're hip, they're cool and they're made by a new cigarette manufacturer you've almost certainly never heard of: the Moonlight Tobacco Company.

But Moonlight Tobacco is not quite the independent upstart it appears. In fact, the name is just a smokescreen for the rather better-known company behind the new brands. It turns out to be R J Reynolds Tobacco, the second biggest cigarette manufacturer in the US.

Moonlight Tobacco takes its cue from recent events in the US beer market, which has seen a surge in demand for 'alternative' beers brewed by small beer companies, or microbreweries. Some big brewers have responded by launching microbrewery-style sub-brands: Philip Morris's Miller Brewing subsidiary, for example, has set up a company called Plank Road Brewery, which sells the trendy Red Dog and Icehouse brands.

RJR, maker of old-established brands like Camel, Winston and Salem, says Moonlight Tobacco was inspired by a belief that an unexploited niche existed for 'alternative' cigarettes.

It quietly started test-marketing seven Moonlight Tobacco brands in New York City, Chicago and Seattle just over a year ago. Since then, it has added two more brands, and the testing has been extended to Cleveland, Ohio; Portland, Oregon; and North Carolina.

The cigarettes themselves are nothing out of the ordinary: they are made in RJR's factories alongside the company's traditional cigarettes. But the packaging, designed by Chicago artist Thomas Van Housen, is quirky and distinctive, as are the names of the cigarettes – B's, Sedona, Politix, Jumbos, Northstar, City, Metro, Planet and Icebox.

The advertising, too, is off-beat. Posters and billboards do not promote individual brands: instead, they feature the Moonlight Tobacco name and slogan against a collage of the package designs. This gives Moonlight Tobacco the flexibility to introduce brands and drop less successful ones, almost at a whim.

Moonlight Tobacco is not expected to produce another big-selling brand but the point is that a cigarette brand does not have to command a large share of the US market to be profitable. With annual cigarette sales totalling $47bn (£29bn), even a market share of 0.5 per cent is worth $235m a year.

Britt Breemer, chairman of America's Research Group, a market research company in Charleston, South Carolina, says that once companies achieve a certain market share with their big brands, they can find it extremely difficult to make further gains. 'So what they are attempting to do by creating these small companies is to pick up an extra few points of market share by having much more specific target marketing.'

Renee Frengut, president of Market Insights, a marketing consultancy in Bronxville, New York, says RJR is aiming the Moonlight Tobacco cigarettes at a new generation of consumers who are looking for alternatives to products from 'big, nasty, American' companies. 'They're trying to say: "We are really not these big bad guys. We are really just small entrepreneurs, people like you, and you should support us because we are hip like you guys are".'

One problem with this approach is that it appears to target younger people at a time when the Clinton administration is trying to crack down on underage smoking. But RJR denies the cigarettes are aimed at younger smokers – or, indeed, at any age group at all. Says Cliff Pennell, RJR's senior vice-president for brands: 'It's for folks who have much more of an alternative, free-spirited lifestyle.

'They don't like the mass or the mainstream. They like to be different, they like to make a statement about themselves. They are the folks who are more comfortable drinking a microbrew than a Budweiser. There's a level of sophistication with this proposition that I don't think is younger adult at all.'

**Ethical issues case study 6.2 *continued***

Clearly, RJR is taking a risk by moonlighting as an iconoclastic underdog, for the ploy could backfire if it were seen as a form of deception. So although RJR's name does not appear in Moonlight Tobacco's advertising, the cigarette packs contain leaflets explaining Moonlight Tobacco's parentage and portraying the company as a whacky, 'independent' outfit operating under the RJR umbrella.

Could RJR's example be worth following? Possibly, though RJR will not give any figures, and the project has yet to progress beyond the test-marketing stage.

It may also be worth remembering the cautionary tale of Quaker Oats, another big US company that tried to be trendy by buying the Snapple Beverage soft drinks manufacturer. The customers rebelled, sales plunged, and the acquisition turned into one of the most disastrous in US corporate history – proof positive, it seems, that just wanting to be hip is not enough.

*Source: Financial Times*, 3 February 1997. Reprinted with permission.

**Questions for ethical issues case study 6.2**

1 In your opinion, which media are acceptable for use in advertising cigarettes?

2 Explain if the 'hip' and 'cool' cigarettes promoted by the Moonlight Tobacco Company will attract under-age smokers or 'folks who have much more of an alternative, free-spirited lifestyle'.

3 Discuss the likelihood of success or failure for the Moonlight Tobacco Company's strategy of developing cigarettes brands for small niche markets.

## EXIT CASE STUDY 6.3

# Raise your glasses to craft beers

**By John Murray Brown**

Starting a microbrewery might seem an ambitious project in a country where Guinness accounts for one of every two pints sold and is seen by many drinkers as a 'craft beer' in its own right.

But Bill Dwan is confident he can carve a niche in the Irish market with his Tipperary-based Dwan beers which, like other craft beers, put the emphasis on taste, quality of ingredients and freshness.

Mr Dwan points out that in the US, where the microbrewery concept was invented, craft beers account for 3.5 per cent of the market. The Irish are big drinkers – only the Czechs drink more beer per head. If the Irish market is worth about I £2bn (£1.7bn), the microbrewing sector could be worth as much as I £70m.

'We obviously don't have the branding power to take on Guinness,' Mr Dwan says. 'But we're about an alternative choice. And I don't think younger drinkers are so hung up with brands. What they want is atmosphere and style.'

He believes microbrewing is handicapped by associations with real ales. In contrast to real ale, which is conditioned and fermented in the cask, and hand tapped, Dwan beers are pressurised and bottled or sold in kegs. Where real ale enthusiasts are concerned with how a beer is stored, microbreweries focus on the quality of the hops.

As consumers become increasingly health conscious, the fact that his beers use no additives or preservatives is another selling point, Mr Dwan believes.

Microbrewing, where typically the beer is brewed and consumed on site in a bar or restaurant, requires a relatively small initial outlay.

Mr Dwan has invested about £1.5m – £600 000 of that being a grant under the government's small business expansion scheme.

He produces two stouts and two lagers – one with a European pilsner flavour and the other lighter coloured and more like an American lager – as well as a traditional red ale.

He bought the brewing kit from Dobbin and Dobbin Contracts, a Manchester-based company which also supplies the recipes. Dobbins helped recruit David Jones, Dwan's master brewer, who was working in northern Sweden on another of Dobbin's projects. Mr Dwan is giving work experience to three apprentices from another of Dobbin's start-ups in San Sebastian in Spain's Basque country.

**Exit case study 6.3 *continued***

The business a re-branding of a soft drinks concern started by his grandfather in the 1920s, comprises the one microbrewery with a pub and restaurant. Mr Dwan is in negotiations to extend the brand to Limerick and Cork.

Last month the company won an award at the Swedish beer festival, and has now appointed a Swedish distributor. Exports are expected to start in the next few weeks.

The small scale of the operation – with Dwan producing in six months what Guinness would make in a day – is critical to the freshness that is one of the beer's distinguishing features.

*Source*: *Financial Times*, 17 November 1998. Reprinted with permission.

**Questions for exit case study 6.3**

1 Consider the marketing activities discussed in this chapter. Explain how you would undertake each of the activities for craft beers.

2 In order to ensure that craft beers remain popular, which of the customer growth options should be followed and why? (You can use more than one of the customer growth options.)

## Short-answer questions

1 Explain the term 'functional convergence'.

2 Define the term 'production era'.

3 Define the term 'sales era'.

4 Define marketing.

5 List the three basic points of marketing, from a marketer's perspective.

6 Explain Maslow's hierarchy of needs.

7 List three basic assumptions underpinning the marketing concept.

8 Name the activities that go to make up the marketing department.

9 State the 4 Ps of marketing.

10 Name the fifth P of marketing.

11 Sketch the product life-cycle curve accurately.

12 In which stage of the product life-cycle does a product make most money?

13 In which stage of the product life-cycle does a product sell most?

14 In which stage of the product life-cycle does a product make virtually no profit and why?

15 Name three strategies that are used to extend the life-cycle of a product.

16 Name the two customer growth options that can be used in combination.

17 Sketch the Boston Consulting Group matrix correctly.

18 List the characteristics of a cash cow product.

19 In what circumstances should a company decide to keep a dog product?

20 Explain the path of an unsuccessful product around the Boston Consulting Group matrix.

## Assignment questions

**1** Consider your own organisation (employing organisation or university/college). Using the marketing tools discussed in this chapter, analyse the marketing that it undertakes. Use the findings of your analysis to recommend what the organisation should currently be doing in terms of market penetration, product development, market development and diversification. Present your findings in a 2000-word report.

**2** You are setting up a small travel company that specialises in outdoor pursuits holidays in the UK. Write a 2000-word report outlining both the marketing and development plans for the venture.

**3** Identify two organisations, one sales oriented and one marketing oriented. Compare and contrast the activities they undertake when dealing with new customers. Present your findings in a 2000-word essay.

**4** You work for a market research agency and are in charge of the latest contract for which the agency is bidding. The contract is to undertake market research into the suitability, acceptability and feasibility of your client developing and opening a *Star Wars* theme park in the UK. The client has asked for a plan of the market research work that the agency would undertake if it secured the contract. Present the plan and its justification in a 2000-word report.

## References


1 Drucker, P (1954) *The Practice of Management*, Harper & Row.
2 Kotler, P, Armstrong, G, Saunders, J and Wong, V (1996) *Principles of Marketing*, Hemel Hempstead: Prentice Hall.
3 Lancaster, G and Massingham, L (1988) *Essentials of Marketing*, Maidenhead: McGraw-Hill.
4 Maslow, A H (1943) 'A theory of human motivation', *Psychological Review*, 50.
5 Borden, N H (1964) 'The concept of the marketing mix', *Journal of Advertising Research*, June, reprinted in (1998) *Management Classics*, 6th edn, Harlow: Allyn & Bacon.
6 McCarthy, E J (1981) *Basic Marketing: A Managerial Approach*, Burr Ridge, Il: Irwin.
7 Ansoff, I (1987) *Corporate Strategy*, London: Penguin Business.
8 Jenkins, M (1997) *The Customer Centred Strategy*, Harlow: Prentice Hall.
9 Ibid.
10 Ibid.
11 Henderson, B (1970) *The Product Portfolio*, Boston: Boston Consulting Group.


## Further reading

The following books and articles all look at what marketing is about:


Blois, K J (1989) 'Marketing in five "simple" questions!', *Journal of Marketing Management*, 5 (2).

Burke, R R (1996) 'Virtual shopping break through in marketing research', *Harvard Business Review*, March/April.

Christopher, M and McDonald, M (1995) *Marketing: an Introductory Text*, Basingstoke: Macmillan.

Iansiti, M (1997) 'Developing product on Internet time', *Harvard Business Review*, September/October.

Keith, R J (1960) 'The marketing revolution', *Journal of Marketing*, January.

Kotler, P and Levey, S J (1969) 'Broadening the concept of marketing', *Journal of Marketing*, January.

Kotler, P, Armstrong, G, Saunders, J and Wong, V (1996) *Principles of Marketing: European Edition*, Harlow: Prentice Hall.

Levitt, T (1960) 'Marketing myopia', *Harvard Business Review*, July/August.

Levitt, T (1975) 'Marketing myopia 1975: retrospective commentary', *Harvard Business Review*, September/October.

The following books and articles all look at marketing tools:

Abell, D F and Hammond, J S (1979) 'Portfolio analysis', an excerpt from *Strategic Marketing Planning*, Harlow: Prentice Hall, reprinted in Quinn, J B, Mintzberg, H and James, R M (1996) *The Strategy Process*, 3rd edn, Hemel Hempstead: Prentice Hall.

Christopher M and McDonald, M (1995) *Marketing: an Introductory Text*, Basingstoke: Macmillan.

Doyle, P (1976) 'The realities of the product life-cycle', *Quarterly Review of Marketing*, Summer.

Fisher, M L (1997) 'What is the right supply chain for your product', *Harvard Business Review*, March/April.

Hofer, C and Schendel, D (1978) 'Portfolio analysis', adapted from *Strategy Formulation: Analytical Concepts*, West, reprinted in de Wit, B and Meyer, R (1994) *Strategy Process, Content, Context*, St Paul, MN: West.

Kotler, P. Armstrong, G, Saunders, J and Wong, V (1996) *Principles of Marketing: European Edition*, Hemel Hempstead: Prentice Hall.

The following articles all look at examples of marketing in action:

Aaker, D A (1997) 'Should you take your brand to where the action is?', *Harvard Business Review*, September/October.

Gwyther, M (1999) 'The big box office bet', *Management Today*, March.

Jones, H (1995) 'Charity coffee aims for a richer blend', *Independent on Sunday*, 12 November.

Mitchell, A (1998) 'The dawn of a cultural revolution', *Management Today*, March.

Parry, R (1996) 'Sites set on a roller-poster success story', *Independent*, 24 June.

Rigby, R (1998) 'Tutti-frutti capitalists', *Management Today*, February.

Shepherd, J (1995) 'Young drinkers targeted in drop of the hard sell', *Independent*, 2 September.

Surowiecki, J (1998) 'The billion dollar blade', *Management Today*, August.

Trapp, R (1996) 'Brands with no barriers', *Independent on Sunday*, 17 November.

Vishwanath, V and Mark, J (1997) 'Your brand's best strategy', *Harvard Business Review*, May/June.

# 7 Operations management

Claire Capon with Andrew Disbury

## Organisational context model

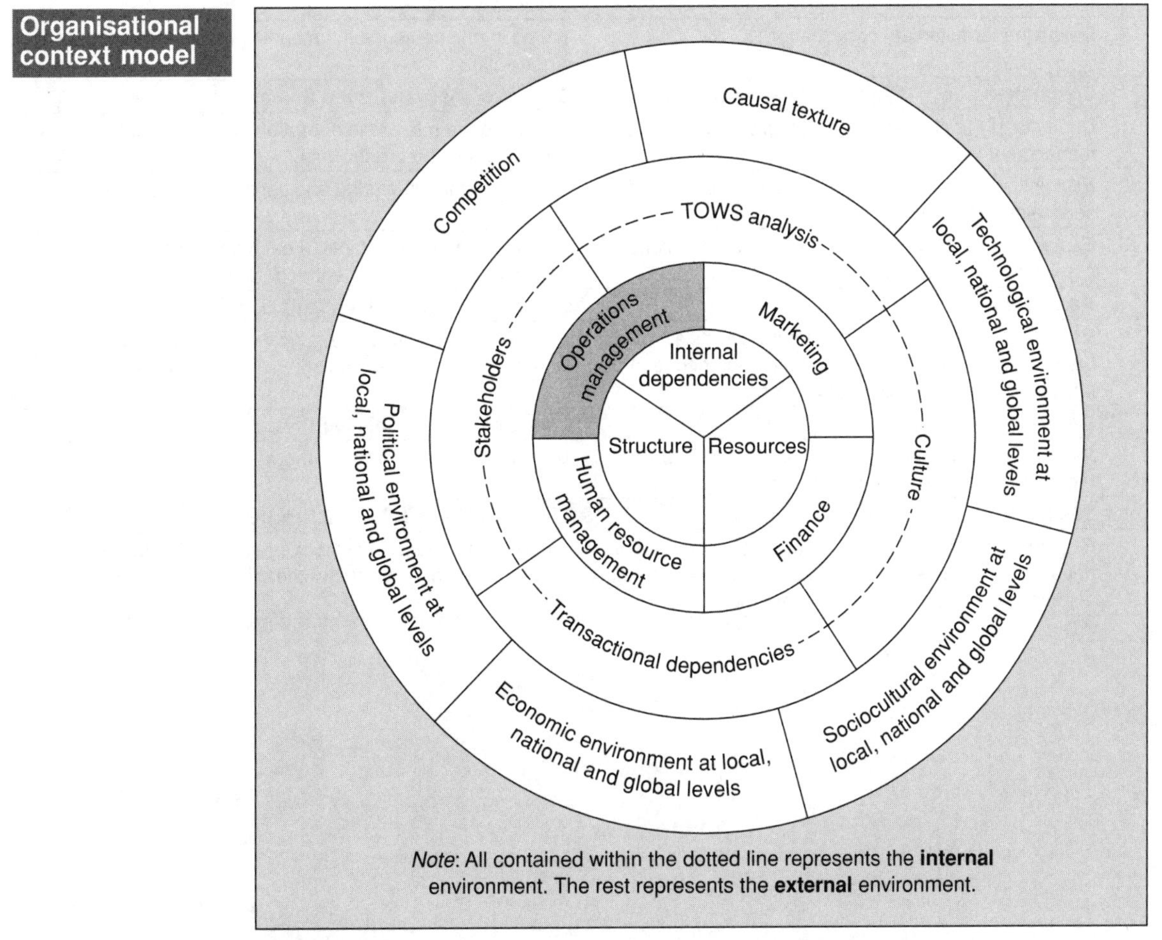


*Note*: All contained within the dotted line represents the **internal** environment. The rest represents the **external** environment.

**Exhibit 7.1**

## Learning outcomes

While reading this chapter and engaging in the activities, bear in mind that there are certain specific outcomes you should be aiming to achieve as set out in Exhibit 7.2 (overleaf).

**Exhibit 7.2 Learning outcomes**

| | Knowledge | Check you have achieved this by |
|---|---|---|
| 1 | Define operations management | stating in one sentence what operations management is |
| 2 | List the activities of the operations management department | naming the tasks the operations management department performs |
| 3 | Define the term 'primary organisation' | stating in one sentence the meaning of the phrase 'primary organisation' |
| 4 | Define the term 'secondary organisation' | stating in one sentence the meaning of the phrase 'secondary organisation' |
| 5 | Define the term 'tertiary organisation' | stating in one sentence the meaning of the phrase 'tertiary organisation' |
| | **Comprehension** | **Check you have achieved this by** |
| 1 | Understand the organisational context of operations management | identifying the role of the operations management department in the organisation |
| 2 | Explain the differences between manufacturing and service organisations | contrasting manufacturing and service organisations |
| 3 | Explain what each of the activities of the operations management department involves | summarising the key points of each activity carried out by the operations management department |
| | **Application** | **Check you have achieved this by** |
| 1 | Explain how operations management activities, which are traditionally associated with manufacturing, can be applied to service-sector organisations | demonstrating, by use of examples, how operations management activities occur in service-sector organisations |
| | **Analysis** | **Check you have achieved this by** |
| 1 | Assess the operations management activities of an organisation with which you are familiar | identifying areas for improvement |
| | **Integration** | **Check you have achieved this by** |
| 2 | Explain the links between the operations management department and the other key departments of marketing, finance and human resource management | summarising and discussing the relationship between operations management and the other key departments |

ENTRY CASE STUDY 7.1 FT

# Lessons in improvement

**By John Griffiths**

Perkins Engines, the UK-based diesel engine maker sold for $1.3bn recently by Lucas Varity Corporation to Caterpillar of the US, invited in a few outsiders just before Christmas to help Perkins tear itself apart.

Large mobile cranes, forklift trucks and anything else that might be needed were placed to hand. Then the management stood back – well, almost – and let them get on with it.

The outsiders came from a diversity of companies; some to which it is a supplier of engines; others in the financial services and accountancy fields. They came from a variety of working backgrounds, with few having close connections with process engineering.

Yet just four days after they had first started walking purposefully towards key areas of Perkins' production lines, to embark on what was formally named a 'Shop-floor Kaizen Breakthrough', the transformations they had achieved in terms of improved efficiency and productivity could be described accurately as radical.

In the four production areas which the outsiders, formed into teams with a mix of Perkins' own workers, tackled, few of the processes in place just four days earlier were immediately recognisable.

A large area devoted to machining of engine connecting rods had been completely reconstructed, including the reshuffling of process machinery weighing several tonnes. The floor space needed for all the processes had been cut by 72 per cent.

Work in progress had been reduced by 93 per cent. The labour force required to carry out the processes had been reduced by 40 per cent, and the number of machine tools required from eight to five.

The results, according to Perkins' general manger and divisional director, Brian Amey, were even better than the company has come to expect from the ongoing series of 'kaizen breakthrough' taskforces it has now sent into action in some 150 areas of the plant.

Until the most recent four-day exercise in 'kaizen' – the Japanese term for continuous improvement – the teams had consisted always of Perkins' own employees, but from different areas of the company from that being targeted. The 'breakthrough' idea is that the teams, comprising usually a dozen or so, bring a fresh and critical eye to long-entrenched, but not necessarily efficient, production processes in the space of a short, highly concentrated exercise lasting just a few days.

This time round, however – an occasion regarded by Perkins as a big first – was to be a public kaizen, opening the company's doors and its processes to wider critical scrutiny.

The productivity and cost-saving gains were made by all teams involved in the exercise, with empirical evidence of the improvements when the four teams of the public kaizen made formal presentations to Perkins management at the end of the exercise.

Another team of a dozen or so which examined the core build area of Perkins' four-litre engine range made sweeping changes.

Much process plant was shifted around, with parts for assembly made to fall more easily to hand by opening up both sides of the assembly line and reducing radically the number of movements required of operators. Result:

- A 41 per cent improvement in operator productivity;
- A 79.5 per cent reduction in inventory;
- Floor space reduced by 45 per cent;
- Cycle time reduced by 25 per cent.

Total distance travelled by employees between process functions for a complete cycle reduced from 350 metres to 50 metres. Previously, the team had found 'workers hopping around the production line like rabbits'.

The process meets no apparent resistance from the shop floor. To the contrary, most of the company's several thousand employees, many of whom have themselves now been involved in internal kaizen breakthrough, say they have found the exercises stimulating as well as surprising.

However, part of the ground rules set down by North Carolina-based TBM Consulting Group, which has overseen Perkins' kaizen activities, is that employees receive pledges of no redundancies arising from the exercises.

In Perkins' case such assurances have not been difficult to give. The company has embarked on a number of expansion programmes expected to result in engine production doubling to around 500 000 units a year by early next decade.

Some of the public kaizen solutions were relatively obvious and clearly had been in the back of managers' minds as needing improvement even before it began.

For most of the teams, indeed, the pre-exercise discussions with plant managers gave some clear pointers as to where some improvements might come. Nonetheless, Perkins managers are adamant that the teams were not merely following up ideas for improvement planted by managers themselves.

Much of the teams' attention was taken up with dismantling 'pipeline' processes – so restricted and sequential that they can only move at the speed of the slowest operator or piece of equipment – and putting in their place a system which could flow around obstructions.

In all operations like this, says Anand Sharmah, TBM's president and chief executive, the idea is 'to lower the water

**Entry case study 7.1 *continued***

around the process to make any efficiencies visual and painful. If you let people continue to hide problems then you cannot manage those problems'.

Mr Amey, who joined Perkins in late 1995 from Nissan, where he had been deeply involved in ongoing improvements of production systems, knew very substantial production increases were in prospect for Perkins when he joined, and one of his first acts was to start kaizen activities.

Factory floor space started appearing as if from nowhere. 'It proved time and time again that we already had the space capability to double production,' recalls Mr Amey.

But can kaizen breakthroughs work for everyone?

Mr Sharmah agrees that this is not the case, but says that what he would regard as companies failing to get a serious grip on kaizen-type improvements represent no more than 5 per cent of the more than 100 US and more than 20 UK companies with which TBM has worked.

Kaizen, stresses Mr Sharmah, is surprisingly easy to make gains initially. 'But it is much more difficult to sustain.'

*Source*: *Financial Times*, 23 February 1998. Reprinted with permission.

## Introduction

Operations management is the term applied to the activities at the core of any organisation's business and is concerned with the way in which the organisation actually puts into practice what it has set out to do. An organisation will undertake operations to make a product, provide a service or perform a combination of the two. Hence Glaxo manufactures pharmaceuticals; BT provides telecommunications services; and Laura Ashley produces and sells clothes. Accordingly, operations management is concerned with managing the way products are made and/or service delivered, which has a direct connection with how the organisation achieves its objectives. The principles of operations management can be applied to any organisation.

## Organisations and operations management

On comparing and contrasting two very different organisations, it would appear as though their operations have few similarities. The operations of Glaxo, for example, would seem very unlike the operations of a chip shop run by its self-employed owner. However, closer examination will reveal surprising similarities. Both organisations have to choose the best location; buy raw materials; forecast demand for their products; calculate the required capacity; arrange resources to meet demand; use the raw materials to make products; sell the products to customers; manage cashflows and human resources; and seek out reliable suppliers. Both Glaxo and the chip shop want to run an efficient operation, with high productivity.

There are two basic ways of categorising organisations and the operations they undertake. The first is to consider organisations as belonging to different sectors: primary, secondary or tertiary; *see* Exhibit 7.3. Primary-sector organisations are concerned with producing raw materials and include oil extraction, coal mining, diamond mining and farming to produce food. Secondary-sector organisations manufacture and produce goods, often from raw materials

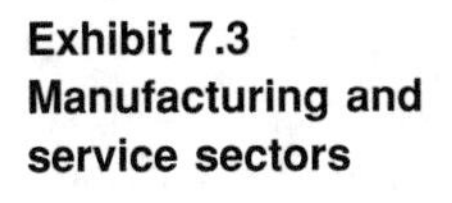

**Exhibit 7.3 Manufacturing and service sectors**

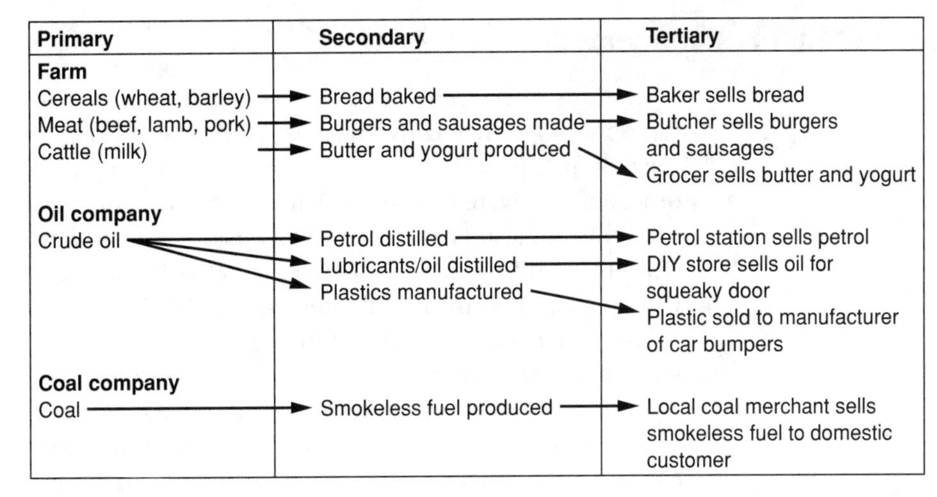


produced by primary-sector organisations. Tertiary-sector organisations sell goods produced by primary and secondary organisations. The tertiary sector also includes service-sector organisations like banks and social services.

An alternative way of viewing organisations is to consider whether the organisation produces goods, provides a service or delivers a mixture of both, and whether it is a private-sector organisation or not; *see* Exhibit 7.4, for more details. There are no public-sector/not-for-profit organisations that manufacture. If a public-sector or not-for-profit organisation is to provide a manufactured product, it is most likely that manufactured goods will be made by a subcontractor from the private sector. For example, local councils provide domestic and commercial council tax payers with wheelie bins, which are not manufactured by the council but bought in bulk via a negotiated contract from a supplier in the private sector.

**Exhibit 7.4 Manufacturing and service organisations**

| | **Not-for-profit organisations** | **Public-sector organisations** | **Private-sector organisations** |
|---|---|---|---|
| **Manufacturing** | | | • Pharmaceuticals (Glaxo)<br>• Cars (Vauxhall)<br>• Food (Northern Foods) |
| **Manufacturing and service** | • Retailing (Oxfam shop/fair trade initiatives) | • Housing associations (build and let homes)<br>• Provision of artificial limbs (NHS) | • Restaurant (Pizza Hut)<br>• Retailing (Laura Ashley)<br>• Carpet shop (supply and fit carpet) |
| **Service** | • Charities (Red Cross)<br>• Religious organisations (Church of England) | • General practitioner (GP – NHS)<br>• Refuse collection (local council)<br>• Education – schools (LEA) | • Banking (Abbey National)<br>• Telecommunications (BT)<br>• Hotels (Hilton) |

## Operations management

Operations management can be considered from the perspective of the organisation as a resource convertor; *see* Chapter 4. It is concerned with forecasting the output required and scheduling the conversion process such that customers' orders are delivered on time. The purchasing and just-in-time management of inputs are also crucial if the conversion or operation process is to happen efficiently and effectively. It is these activities, along with a few more, that will be examined in this chapter; *see* Exhibit 7.5.

The principles of operations management examined in this chapter can be applied to organisations providing a product, service or mixture of both. This section considers the characteristics of all three types of organisations: an organisation delivering a service with the example of a bank providing financial advice to a customer; a product with the example of a car company producing a car that has been ordered by a garage for a customer; and a mixture of product and a service, a pizza restaurant.

First, the service and product organisations are considered. The most basic difference between a service and a product are that a product is tangible – the car can be touched and driven by the customer – whereas a service is intangible – the financial advice cannot be seen and touched. The latter is delivered by the financial adviser and assimilated by the customer simultaneously and cannot be stored to be repeated another day. This contrasts with a product, which is able to be stored, highlighting the delay between manufacture and consumption. For example, the car is built in the factory and there will be a delay of at least a few days, maybe longer, before it finally reaches the customer who is going to own and drive the car.

The level of contact that occurs between a service provider and customer and a manufacturer and customer is also very different. In delivering a service there is significant contact between the service provider, the financial adviser and the customer; in contrast, the buyer of a good, such as the purchaser of a car, and its manufacturer are very unlikely to have any contact at all. This is because in providing a service the customer is part of the process of its delivery: the customer has to be there to receive the financial advice. Therefore

**Exhibit 7.5 Operations management: an overview**

| | | | |
|---|---|---|---|
| Location | Product development | Forecasting | Layout of facilities |
| Process and system performance | Inventory management | Material requirements planning | Just in time |
| Quality | Scheduling | Purchasing | Maintenance |

the facilities are located close to the customer, e.g. the bank's office will be on the local high street and accessible to the individual receiving the financial advice. In contrast, the customer will not participate in the manufacture of their car and the factory is likely to be located some distance from the end user, maybe even in another country. Finally, in general services are labour intensive and production is automated.

An organisation that both provides a service and delivers a product will assume characteristics of a service provider and/or a manufacturer. Taking the example of a pizza restaurant, the food is a tangible product, but cooking and serving the food are intangible services. The food may have been stored in the restaurant's fridge or freezer before being used to produce a pizza. Serving a meal is a service that cannot be stored, and indicates the simultaneous nature of service provision: the food is served hot as soon as it has been cooked and is eaten as soon as it is served. Showing the diners where to sit, giving them menus, taking their orders, serving the food and taking payment are all service provision and will therefore involve contact between the waiting staff and customers in the restaurant. In contrast, there will be limited or no contact between diners and the kitchen and cooking staff who produce the pizza. Again, as a service is being provided the location will be easily accessible to diners; pizza restaurants are on the high street in most towns in the UK. The dining area of the restaurant will be the section of the premises most accessible and used by diners; the storage areas, kitchens and bins will be towards the back of the premises and rarely accessible to customers. A restaurant is one example where providing the service, done by the waiting staff, and production, food preparation and cooking, are both labour intensive.

The examples discussed in the section illustrate that the scope of operations management is wide ranging and applicable to organisations undertaking different operations. The rest of this chapter goes on to examine the activities of operations management and how they apply to all types of organisations; *see* Exhibit 7.5.

## Location

An organisation deciding on a location will have to consider a number of alternatives. The best location for a manufacturing organisation may be one where the overall costs are minimised. In a service organisation the customer is directly involved in the supply process, therefore issues such as ease of access and speed of delivery have to be considered along with costs.

### Location strategies

Naylor[1] identifies three location strategies. The first is product-based location and is commonly used by large organisations. A large organisation using this location strategy takes into account that it has different divisions, each responsible for their own product ranges. Therefore several divisions with similarities or differences can be in the same area or on neighbouring sites. This separation based on product range allows each division to adopt and utilise the appropriate resources for its business. Locating all divisions in a large organisation together is likely to cause problems of focus and control.

An alternative to the product-based location is the market-based location. This strategy reflects the geographic divisions of the organisation and locates facilities in a location convenient to its geographic markets. For example, new supermarkets are usually built close to residential areas in out-of-town locations, which means that they are close to the customers – most customers will not want to travel very far to do supermarket shopping, as it is something that is done fairly frequently, for example every week.

Finally, a vertically differentiated location strategy is when separate stages of the supply process are in different locations. Some industries have vertically integrated firms who combine several stages of the manufacturing cycle. Rather than locate the whole operation on one site, location decisions are made for each stage.

### Push and pull factors

The decision to relocate is often made by small and medium-sized enterprises (SMEs) for a variety of reasons, which include the need for more space; increase in the scale of operations; and a reduction in unit costs. Larger organisations often relocate if more locations are required.[2] The factors causing organisations to relocate can be categorised as push or pull factors. Push factors result from dissatisfaction with existing locations, hence causing the organisation to consider changing location. Push factors originate from a wide variety of issues, some of which are presented below.

1 Current location is inconvenient for current customers and makes providing a good service cumbersome.
2 Competitors have locations that offer competitive advantage.
3 Different facilities are required for changing product/service range.
4 Regulatory authorities impose constraints related to health and safety, effluent disposal or noise.
5 Shortage of appropriately skilled labour.
6 Increasing cost of current site, for example due to rising rent.

In contrast, rather than pushing an organisation out of an old location, the pull factors attract or pull it towards a new location. For instance, an organisation may be pulled to locate in a particular region or country due to the availability of cheap skilled labour. Many consumer electronics companies have located assembly facilities in Malaysia as there are educated workers available, requiring wages that are a fraction of those paid in western economies. Sometimes a combination of factors will pull an organisation to a particular location. The Japanese car companies Nissan and Toyota located in Sunderland and Derby respectively as there were availability of labour and government grants. The location of the North of England also allowed both companies access to the European market and a location geographically close to mainland Europe, making selling cars in Europe much easier than importing directly from Japan.

## Product development

Product development and forecasting are both activities that occur early on the operations management process. The commercial evaluation of a new product will include assessing or forecasting likely demand. The process of product development is discussed in this section and forecasting in the next section.

To be successful, an organisation has to manufacture the products that customers desire. Therefore it must discover the kind of products that customers require and continue to supply them. To do this an organisation has to introduce new products and update or withdraw old products from the market; *see* Chapter 6. The development and introduction of a new product are expensive activities, hence careful planning is essential. The development of the new Rover 75 will have taken several years and cost several million pounds before it was launched on the market in the summer of 1999.

There are a number of steps in the product development process, shown in Exhibit 7.6.

**Generation of ideas**

The first step, the generation of ideas, relies on a number of sources of ideas for new products or services. The results of research and development may lead to a new product, for example the drug Viagra. Alternatively, sales staff out in the field may report customer demand for a new product and/or customers themselves may contact the organisation to suggest new products. The operations management department itself may come up with ideas for new or better products. Finally, the competition may be a source of ideas. Cadbury developed the textured chocolate bar Wispa in the 1980s to compete with the successful Aero bar produced by its main rival at that time, Rowntree.

**Exhibit 7.6 Product development process**

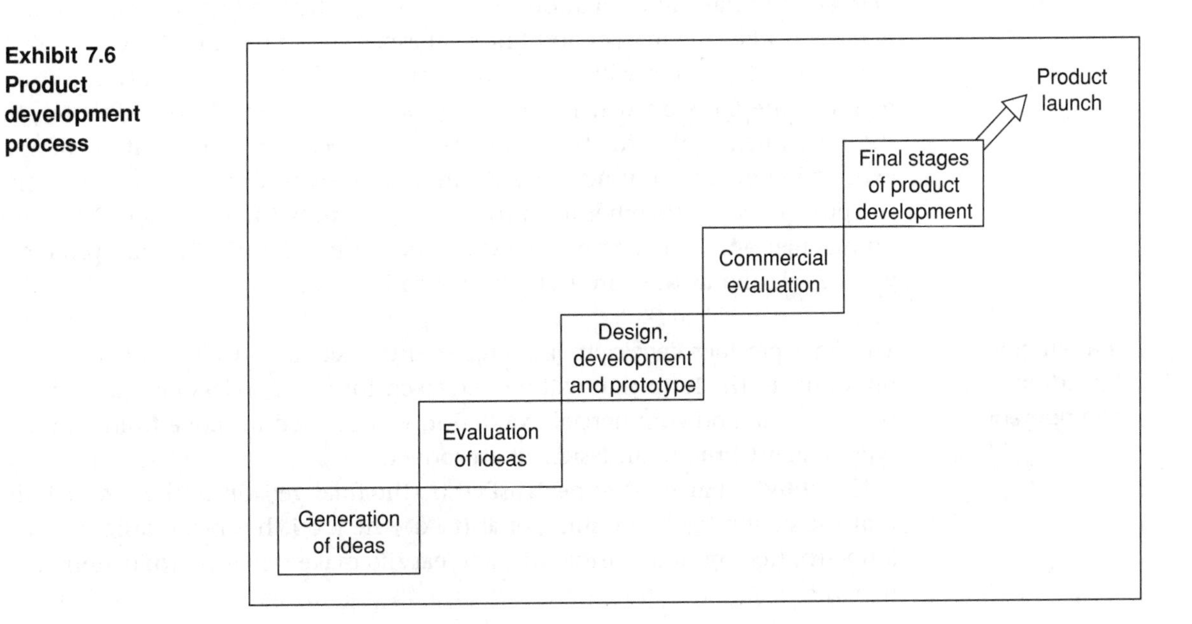

**Evaluation of ideas**

The evaluation of all ideas is necessary to filter out those with obvious deficiencies and weaknesses. The sources of deficiency and weaknesses in an idea for a new product or service can arise from a number of areas. For example, the manufacture of the product may be technically impossible at present. Alternatively, the manufacture of the product may require skills and abilities that the organisation does not possess or the product does not fit into the organisation's current product ranges.

This initial evaluation needs to be done by a team containing a range of people from different areas of the organisation, so that all relevant points of view are considered. Such a team includes representation from marketing, finance and operations. The ideas that are appraised as worthy of further consideration are taken forward to the next stage of the product development process, design, development and prototype. In that stage the technical specification required to deliver a successful final version of the product is considered.

**Design, development and prototype**

Initial basic questions concern whether: (a) it is possible to make the product; (b) there are any patent problems; (c) there are any competitors to consider; or (d) there are any current technical developments likely to overtake the product and render it obsolete by the time it arrives in the marketplace. Next, the following questions need to be addressed. Is the current design technically feasible for the organisation to manufacture and does the organisation have the necessary technology and skills? Does the new product complement the organisation's current products? The production and testing of a prototype will often help answer such questions.

**Commercial evaluation**

Products with a satisfactory technical evaluation move on to undergoing market and financial evaluation to determine if the product will make a profit. This studies the market and financial aspects with the aim of determining how well the product will sell. A commercial evaluation will study the competition; the basis on which the new product will compete (*see* Chapter 3); the investment required; the return to be expected; and estimation of the expected sales. The commercial evaluation and the technical evaluation from the previous stage together form the feasibility study for a product. Therefore on completion of the commercial evaluation some technically valid products will be rejected as they are not commercially valid.

**Final stages of product development**

The final product development stage sees the feasible product move from a prototype to the form that will be sold to customers. The lessons arising from the technical and commercial evaluation will be used to move from a prototype design to the final design of a product.

The criteria that need to be satisfied by the final version of the product are that the design has to be functional (to do the job it has been designed for); attractive (to appeal to customers); and easy to make (to keep production costs down).[3]

**Product launch**

Finally, the new product is launched on to the marketplace and its depends on a number of factors that customers are likely to consider. First price: some customers are price sensitive, other are not (*see* Chapter 3). Next, the quality of the product or service: the quality of food purchased from Marks & Spencer is different from that sold in Netto; and a consumer's response to quality is often directly related to their response to price. Accordingly, consumers who are extremely price sensitive will shop at Netto and those who are very quality sensitive will shop at Marks & Spencer. Finally, ease of access for the customer: a home computer that can be taken home from the shop today will sell better than a model for which there is a 10-week waiting list.

The operations management department of an organisation will attempt to simplify and standardise the design of its products. This makes manufacture easier, as fewer components are used. In addition, the manufacturing process contains fewer steps, is cheaper and waste is minimised. The same principles can be applied to the delivery of a service. An organisation delivering a service will seek to reduce the number of steps a customer has to move through and to minimise queuing time.

## Forecasting

Forecasts become effective at some point in the future when a decision is made concerning, for example, demand for a product. Hence forecasts need to be based on the likely conditions in the future. In the example of a company trying to predict demand for a product, conditions such as the amount of disposable income consumers will have and the competition's activity will affect the forecast. For instance, a company manufacturing ice cream making a forecast in December concerning demand for ice cream in the following June, July and August will consider the likely weather conditions and the new varieties of ice cream to be launched by its competitors for the summer.

There are a number of different ways in which forecasting can be done. One key criterion for a forecast is the time period in the future that it covers. Long-term forecasting looks ahead over a number of years. The types of projects that will be affected and influenced by long-term forecasts are, for example, capital expenditure projects, such as the building of office blocks or shopping centres. The decision to build a new out-of-town shopping centre will have been influenced by forecasts covering factors such as the number of people likely to visit at different times of the year, Christmas, school holidays and wet Tuesday afternoons in February; the size of the surrounding population; and the distance people will be prepared to travel to get to the shopping centre.

Medium-term forecasts are defined as covering a number of months, from say three to twenty-four months. The types of decisions that will be influenced by medium-term forecasts are the launch of new products and fashions on the market. Medium-term forecasts tell retailers and producers which toys will be popular with children in 12 to 18 months' time and what will be in greatest demand from the retailers at Christmas, for example. This will depend on what toys are likely to be made fashionable by the latest films, cartoons, TV programmes and bands. Toy manufacturers forecast that *Stars Wars* games and figures would be popular for Christmas 1999, due to the release of *Star Wars – Episode 1* in the summer of 1999.

In contrast to long- and medium-term forecasts, short-term forecasts cover a few weeks and often have a direct operational effect on the factory. Returning to the company producing ice cream and expecting a good summer in June, July and August, in the UK a wet summer and a forecast of more rain to come at the end of July, as the schools in England break up for the summer holiday, will have a very rapid and direct effect on sales and the amount of ice cream that needs to be produced by the ice cream factory.

## Quantitative forecasting

Forecasting can be either qualitative or quantitative in nature. Quantitative or numerical forecasting is feasible if the company is already producing the product or providing the service, as historical data already exists concerning the demand for a product or service and the factors affecting demand.

Quantitative data can be used in two ways to forecast future demand. First, projective methods examine the patterns of previous demand and extend the pattern into the future. For example, if a supermarket has purchased the following numbers of tubs of ice cream over the last four weeks in June, 100, 150, 200 and 250, it could be projected that demand in the first week of July will be for around 300 tubs. Secondly, causal methods examine the impact of external influences and use them to forecast future demand or activity. The productivity of the ice-cream factory might depend on the bonus rate paid to employees over the busy summer months. In this situation it would be more accurate to use the bonus rate than demand from shops and supermarkets to forecast productivity.

Both projective and causal methods of forecasting depend on accurate data and figures being available. This will not be the case if the company is launching a new product for which no data exists. In this situation it is not possible to use quantitative methods of forecasting, therefore qualitative methods have to be used. Qualitative methods of forecasting rely on the views and opinions of different stakeholder groups.

## Qualitative forecasting

There are five commonly used methods of qualitative forecasting:[4] personal insight; panel consensus; market surveys; historical analogy; and the Delphi method.

### Personal insight

Personal insight is a frequently used method of forecasting and should be avoided by managers when making critical decisions. Personal insight is simply when a manager who is familiar with the situation produces a forecast based on their own views and opinions. This means the individual's views and opinions are taken into consideration along with their individual prejudices and misconceptions concerning the situation. Hence personal insight in a method of forecasting that is unreliable.

### Panel consensus

The panel consensus is an attempt to dilute the prejudices and misconceptions of an individual. The panel, assuming it is able to talk openly and freely, should produce a more credible agreement. The disadvantages of a panel will occur if the views of the panel members are too wide ranging to come to a consensus. Also some members of the panel may not perform well in a group and fail to get their views across, hence leaving those who are loudest and most forceful to win through and falsely represent the group.

### Market surveys

Market surveys collect data and information from a sample of customers and potential customers. The data and information are analysed and inferences made about the population at large. However, market surveys can be expensive and time consuming and rely on the following being the case, if they are to provide reliable information: a valid sample of customers accurately representing the population; unbiased questions being asked; customers giving honest answers; correct analyses of the answers; and valid conclusions being drawn from the results.

### Historical analogy

The product life-cycle has periods of introduction, growth, maturity, saturation and decline (*see* Chapter 6). If a new product is being launched, it may be valid for the organisation to assume that demand for the new product will follow the same pattern as that for a similar product already on the market. For example, a publisher launching a new book is able to forecast demand based on the actual demand for a similar book that it published earlier. In the summer of 1999 the publishers of the popular Harry Potter children's books correctly forecast that demand for tales of Harry Potter's latest adventures would be very large. This was based on the fact that the previous two volumes of Harry Potter tales had sold extremely well.

### Delphi method

This is more formal than the other qualitative methods that have been discussed. The Delphi method follows a well-defined set of procedures in which a number of experts are asked to complete and return a questionnaire by post. The replies from the experts are analysed, the results summarised and posted back to the experts. They are asked to amend their previous replies in light of the summarised results. The replies to the questionnaires are anonymous and the experts do not know who the other experts are. Therefore the problems of face-to-face discussion, mentioned in the section on panel consensus, are avoided. The amending of replies is repeated several times, up to about six occasions. This should allow a range of opinions to emerge that is narrow enough to aid the decision-making process.

## Layout of facilities

Facility layout is concerned with the physical arrangement of resources in the organisation's premises. It covers all types of organisations, for example factories, offices, schools, shops and hospitals. The location of resources and their location with respect to other resources is important, as if it is done well the flow of work is smooth and efficient, in contrast to poorly laid-out facilities that disrupt operations and reduce efficiency.

The layout of resources in an organisation therefore has two clearly linked aims: to organise the resources and facilities so that the desired output of product or throughput of customers is achieved using minimum resources; and to ensure that the physical arrangement of resources allows maximum output or throughput. Consequently the layout and design of an organisation's premises should allow operations to run efficiently.

Take the example of retail premises, such as a supermarket like Tesco or Sainsbury's, where the goods are organised in parallel aisles. The layout is designed to allow a steady throughput of customers, even on a busy Saturday morning, and to encourage shoppers to purchase particular items in addition to those they need. For example, staple products such as bread and milk are located around the outside of the store some distance from each other, so the shopper has to walk past aisles of other goods to get to them, hence providing an opportunity for other goods to be promoted to the customer in-store. This includes stocking particular goods on the ends of aisles that customers will see as they search for more staple products.

## Layout policies for manufacturing

There are a number of layout policies that manufacturing organisations can follow: the process layout, the product layout and the fixed position layout.

### Process layout

A process layout involves similar equipment and machinery being located together. In a factory manufacturing armchairs and sofas, the process layout would mean: all sawing machines would be located in one area; all drilling machines in one area; all equipment used to assemble the frames in another area; and all equipment to upholster the chairs and sofas in another. The process layout works best when a range of products is manufactured on the same general-purpose equipment, as this is less expensive than specialised equipment.

One benefit of the process layout is the ease with which specific orders and variable demand can be met. However, this can mean low utilisation rates of the equipment and a high unit cost if the batch size for production is small. Nevertheless, the process layout does allow operations to continue if some equipment is unavailable because of breakdown or planned maintenance. For example, if the upholstery equipment is broken, the factory can continue producing frames. Consequently, the scheduling and controlling of the work in the process layout have to be carefully managed otherwise queues and backlogs of work can occur, resulting in very large stocks of work in progress. Finally, in the process layout people are grouped together according to their work and skills, which can lead to high morale and productivity. Hence splitting such groups up can be difficult when reorganisation and changes occur.

**Product layout**

A product layout puts all the equipment required to manufacture one particular product together in one location. This forms the basis of a traditional production line, where all machinery is lined up and each unit passes from one piece of equipment to another. A good example of the product layout is the production line in a car plant. The car body moves down the production line with different bits being added as the car moves along the assembly line, for example the engine, the doors, the seats, the brakes, the lights, the windscreen and windows and the wipers, so at the end of the production line the car is complete. The process uses dedicated equipment that is laid out so that the product can move through in a steady flow.

The production line layout results in a high rate of output and high levels of equipment utilisation and low unit cost. In most production lines the unit cost is further reduced by the use of automation and different methods of inventory control, such as JIT; *see* later in this chapter. The implementation of an appropriate system of inventory control will reduce levels of materials, components and work in progress. On a production line scheduling and controlling operations are easy and it is possible to achieve high and consistent quality.

The disadvantages of the production line layout include the inflexible nature of the operations, as it is difficult to adapt a production line to make another product. For example, adapting a car production line to produce washing machines is impossible without a major refit of the premises. Hence production lines are dedicated and expensive, with failure in one part of the production likely to disrupt the whole production line.

**Hybrid layout**

If neither a product nor process layout is suitable, then it is possible to combine them. For example, this allows a product to be assembled from two components, one being manufactured on a production line and the other in another part of the factory in a job shop using the process layout.

**Fixed position layout**

A fixed position layout occurs when the product is to big or heavy to move, as in shipbuilding, airplane assembly and oilrig construction. All the operations are carried out on one site around the static product. The difficulties of the fixed position layout are that materials, components and workforce all have to be moved on to the site and this will be difficult if there is limited space on site. In the fixed position layout careful management is required to ensure that the schedule of work is maintained, otherwise completion dates will be in jeopardy. Factors such as weather conditions may also affect operations and completion dates.

## Process and system performance

All organisations have a finite capacity: a factory can only manufacture so many TV sets in a month, and a school can only accept a finite number of new pupils into year 1 every September. Therefore consideration at the design stage of the process system is needed to determine the capacity required in order that products can be made, or services can be offered, to meet the demand of customers

System capacity involves a significant capital investment, hence careful planning should be undertaken to optimise the utilisation of financial resources and meet demand. This can be crucial, as customers can be lost quickly if a firm's capacity is insufficient to meet demand. Alternatively, under-utilised capacity can be very costly. For example, a local education authority will close down a school if pupil numbers fall significantly, as maintaining school buildings and employing staff are costly activities.

## Defining capacity and measuring performance

In theory, an organisation examines the forecast demand for a product and from this determines the capacity needed to satisfy that demand. However, in practice factors other than forecasted demand affect capacity, for example how hard people work, the number of disruptions, the quality of products manufactured and the effectiveness of equipment.

Capacity is a basic measure of performance. If a system is operating to capacity then it is producing the maximum amount of a product in a specified time. Decisions concerning capacity are made at the location and process design stage of an organisation's operations management activities. Ideally, an organisation should aim for the capacity of the process to match the forecast demand for products. Mismatches between capacity and demand will result in unsatisfied customers or under-utilised resources. If capacity is less than demand, the organisation cannot meet all the demand and it loses potential customers. Alternatively, if capacity is greater than demand, then the demand is met, but spare capacity and under-utilised resources result.

In contrast, if the capacity utilisation hovers around 100 per cent during certain time periods, then on those occasions bottlenecks or queues will occur. A common example of capacity being less than demand is when you are left standing in a long queue in a sandwich bar at lunchtime. You may exercise your consumer choice and go to another sandwich bar with no queues and many staff waiting to serve you. Here capacity is greater than demand, but the cost of paying these under-employed staff will be reflected in your bill.

Utilisation and productivity are directly related to capacity. Utilisation measures the percentage of available capacity that is actually used, and productivity is the quantity manufactured in relation to one or more of the resources used. Take the example of the lunchtime sandwich bar, which makes up sandwiches to order. It has five staff serving at lunchtime and its full capacity is 150 sandwiches per hour. If one member of staff phones in sick with food poisoning on Thursday morning, then utilisation of staff, a key resource in the sandwich bar business, is 4/5 or 80 per cent. If it takes on average two minutes to serve a customer and make their sandwich up to order, then one server has the productivity of 30 sandwiches per hour.

Another measure of how well operations are proceeding is efficiency. Efficiency is the ratio of actual output to possible output, usually expressed as a percentage. Returning to the sandwich shop, where a long-term member of staff and experienced sandwich maker has left and been replaced by a trainee 16-year-old school leaver with no catering experience, on their first day the new staff member can only manage to make 20 sandwiches an hour and is therefore operating at 20/30 or 67 per cent efficiency.

Efficiency should not be confused with effectiveness. Effectiveness is how well an organisation sets and achieves its goals. For example, the sandwich shop may not be 100 per cent efficient while the new member of staff is in training, but it can still remain effective if it achieves its goals of serving sandwiches made from fresh and organic ingredients to its customers.

In considering capacity, utilisation, productivity, efficiency and effectiveness, thought should be given to how these measures combine. For example, high productivity is of no use if the quality of products produced is poor or if the finished products remain in a warehouse because there is no demand for them.

## Process flow charts

The activities, their order and relationship between activities can be shown in a process flow chart. For example, the process a customer goes through when visiting the hairdresser is examined. The operations carried out at the hairdresser's might be described as:

- junior sweeps up hair clippings;
- pay receptionist;
- arrive on time and tell receptionist you've arrived for appointment;
- junior makes you a cup of tea or coffee;
- make next appointment;
- hair is cut by stylist;
- hair is washed by junior;
- you look in mirror and confirm you are happy with haircut;
- wait for stylist to finish cutting previous client's hair;
- sit down and read magazine until called;
- hair is dried by junior.

The following steps are gone through to complete a process flow chart form; *see* Exhibit 7.7:

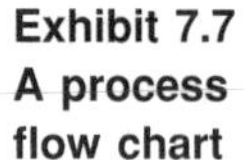

**Exhibit 7.7 A process flow chart**

| Symbol | Activity | Time | Cumulative time | No of occasions activity can occur in one hour |
|---|---|---|---|---|
| ● ➡ ■ ◗ ▼ | | | | |
| ● ➡ ■ ◗ ▼ | | | | |
| ● ➡ ■ ◗ ▼ | | | | |
| ● ➡ ■ ◗ ▼ | | | | |
| ● ➡ ■ ◗ ▼ | | | | |
| ● ➡ ■ ◗ ▼ | | | | |
| ● ➡ ■ ◗ ▼ | | | | |
| ● ➡ ■ ◗ ▼ | | | | |
| ● ➡ ■ ◗ ▼ | | | | |
| ● ➡ ■ ◗ ▼ | | | | |

Key ● Operation ➡ Transport ■ Inspection ◗ Delay ▼ Store

**Exhibit 7.8 A process flow chart: visit to hairdressers**

| Symbol | Activity | Time (mins) | Cumulative time (mins) | No of occasions activity can occur in one hour |
|---|---|---|---|---|
| ● ➡ ■ ◗ ▼ | Arrive on time and tell receptionist you've arrived for appointment | 0.5 | 0.5 | 120 |
| ● ➡ ■ ◗ ▼ | Sit down and read magazine until called | 20 | 20.5 | 3 |
| ● ➡ ■ ◗ ▼ | Junior makes you a cup of coffee | 1.5 | 22 | 40 |
| ● ➡ ■ ◗ ▼ | Hair is washed by junior | 8 | 30 | 7.5 |
| ● ➡ ■ ◗ ▼ | Wait for stylist to finish cutting last client's hair | 5 | 35 | 12 |
| ● ➡ ■ ◗ ▼ | Hair is cut by stylist | 14.5 | 49.5 | 4.1 |
| ● ➡ ■ ◗ ▼ | Junior sweeps up hair clippings | 1 | 50.5 | 60 |
| ● ➡ ■ ◗ ▼ | Hair is dried by junior | 10 | 60.5 | 6 |
| ● ➡ ■ ◗ ▼ | You look in mirror and confirm you are happy with haircut | 0.5 | 61 | 120 |
| ● ➡ ■ ◗ ▼ | Pay receptionist | 1 | 62 | 60 |
| ● ➡ ■ ◗ ▼ | Make next appointment | 1 | 63 | 60 |

Key ● Operation ➡ Transport ■ Inspection ◗ Delay ▼ Store

1 Look at the processes and list all the activities in their proper order on the process flow chart. This is shown in the column headed 'activity' in Exhibit 7.8.
2 Classify each activity using the symbols shown in Exhibit 7.9. This is shown in the column headed 'symbol' on the process flow chart.
3 Find the time taken for each activity and record it in the column headed 'time'. Also record the 'cumulative time' and the 'number of times an activity can be carried out in one hour'.
4 Summarise the process by adding up the number of each type of activity and the total time.

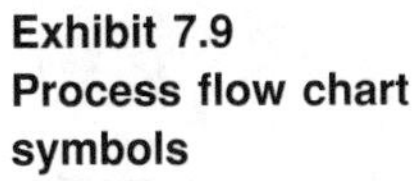
**Exhibit 7.9**
**Process flow chart symbols**

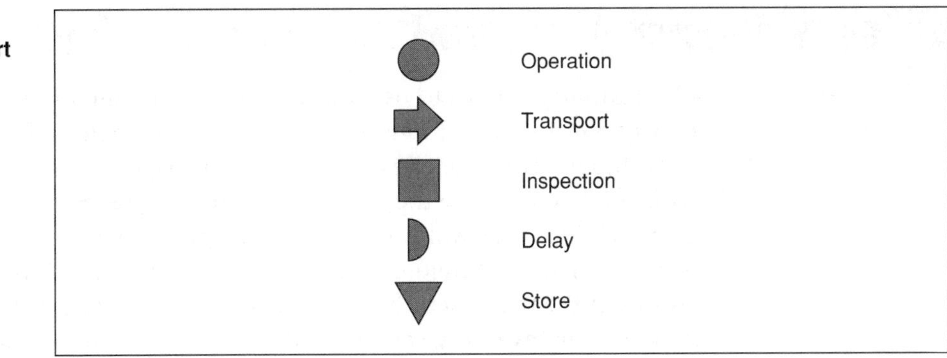


For the visit to the hairdresser, the following time is spent:

| | *No.* | *Time* |
|---|---|---|
| Operations | 7 | 37 |
| Delay | 2 | 25 |
| Transport | 1 | 0.5 |
| Inspection | 1 | 0.5 |

Drawing up a process flow chart will help answer the following questions:

- What operations are performed?
- What is the sequence of these?
- Which operations cannot be started until others have finished?
- How long does each operation take?
- Is the system being used to full capacity?
- Are products being moved?

Once the process flow chart for a product or service has been drawn up and the basic questions above answered, then areas for improvement in the process can be looked for and examined.

In the example of the visit to the hairdresser, finishing the longest activity takes 20 minutes, therefore at the moment the maximum number of people that can be processed in one hour is three. However, the longest activity is waiting, therefore this indicates that the appointments system is one area for improvement. The activity that takes the next longest amount of time is cutting hair and up to four people can be processed by one stylist in one hour. If demand is greater than four haircuts per hour, then the number of stylists will need to be increased. An increase in the number of stylists may only be needed on the busiest days, for example Friday and Saturday.

The first three steps give a description of the procedure for drawing up a process flow chart and step four provides some indication of the types of issues looked at if improvement is sought.

Operations management should aim for fewer operations and shorter times, while still ensuring that each operation gives the output required by the customer. If bottlenecks occur, the process and/or equipment need to be adjusted so that the process improves.

## Inventory management

All organisations have to use raw materials, components and/or consumables to carry out their operations and meet forecast demand. Insurance companies and council offices use consumables such as paper, pens, computer discs and stock enough to ensure they do not run out of these items. In contrast, a shoe shop such as Clarks will hold stocks of finished goods in the form of pairs of men's, women's and children's shoes in different styles and sizes. Equally, organisations in the manufacturing sector hold inventory or stock of different types of items. The inventory can be raw materials, for example paper pulp, wheat, coal and crude oil. Inventory can also be components, for example a car production plant will buy in certain items of inventory in component form, such as tyres, lights and assorted engine parts.

### Raw materials and component inventory

Raw materials and components are held as stock by manufacturing companies in case supplies of required raw materials and components cannot be supplied on demand. Consequently, holding a certain level of inventory allows for production planning to continue. Anticipation of increases in the price of raw materials such as coal, cocoa, crude oil or wheat may mean that organisations choose to purchase larger than normal amounts, possibly via futures contracts. Alternatively, large quantities of raw materials or components may be purchased to take advantage of a lower unit price or reduced transportation costs. For example, coal-fired power stations negotiate rates for the coal they use based on the large amount purchased and its delivery by the trainload.

### Work-in-progress inventory

Components and raw materials that have been partly processed by the manufacturing operation have a value to the company and hence are counted as inventory and referred to as work in progress or WIP. There are two points of view on work-in-progress inventory. The first is that if production rates are uneven, work-in-progress inventory ensures that the system always has work to carry out and provides flexibility. The opposite point of view is that work-in-progress inventory merely creates queues and bottlenecks. Consequently the production rate should be balanced, allowing a smooth flow of work right through the process, with no queues.

### Finished goods inventory

The final type of inventory is goods that have completed the manufacturing process and are finished goods ready to be passed on to the customer. For many manufacturing organisations the customer is not the end consumer, but a manufacturer buying components, for example the car manufacturer buying light fittings, or a distributor of goods, such as a car showroom, acquiring cars to sell.

Retailers such as Marks & Spencer, Next and Debenhams hold an inventory or stock of goods for a variety of reasons. Forecasts for goods are not always completely accurate and the extra inventory allows consumer satisfaction to be met rapidly, rather than the customer purchasing the goods at a competing shop. Alternatively, the distributor may offer a significant discount if finished products are purchased in bulk, hence making it more economic to take advantage of a lower unit price and store what is not needed immediately.

## Types of inventory

Inventory can be categorised into two broad types of stock. Independent demand inventory is items that are not dependent on other components, i.e. they are finished goods, like cars or shoes. Demand for such goods is directly dependent on consumer demand and to manage this inventory requires the use of forecasts of consumer demand (*see* earlier section in this chapter on forecasting).

The other category of inventory is dependent demand inventory, which covers items or components used in the assembly of a final product. For example, manufacturing a child's tricycle requires three wheels, one frame, one pair of pedals and one saddle. Hence demand for the component parts depends on the demand for the product. This is the dependency and it can be managed using MRP, discussed below. If 1000 tricycles are to be produced next month, 1000 frames are needed, which in turn means 1000 saddles, 2000 pedals and 3000 wheels.

## The cost of holding inventory

Inventory is of value to an organisation and costs money to store. The costs associated with inventory are carrying costs, ordering costs and stockout costs. Inventory carrying costs are the expenses associated with storing stock, borrowing money to purchase the stock, the opportunity costs, the purchase or rental of warehouse premises, which can be expensive if the goods are perishable like food or cut flowers and require refrigerated facilities, insurance, obsolescence and security costs.

### Inventory ordering costs

Inventory purchased from an external source incurs the cost of the salary of purchasing staff; preparation and dispatch of the order; salary of accounting staff involved in processing the necessary invoices and making payment; communication including postage, telephone, fax and electronic mail; expediting of the goods if they do not arrive on schedule; and receiving, handling, classifying and inspecting of incoming goods. In summary, inventory ordering costs are the expenses of procurement of the inventory and do not include the cost of purchasing the stock.

### Inventory stockout costs

Inventory stockout costs are those costs incurred when inventory is too low to satisfy customer demand. These costs can be difficult to quantify, but include the profit lost on not making a sale and potentially the cost of a client moving all their business to another supplier.

**Exhibit 7.10 Stock levels using MRP**

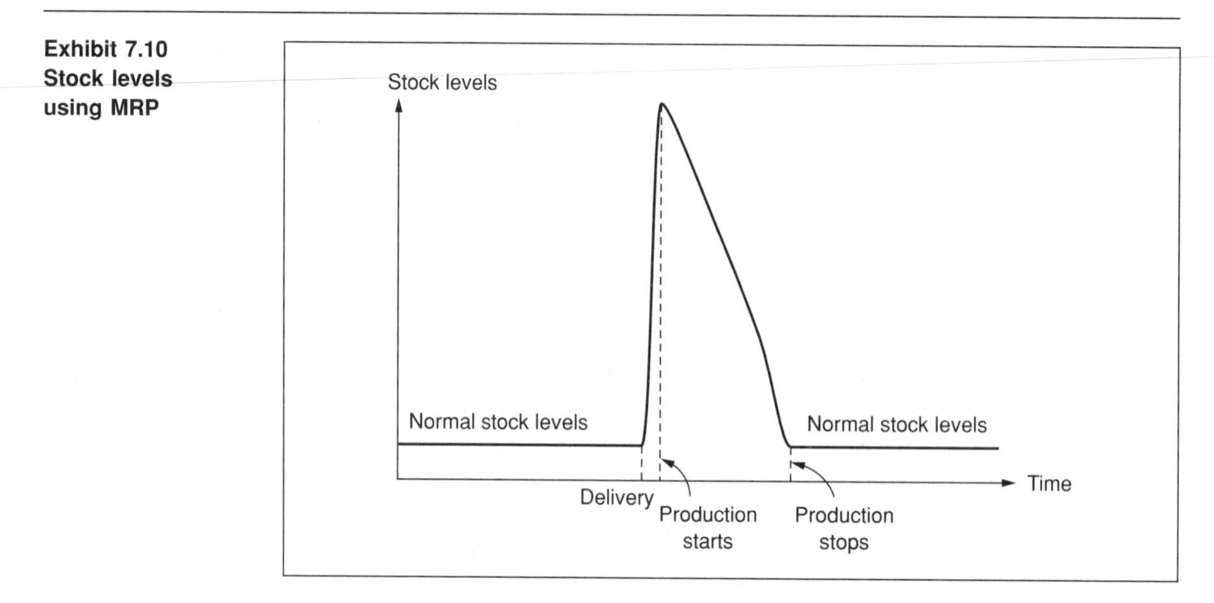


## Materials requirement planning

The dependent demand inventory system can be managed by use of materials requirement planning or MRP. MRP relies on production plans to propose a timetable for when materials orders are required. Consequently the resulting stocks of materials depend directly on a known demand. The alternative is an independent demand inventory system, which means that large enough stocks of materials to cover any probable demand are held.

A hotel coffee shop using an independent demand system would look at the ingredients used last week and ensure that there is enough of the same ingredients in stock to cover likely demand. In contrast, if a MRP system were in use the number of meals and snacks to be served each day would be assessed and this information used to determine the food required and the time and day of delivery. Hence with an MRP system overall stocks are lower, as only the ingredients and goods needed are ordered and are delivered just before production commences; *see* Exhibit 7.10. In contrast, with independent demand systems the stocks are not related to production plans, so they are kept higher to cover any level of expected demand and when consumed are replenished to maintain levels to cover any demand; *see* Exhibit 7.11.

**The MRP procedure**

The MRP inventory control system requires a great deal of information about components and products and is therefore computerised. The main information comes from three files:

- master production schedule;
- bill of materials;
- inventory records.

**Exhibit 7.11 Stock levels for an independent demand inventory system**

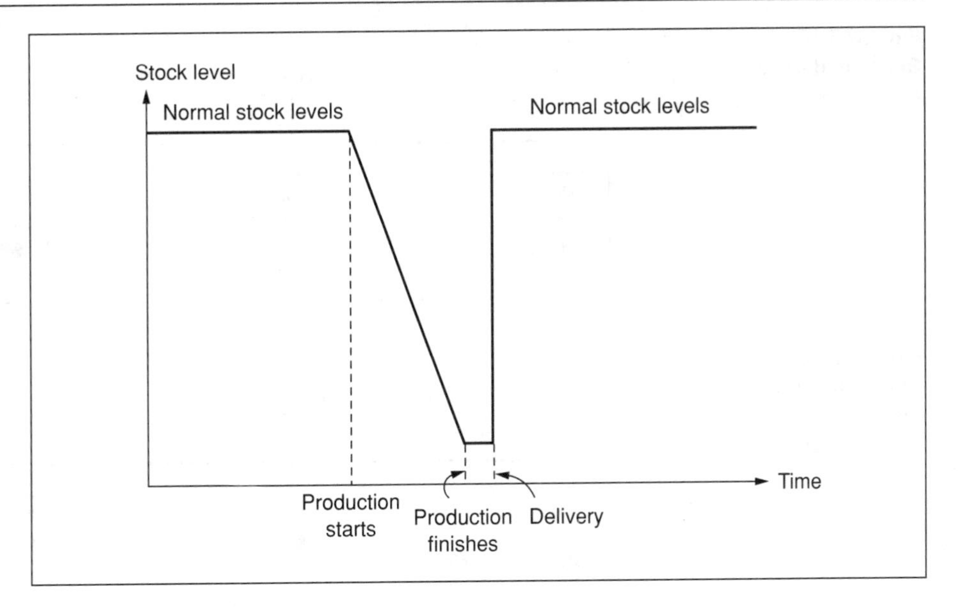


The MRP procedure starts with the master production schedule, which indicates the number of each product to be made in each period. The bill of materials is prepared by the designer or production engineer and is the result of the MRP being broken down to show the materials and components needed to manufacture a product and the order in which they are used. The bill of materials for the desk shown in Exhibit 7.12 is shown in Exhibit 7.13. The figures shown in the circles are the numbers needed to make each desk and every item is given a level number that shows where it fits into the process. The finished product is always level 0, with level 1 items used directly to make the level 0 item, the level 2 items used to make the level 1 items and so on.

**Exhibit 7.12 Desk**

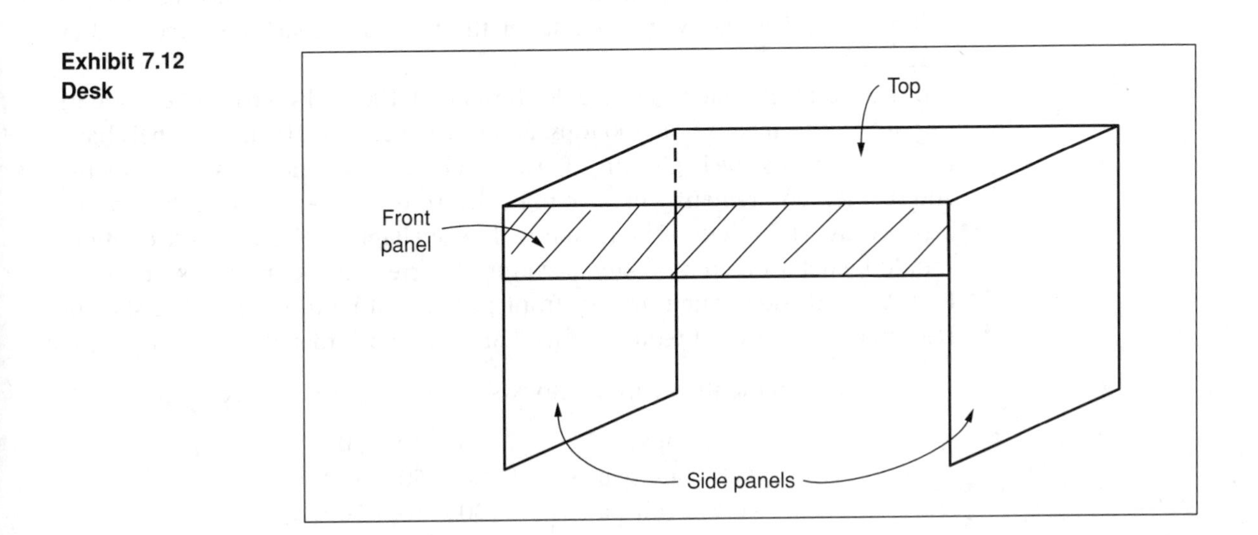

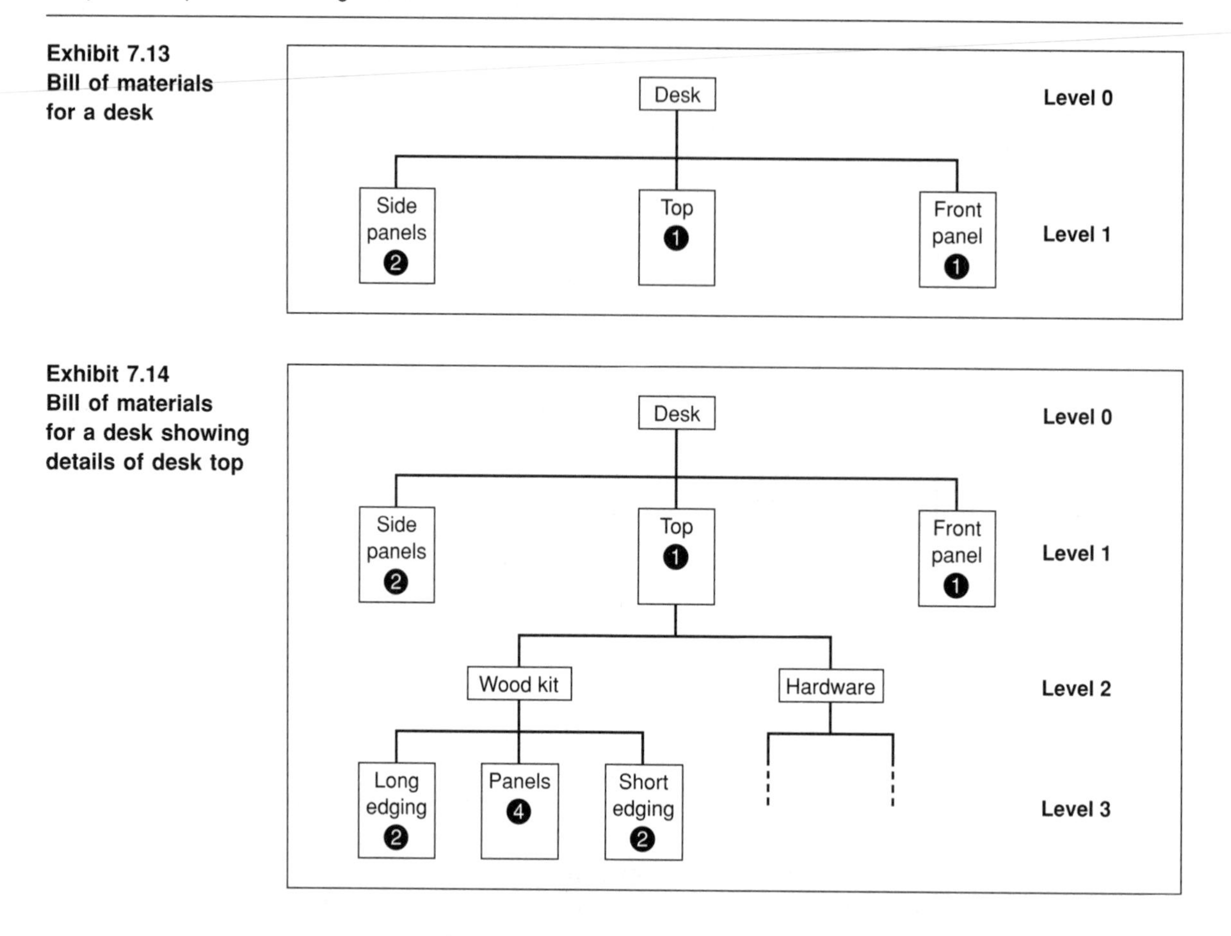


**Exhibit 7.13 Bill of materials for a desk**

**Exhibit 7.14 Bill of materials for a desk showing details of desk top**

Closer examination shows that the desktop is made from a wood kit and hardware. The wood kit contains four panels 150cm long, 15cm wide and 2cm thick, two pieces of edging 150cm long and two short pieces of edging each 60cm long. The next level of the bill of materials for the desktop is shown in Exhibit 7.14. The hardware consists of the tools and machinery required at that stage.

If the master production schedule shows that 100 desks have to be made in December, this means that 100 tops, 200 side panels and 100 front panels have to be available by the beginning of December. These are the gross requirements and the net requirement will have to be determined by examining the inventory records. The stocks of parts that will be available at the start of December should be determined by taking account of current stock and stock on order. If 20 tops, 50 side panels and 10 front panels will be in stock at the start of December, then the net requirements that need to be ordered are shown below.

Gross requirements – current stock/stock on order = net requirements

| | |
|---|---|
| Tops | 100 – 20 = 80 |
| Side panels | 200 – 50 = 150 |
| Front panels | 100 – 10 = 90 |

Therefore 80 top panels, 150 side panels and 90 front panels need to be ordered, with delivery scheduled for the end of November. For example, if tops take six weeks to arrive (sometimes called lead time) then the order needs to be placed by mid-October at the latest. Information from the supplier also has to be considered when placing the orders, for example if front panels are only delivered in batches of 50, then 100 front panels will have to be ordered and the extra 10 kept in stock until the next batch of desks is made. Alternatively, the supplier may offer a 15 per cent discount if side panels are purchased in batches of 100. Therefore careful calculation and consideration have to be given to deciding if it is worthwhile buying 200 side panels instead of the 150 required. Once the stock and order situations have been clarified, a timetable for production in December can be produced.

## Just in time

This chapter has looked at independent inventory demand systems and dependent demand inventory systems, which both operate with the main aim of managing mismatches between supply and demand for stock. Independent demand inventory systems manage the mismatch by ensuring that stocks are high enough to cover any expected demand. In comparison, dependent demand inventory systems using MRP overcome the stock mismatch by using a master schedule to match the supply of materials approximately to demand. The closer the match of supply to demand, the lower the stock levels needed to cover any mismatches. The just-in-time system (JIT) takes things a stage further and attempts to eliminate the stock mismatch altogether. A just-in-time system is organised so that stock arrives just as it is needed. Accordingly, the immediate nature of JIT systems depends on suppliers and customers working together to achieve the common objective of supplies arriving on time.

JIT systems operate in car plants, such as the Nissan plant in Sunderland. The car body moves down the production line to a work station and the doors arrive at the same point at the same time and are fitted. This is repeated all the way down the production line for seats, engines, windows and other parts, until the finished car emerges from the end of the production line.

JIT is commonly seen as a way of reducing stock levels to zero in the organisation. However, it does influence other aspects of operations management. Traditionally, organisations have set arbitrary quality levels, such as a maximum of two defective items per 100 produced. However, the JIT system takes the view that any defect costs the organisation and it is cheaper in the long run to prevent defects happening in the first place than paying to correct them later. Hence JIT supports the principles of total quality management; *see* later in this chapter. The principle of no defects also extends to the reliability of equipment. The JIT system of supply of materials and components works on the basis of operations being continuous. Hence if a piece of equipment is unreliable and breaks down, managers have to discover why this happened and take action to ensure that this failure does not re-occur. This may include reviewing maintenance procedures and policies; *see* later in this chapter.

Accordingly, JIT is not only a means of reducing inventories, but a way of viewing the whole range of activities in which an operations management department is involved and minimising delays of all kinds, including stock-outs, breakdowns and defects.

## Quality

Quality can be defined as the ability of a product or service to meet and preferably exceed customer expectations. For example, a breathable hill-walking jacket that keeps the rain and wind out and the wearer comfortable fulfils its quality expectation, as does a meal in a restaurant to celebrate your birthday, if good food, wine and ambience are in evidence. This illustrates the importance of quality. Quality contributes to helping an organisation remain competitive by producing goods and services of the quality demanded by customers. If the quality fails to meet the quality needs and wants of customers, then market share and profits will be lost. Hence managers and organisations invest significant effort into quality management, which is concerned with all aspects of product or service quality.

Quality management is affected and influenced by improvements in technology that ensure greater accuracy in the manufacturing processes, resulting in consistently high-quality products. This consistent quality is used by organisations as a competitive tool to gain competitive advantage (*see* Chapter 3). Therefore organisations have to pursue the manufacture of high-quality products, as consumers have come to expect them and have become reluctant to tolerate anything less. Hence if the demand for high-quality products is ignored, an organisation will lose out to competitors that can meet customers' quality demands.

Now you can see why organisations must make high-quality products. Any organisation that ignores the demand for high quality will lose out to competitors that can meet customer expectations. Although high quality will not guarantee success for a product, low quality will guarantee failure. There are clear benefits to an organisation from manufacturing high-quality products, such as improved reputation, competitiveness, sales and productivity, along with reduced marketing, defects and costs.

The quality of an organisation's products or services can be viewed from two basic points, inside and outside the organisation. The inside or organisational viewpoint of quality is that the performance of a product or service meets its design specifications exactly. The external or customer viewpoint is defined as how well a product or service does the job for which it was purchased. Hence there are two types of quality. The first is designed quality, which is the level of quality designed into a product or service. For example, a Fiesta has a different level of designed quality to an Audi. Second is achieved quality, which illustrates how closely a product achieves the designed quality. For example, at most British main railway stations there is a board showing the targets or designed quality for trains running on time, say 97.5 per cent, and alongside these figures the achieved quality, i.e. the actual trains running on time, say 84 per cent. Accordingly, achieved quality is often lower than the designed quality.

**Quality costs**

The management of quality will both incur and save costs for an organisation. Suppose that a faulty computer games system is sold to a customer buying a Christmas present for their child. The customer complains and the manufacturer arranges for the system to be repaired. However, money could have been saved if the manufacturer had found the fault prior to the games system leaving the factory, and even more money could have been saved by producing a games system that was fault free in the first instance. These costs are known as external failure costs.

There are three other categories of costs associated with quality: design costs, appraisal costs and internal failure costs. Design costs cover the expense of designing a good-quality product. This involves employing appropriate design staff, considering the type and cost of materials, the number of components, the manufacturing time, the ease of production, the amount of automation used, and the skill level required by the workforce. Appraisal costs cover verifying that the designed quality is the same as the achieved quality, which includes quality control costs such as those for sampling and inspecting the goods and work in progress. Finally, internal failure costs cover the cost of any items not meeting the designed quality. This can cover scrapping the item or returning it to an earlier point in the process to be reworked. There is also the cost of the work carried out on the defective item before it was detected. Therefore defects should ideally be found as early as possible in the process.

Savings are made when a quality system eliminates defective items from the system. For example, if a manufacturer sells goods, 10 per cent of which are defective, and replaces them under guarantee when reported by customers, then there is great inefficiency. The inefficiency arises as the manufacturer has to increase production by 10 per cent to cover the faulty goods. A system for dealing with complaining customers and faulty goods also has to be maintained, which incurs costs for the organisation. It is the removal of these inefficiencies that will save money. If the manufacturing process were to eliminate the manufacture of defective goods, productivity would rise by 10 per cent, unit costs would fall, there would be no more customer complaints, and the cost of dealing with them would no longer be incurred.

**Total quality management**

Total quality management is defined as the whole organisation working together to improve product or service quality. The aim of total quality management is zero defects.

Organisations typically had separate quality control and production departments. These two departments would often have conflicting objectives, with production departments aiming to manufacture products as quickly as possible, and quality control slowing down production by inspecting products, removing defective products and asking for them to be reworked. In this arrangement it is easy for quality control and production to forget that both departments have the overall common objective of customer satisfaction.

Throughout the 1980s and 1990s in the West, quality management and production have become integrated into the operations management department. The role of quality staff is to help and ensure that quality is built in, not inspected out. The production and quality staff look for ways of working with each other, customers, marketing staff, engineers and anyone else with a role to play in ensuring high quality.

## Scheduling

Scheduling involves drawing up a timetable of work that will ensure that customer needs and wants are met. Scheduling is critical in making certain that the utilisation of labour and equipment is optimal and that bottlenecks in the process are avoided. Scheduling generally deals with activities that are normally repetitive and short term in nature. Examples of timetables or schedules are shown in Exhibit 7.15.

Scheduling aims to meet the master production schedule and achieve low costs and high utilisation of equipment. This may appear to be straightforward and easy, but schedules must take account of many different factors. Take the example of drawing up timetables for first-year university students. The availability and requirements of staff, students, subjects and rooms all have to be balanced. The following problems will arise in drawing up a timetable for a large cohort of first-year students. The availability of staff and students

**Exhibit 7.15 Examples of schedules**

| |
|---|
| **Railways:**<br>Railway timetables for trains, drivers, guards, ticket inspectors, catering staff and passengers |
| **Hospitals:**<br>Hospital schedules for operations, patients, nurses, surgeons, beds and operating theatres |
| **Chocolate manufacturer:**<br>Chocolate manufacturer producing handmade chocolates has a schedule for customer orders, employees, equipment, raw materials delivery (cocoa, butter, cream and flavourings) and delivery of completed orders |

has to be considered, as they can only be in one class at a time and rooms of the right size and type have to be allocated to a subject at a time when both staff member and students can attend. This will be difficult if the number of rooms is limited or if special facilities such as a computer lab, language lab or science lab are required for the class.

There are four different ways for scheduling of services[5] and these are discussed below. Scheduling jobs in services is essentially the same as scheduling in manufacturing, but there are some differences. In service industries the customer is personally involved in the process, often being asked to wait in a queue, so queuing times are critical. Services contrast with manufacturing in that they cannot be produced during quiet times and held in stock for busy times, they have to be provided as required. Hence the capacity of a service process has to be organised to meet peak-time demands and cope with uneven patterns of demand. If there are large differences between peak-time demand and the lowest level of demand, then equipment will have a relatively low utilisation rate. In this situation the employment of staff can be dealt with by employing extra part-time staff during the busy period and not in the quiet times.

**First come, first served**

First come, first served scheduling is what most of us encounter on a daily basis, for example in the supermarket or queuing for our lunch. It is simple and straightforward: customers are served in the order they arrive.

**Fixed schedule system**

A fixed schedule system arises when a service is delivered to many customers at once. The timetable or schedules are generally known in advance by customers as the information has been made publicly available. Examples include bus, tram, train and airline timetables.

**Appointment system**

Appointment systems are commonly used by doctors, dentists, lawyers and hairdressers and require the customer to make an advance appointment. The aim of using an appointment system is to ensure the best utilisation of resources and good customer service, such that neither the customer nor practitioner is kept waiting. The problems arise if the appointment systems do not run to schedule, with customers kept waiting beyond their appointed time, or if the customer cancels and the practitioner has to wait for the next appointed customer.

**Delayed delivery**

The delivery is deliberately delayed in situations when the customer will not be significantly inconvenienced. Dry cleaners are good examples of organisations that delay the delivery to enable them to match capacity and workload. A dry cleaner's that has enough work to operate to full capacity today will still accept clothes for dry cleaning. However, they will not be dealt with until tomorrow and the owner will be told to return the following afternoon or the day after next to collect the clothes. Repair shops such as shoe repairs are also good examples of businesses that operate a delayed delivery scheduling system.

## Purchasing

The purchasing activity for organisations can be centralised, decentralised or a combination of both.

**Centralised purchasing**

Centralised purchasing is when the procurement of all purchased items for the whole organisation is arranged and controlled via one department. This allows bulk buying, which usually means that better prices and service can be obtained from suppliers (*see* Chapter 3). Centralised purchasing also yields a consistent standard and quality of purchased products for the whole organisation; reduced administrative costs; streamlined relations with suppliers; and a reduction in transport costs, since orders are delivered in larger quantities.

**Decentralised purchasing**

Decentralised purchasing occurs when every division or department of the organisation makes its own purchasing decisions, which is less bureaucratic than a centralised purchasing system. In addition, if the divisions or departments of a large organisation are buying from local suppliers who are responsive to their individual needs, it may be more cost effective than centralised purchasing.

**Combination of purchasing functions**

If neither a centralised or decentralised purchasing system is completely appropriate for the organisation, then a combination of centralised and decentralised purchasing may be more suitable. If a combination of systems is used, then responsibility for certain items, often of a relatively low value, rests with the decentralised system, which is managed by the division or department. In contrast, the centralised part of the system is used for relatively expensive items and infrequent purchases, such as capital expenditure, which may have to be approved by the board of directors.

## Distribution

Distribution is concerned with moving finished goods from the manufacturer to customers. A normal distribution system involves finished goods being moved from the manufacturer's premises to the distributor's warehouse until they are allocated to customers. This type of distribution system allows manufacturers to achieve economies of scale by concentrating operations in central locations, which in turn means that distribution costs are reduced as large orders are moved from manufacturer to wholesaler, rather than small orders being moved directly to retailers or consumers. This also means that the manufacturer does not need to keep large stocks of finished goods. Wholesalers placing large orders with manufacturers will negotiate a reduced unit price and will also stock a range of goods from many suppliers, hence allowing the retailers a choice of goods. If wholesalers offer short lead times and reliable delivery in addition to a good range of stock, then retailers can carry less stock and still offer the consumer a wide range of goods.

## Maintenance

Maintenance activity supports the operations management department by helping ensure that its equipment and facilities are kept in working order. Therefore an organisation's policy on maintenance is integrated with operations policy. This is important as any unplanned shutdown can have a significant effect on production systems, particularly if other carefully planned systems also support operations, such as a JIT inventory management system as discussed earlier in this chapter.

Maintenance has two key aims: to reduce both the frequency and impact of failures. The frequency of failure can be reduced by proper installation of the correct equipment along with a programme of preventative maintenance and replacement of items that are wearing out. The impact of maintenance can be lessened by its being planned for quiet times and/or minimising downtime and repair times.

There are two types of maintenance policy, run to breakdown and preventive maintenance. If the consequences of failure are limited and the equipment is easily replaceable, then run to breakdown is the sensible option. There are two ways to respond to a breakdown: emergency action if the breakdown has serious effects, or corrective action at a point in the future if the impact is limited.

Preventive maintenance is carried out on a planned basis. The intervals between maintenance work are established by experience, manufacturers or external authorities. Inspection is an important part of maintenance, especially for items that are expensive to replace or repair.

## Conclusion

This chapter gave an overview of the operations management department of any organisation, product or service, public or private, and examined how operations management activities relate to one another and the other major departments of any organisation. This should have furnished the reader with broad general knowledge concerning operations management.

ETHICAL ISSUES CASE STUDY 7.2

FT

# Microsoft 'bullied Apple into alliance'

**By Louise Kehoe**

Microsoft bullied Apple Computer into the 'grand alliance' that the two companies announced with great fanfare last year, a senior Apple executive has charged.

The partnership, which appeared to end years of animosity between the companies, was forced upon Apple by Microsoft's strong-arm tactics, according to written testimony to be presented in the Microsoft anti-trust trial today.

Avie Tevanian, Apple senior vice-president of software engineering, says in his testimony that Microsoft attempted to 'sabotage' Apple's program for Internet multimedia applications, called Quicktime, by causing misleading error messages to appear when the program ran on Windows – Microsoft's operating system.

Microsoft had also proposed that it split the multimedia software market with Apple, according to Mr Tevanian, in a charge that echoed allegations levelled by Netscape Communications in court last week.

'Microsoft does not hesitate to use its operating system monopoly power and application program dominance to try to eliminate competition, acquire control of new markets and block innovation that could challenge its position,' Mr Tevanian said in his written testimony.

The stinging testimony drew sharp comments from Microsoft, which says it was mistakes by Apple's engineers that caused the Quicktime error messages. The company also denied any move to persuade Apple to share the Internet multimedia market.

The testimony was 'another example of how the government deliberately twists facts to support its distorted allegation,' Microsoft said.

Mr Tevanian will be cross-examined by Microsoft's attorneys in court today. He will no doubt be called upon to explain why Apple's previous public comments about its new relationship with Microsoft have been very positive.

Mr Tevanian is hailed in the software world as one of the principal developers of an advanced computer operating system called Mach, elements of which have been widely adopted throughout the computer industry over the past 10 years.

He once turned down a job offer from Bill Gates, Microsoft's chairman, and instead he chose to work for Steve Jobs, Apple Computer co-founder.

Now, his tricky position in the trial stems from the pact under which Microsoft acquired a small stake in Apple and agreed to pay royalties to settle a dispute over multimedia software intellectual property rights.

As part of the deal, Apple adopted Microsoft's Internet browser software, dropping a similar arrangement with Netscape Communications. Evidence presented by the government last week included a note written by Apple's chief financial officer which implied that Microsoft had threatened to drop the development of office applications programs for Apple's Macintosh PCs.

If Apple's role in the court case sours its new-found friendship with Microsoft, it could be a setback for the PC maker.

Apple is planning to launch a new operating system next year and needs the continued support of software developers, and Microsoft in particular.

But it is precisely because it is not in Apple's interests to disparage Microsoft that Mr Tevanian could be a powerful witness against the company.

*Source*: *Financial Times*, 2 November 1998. Reprinted with permission.

**Questions for ethical issues case study 7.2**

1 Do you agree that Microsoft 'bullied' Apple into an alliance? Give your reasons.

2 Comment on the effect of alliances such as that between Microsoft and Apple on competition in an industry.

EXIT CASE STUDY 7.3 FT

# Boeing, boeing, bong

**By Michael Skapinker**

Everett, where the company assembles the 747, 777 and 767 aircraft, is the largest building on earth. Viewed from the platform set up for the 130 000 annual visitors, the scene is one of bustling activity. Thousands of workers scurry to fit wings and cables to rows of glinting aircraft fuselages.

It looks like a boom time for Boeing. It is not. Last week, the group announced a net loss for 1997 of $178m, the company's first for 50 years. 'It's a big disappointment. No question,' said Philip Condit, Boeing's chairman in an interview. What is going wrong at the world's largest aircraft maker?

The problem is not that the aircraft-building business is turning down. Far from it. Boeing has never been busier. Airline orders have risen sharply and the company has had to increase production to keep pace.

By the second quarter of this year, its factories in Seattle and Long Beach, California, will be turning out 47 aircraft a month. In mid-1996, the monthly total was just 18.

Nor has the problem much to do with the vast acquisition that Boeing made last year when it took over McDonnell Douglas. That $16bn deal doubled Boeing's size, helped increase its revenues from $22.7bn in 1996 to $45.8bn last year, and boosted its workforce from 112 000 to 238 000. The acquisition turned the world's leading manufacturer of commercial jets into the biggest maker of military aircraft too. (Boeing is also a substantial manufacturer of space equipment, accounting for 60 per cent of Nasa's budget.)

True, some of last year's losses resulted from indigestion. Boeing made a $1.4bn pre-tax provision to take account of its decision to run down some of McDonnell Douglas's civil aircraft production. Boeing is to phase out the company's MD-80 and MD-90 aircraft.

Yet, given the scale of the task, the merger has gone reasonably well. Harry Stonecipher, former chief executive of McDonnell Douglas and now president of Boeing, says some employees initially found it difficult to make eye contact with people they had regarded as enemies. But soon, he says, the two sides realised how similar they really were.

Staff from both companies confirm this. Ignore the remaining McDonnell Douglas signs that have still to be removed at some Long Beach plants, and it is impossible to tell who came from which company.

Morale has been aided by the fact that the two companies' activities were largely complementary, rather than overlapping. This means few programmes will be discontinued, although some factories will close.

Even the McDonnell Douglas civil aircraft workers have been cheered by Boeing's decision to proceed with their planned 100-seat MD-95 aircraft, now renamed the Boeing 717. Boeing will also continue to make the trijet MD-11 as a freight aircraft. Many McDonnell Douglas workers had feared the MD-11 was doomed.

No, where Boeing stumbled last year was not in its execution of the merger. Rather, the problem has arisen in the activity in which the company has long believed it led the world: the manufacture of commercial aircraft. Faced with what it called 'the steepest production increases since the dawn of the jet age', Boeing's factories seized up under the strain.

For a month last year, the company had to halt the Boeing 747 and 737 assembly lines. This did not mean, as widely reported at the time, that all work on the jets ceased. But managers stopped moving the aircraft along the assembly lines, leaving them in place so that workers could find missing parts and catch up with uncompleted work. The company had to make an additional $1.6bn provision to pay for its production problems.

To add to Boeing's woes, the price of aircraft plunged, in spite of the high level of demand. Some analysts say Boeing's obsession with selling more aircraft than Airbus Industrie, its European rival, resulted in price cuts, as sales staff struggled to win customers.

Boeing will not comment on a statement last month by Manfred Bischoff, chief executive of Daimler-Benz Aerospace, an Airbus partner, that competition between the two manufacturers had forced aircraft prices down by a fifth over the past two years. But operating margins on Boeing's commercial aircraft business dropped to less than 3 per cent in 1997 from 10 per cent in the previous two years.

Mr Condit says the reason prices are low at a time of strong demand is that aircraft manufacturers have to think about how to retain customers over 10 to 15 years. He cites the example of Southwest Airlines. When Boeing was approached by the carrier in the 1970s it did not insist on Southwest paying high prices for its aircraft. Not only is Southwest today the world's most successful low-cost carrier. It is also the biggest buyer of Boeing 737s.

'The decision to support a rag-tag group which came and said we want to buy some used 737s and start an airline in Texas has produced a phenomenal number of sales,' Mr Condit says.

Mr Stonecipher says he does not expect the price competition with Airbus to ease. 'There's no dynamic that indicates that price is going to change any time soon,' he says. For Boeing to raise margins, it needs to cut costs. And to cut costs, Mr Stonecipher says, the company will have to change the way it makes aircraft. 'The hardest way

**Exit case study 7.3 *continued***

to cut costs is by trying to do the same things better. You have to do them differently.'

Some of the changes have already been made. Boeing was half-way through a $1bn programme of updating its manufacturing when it was hit by the surge in orders. The reason it did not cope was because its system of ordering and handling parts, and the manufacture of several of its aircraft models, were still too inefficient and old fashioned.

'You put a tremendous strain on your own system and your suppliers' system when you order some parts that you don't need or you fail to order a part that you do need,' Mr Condit says. Had Boeing completed the transformation of its manufacturing systems by last year, he says, 'I think a lot of the problems would have been avoided.'

The transformation of Boeing's production goes by the ungainly title of Define and Control Airplane Configuration/Manufacturing Resource Management or DCAC/MRM for not-very-short. The overall aim is to bring Boeing up to the manufacturing standards of the motor industry.

Central to the programme is greater standardisation. Aircraft are hugely complex to make: the Boeing 747 has 6m parts. Boeing has traditionally allowed airlines to choose how they want many of those parts arranged. There are, for example, 20 different types of clipboard that pilots can order for their cockpits.

Boeing does not plan to deprive its customers of choice. But those choices will, in future, come from a Boeing menu. Airlines opting for greater variation will have to pay more for the privilege. Bob Dryden, executive vice-president for production, says he expects 85 per cent of aircraft parts to be standard, with airlines specifying the rest.

Boeing also wants to transform the way it handles aircraft components. Until it began changing its manufacturing processes, Boeing kept track of parts through a mass of papers and 400 separate computer systems. DCAC/MRM involves putting all those parts on a single computer system.

The changes should allow Boeing to cut down on the number of parts it holds as inventory. Boeing turns over its inventory 2.5 times a year. Mr Condit says that is higher than the US average but low compared with the Japanese car manufacturer Toyota or with competitors such as British Aerospace. Mr Dryden says he would like to see Boeing turn over its inventory 12 times a year by 2005. As the company has nearly $9bn tied up in inventory, the savings could be substantial.

Boeing has also begun asking its suppliers to design parts on computers so that its aircraft are easier to assemble.

The Boeing 777, which began service in 1995, and the new generation of 737 were computer designed. The 747 was not and the parts did not always fit together easily. 'We were shaving bits off or we had to reject them,' says Mr Dryden. Northrop Grumman, the US manufacturer which makes fuselage sections for the 747, has computerised its production process, which makes assembly easier.

Inspired by Japanese practice, workers now receive their parts in colour-coded boxes, packed, counted and ready for assembly, rather than having to fetch them when needed. Specialists who sort out production problems now sit on the factory floor, not in a separate building.

The DCAC/MRM programme should be completed by next year. Whether it will be enough to solve Boeing's problems is another matter. The company will make 550 aircraft in 1998, compared with 374 in 1997. Production problems will continue to depress earnings until the middle of the year.

Boeing will also continue to face fierce competition from Airbus. The European consortium raised its output by 44 per cent to 182 aircraft last year – without experiencing any production problems. It plans another 30 per cent increase this year.

Holding off Airbus, while keeping aircraft prices at competitive levels will be a difficult task. And getting through 1998 without another production breakdown will be the test of what Mr Condit has achieved so far.



**Questions for exit case study 7.3**

1. Summarise the problems faced by Boeing in 1998.
2. Draw up an outline plan of the operations management activities that need to be dealt with if Boeing's problems are to be resolved.

## Short-answer questions

1. Define operations management.
2. Define primary, secondary and tertiary organisations and give examples of each.
3. Name and briefly describe three different location strategies.
4. Name six stages of the product development process.
5. List five measures of process and system performance.
6. Name two different types of forecasting.
7. List five different methods of qualitative forecasting.
8. Name four different types of layout.
9. Define inventory management.
10. Briefly explain the term 'JIT'.
11. Briefly explain the term 'MRP'.
12. Briefly explain the difference between JIT and MRP.
13. State the key aims of scheduling.
14. Name four different methods of scheduling.
15. Briefly explain the difference between centralised and decentralised purchasing.
16. Define quality.
17. Briefly state the difference between designed quality and achieved quality.
18. Define total quality management.
19. State the key aim of maintenance.
20. Name two different types of maintenance.

## Assignment questions

1. Choose a manufacturing and service organisation of your choice. Research your chosen organisations and compare and contrast the operations management procedures they undertake. Present your findings in a 2000-word report.
2. Complete the process flow chart for doing a business studies assignment (Exhibit 7.16). What mechanisms are there for you to monitor and improve the quality of your work and how effective do you think each one is? What constraints affect the way in which you plan and execute your work?

**Exhibit 7.16 Process flow chart for doing a business studies assignment**

| Order | Activity | Done by whom? | Symbol |
|---|---|---|---|
| | Make rough notes | | |
| | Draw up contingency plan | | |
| | Type up final answers/report | | |
| | Put in correct references | | |
| | Hand report into school office | | |
| | Write questions and assessment criteria/learning outcomes | | |
| | Go to library | | |
| | Collect information | | |
| | Read questions carefully | | |
| | Prepare draft answers to questions | | |
| | Collect information | | |
| | Return feedback and mark for report | | |
| | Buy newspapers | | |
| | Select organisation | | |
| | Mark reports | | |

Key ● Operation ◗ Delay
➡ Transport ■ Inspection
▼ Store

Unit leaders
Unit tutors
Students

## References

1 Naylor, J (1996) *Operations Management*, Financial Times Pitman Publishing.
2 Ibid.
3 Waters, D (1996) *Operations Management: Producing Goods and Services*, Addison-Wesley.
4 Ibid.
5 Ibid.

## Further reading

Bank, J (1992) *The Essence of Total Quality Management*, Harlow: Prentice Hall.

Blattbery, P C and Deighton, J (1996) 'Manage marketing by the customer equity test', *Harvard Business Review*, July/August.

Harrison, M (1996) *Principles of Operations Management*, London: Financial Times Pitman Publishing.

Naylor, J (1996) *Operations Management*, London: Financial Times Pitman Publishing.

Slack, N, Chambers, S, Harland, C, Harrison, A and Johnson, R (1997) *Operations Management*, 2nd edn, London: Financial Times Pitman Publishing.

van Biema, M and Greenwood, B (1997) 'Managing our way to higher service-sector productivity', *Harvard Business Review*, July/August.

Waller, D L (1999) *Operations Management: A Supply Chain Approach*, London: International Thomson Business Press.

Waters, D (1996) *Operations Management: Producing Goods and Services*, Harlow: Addison-Wesley.

# 8 Finance

**Organisational context model**

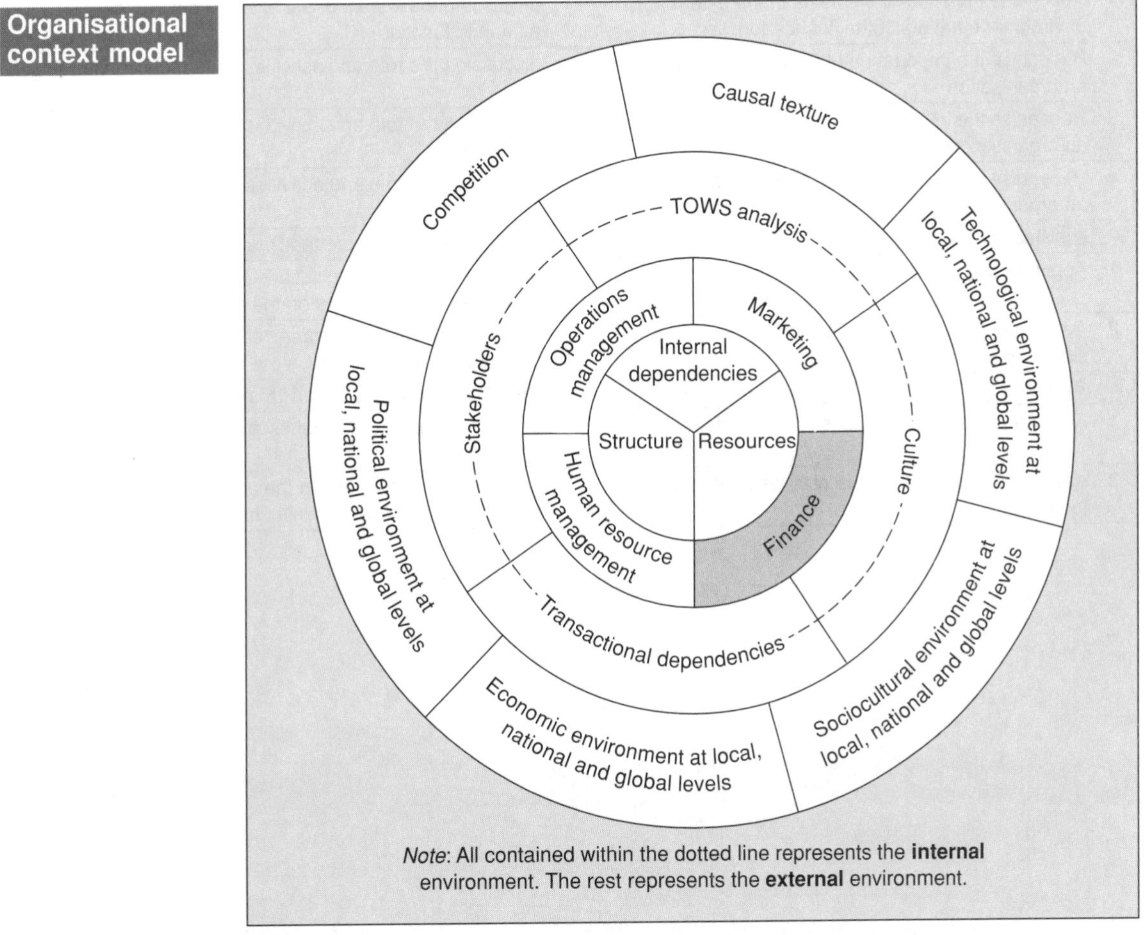


*Note*: All contained within the dotted line represents the **internal** environment. The rest represents the **external** environment.

**Exhibit 8.1**

**Learning outcomes**

While reading this chapter and engaging in the activities, bear in mind that there are certain specific outcomes you should be aiming to achieve as set out in Exhibit 8.2 (overleaf).

**Exhibit 8.2 Learning outcomes**

| | Knowledge | Check you have achieved this by |
|---|---|---|
| 1 | Define the finance function | stating in one sentence what a finance function does |
| 2 | List the key areas of the finance function | reviewing elements of the finance function from memory |
| 3 | Name the financial stakeholders in an organisation | identifying financial stakeholders in generic and specific organisations |
| | **Comprehension** | **Check you have achieved this by** |
| 1 | Recognise the specific elements of the external environment that affect the finance function | discussing the impact of the external environment on a finance function |
| 2 | Recognise the importance of financial management to an organisation | discussing the role and activities of financial management |
| 3 | Recognise the importance of management accounting to an organisation | discussing the role and activities of management accounting |
| 4 | Recognise the importance of financial reporting to an organisation | discussing the role and activities of financial reporting |
| | **Application** | **Check you have achieved this by** |
| 1 | Apply costing techniques to a variety of situations | performing costing calculations and making recommendations based on the answers |
| 2 | Apply investment appraisal techniques to a variety of situations | performing investment appraisal calculations and making recommendations based on the answers |
| | **Analysis** | **Check you have achieved this by** |
| 1 | Appraise the appropriateness of costing techniques | comparing costing techniques and their usefulness for different situations |
| 2 | Assess the appropriateness of investment appraisal techniques | comparing and contrasting the advantages and disadvantages of various investment appraisal techniques |

ENTRY CASE STUDY 8.1

FT

# SmithKline's well-sugared pill

**By Daniel Green**

Have you recently been on a tropical holiday? Given up smoking? Conquered depression? If so, you could have contributed to the record share price performance this week from SmithKline Beecham, Britain's second biggest drugs company.

Its 1996 results, published on Tuesday, showed sales and profits up 13 per cent and 14 per cent respectively. The company is rewarding its shareholders with a 25 per cent dividend increase.

Life was not so comfortable a year ago. SmithKline had been through 18 months of acquisitions and disposals and had to take restructuring charges and provisions for legal action against it.

Moreover, it was hit by the collapse in sales of its best selling product, Tagamet, an ulcer drug, following the expiry of its patents. And while most drug companies generate plenty of spare cash, debt was rising at SmithKline. Not surprisingly, the first half of 1996 was a period of poor share price performance.

The transformation of the company, run by charismatic former tennis star Jan Leschly, is due largely to the performance of the company's new product portfolio. Sales of Seroxat, an anti-depressant launched in 1992, reached £706m last year.

They will continue to grow rapidly as it chases Prozac, which made $2.4bn in sales for its maker Eli Lilly of the US. Indeed, SmithKline gleefully produced figures this week showing that Seroxat sales were growing three times as quickly as those of Prozac.

Sales of Havrix, a hepatitis A vaccine used by travellers which also was launched in 1992, may be more modest than Seroxat's. But any product showing a year-on-year growth of 29 per cent, with sales of £166m, is welcome.

Medications do not have to be high-tech to be money-spinners. Sales of SmithKline's newly launched nicotine chewing gum rose from £22m in the first quarter of last year to £78m in the last quarter.

Then, too, one of its most successful products is the toothpaste Aquafresh. Sales of this grew 24 per cent last year, to £258m, thanks to the launch of the brand in new markets.

Indeed, SmithKline now claims to be the market leader in oral health in western Europe, having overtaken Colgate and Unilever in the past three years.

Leschly also lifted the curtain a little this week on what some stock market analysts regard as the company's most serious mistake of recent years – the $2.3bn acquisition in 1993 of Diversified Pharmaceuticals Services.

Among other things, DPS manages drug purchasing in the US for clients such as hospital chains. Several related businesses were bought by other drug companies in 1994. Unfortunately, a few months later, the US Federal Trade Commission told them they could not use their new acquisitions to boost sales of their products.

In spite of this, DPS is turning into the cornerstone of Leschly's grand vision – which is for SmithKline to become a healthcare company, not just one selling specific products. SmithKline is using DPS to forge relationships with some of the most important healthcare buyers in the US and to gather data on patterns of drug consumption.

That, in turn, is helping what has for many years been SmithKline's weakest link: Clinical Laboratories, which conducts tests for hospitals. New technologies and heavy competition have helped to cut prices in this area of business but profits grew in 1996 for the first time since 1993.

The future for more conventional medications is promising. Leschly gave details of five drugs in the final stages of development. They range from Memric, for Alzheimer's disease, to one for irritable bowel syndrome, code-numbered SB207266. The launches are planned for 1999 to 2000, provided the clinical trials go well.

There is still plenty that could go wrong for SmithKline. The risks in drug development are exemplified by the company's decision to suspend work on Ultrair, an asthma drug, in the late stages of testing. Then, too, the US Food and Drug Administration is hesitating over Coreg, a drug for heart failure. A decision on whether to approve it is expected within weeks.

DPS remains a problem in financial terms. The business might be opening doors for SmithKline and sales are growing well, but Leschly is careful not to give profits figures away.

Nor are SmithKline's shares a bargain for investors. The price is now so high that even the company's own stockbroker, NatWest Markets, says Zeneca's are probably better value.

Even so, SmithKline embodies what has made the drugs sector an attractive investment for the past two decades. Good medicines give a company the security of a utility: the customers must buy its products even if times get hard.

At the same time, top-class research and development, plus the slice of luck that any research needs, make medicines into high-tech, high-growth products comparable with computer chips and software programmes. Historically, pharmaceuticals is one of the world's most profitable industries. At the moment, SmithKline is living up to that reputation.

*Source*: *Financial Times*, 22 February 1997. Reprinted with permission.

## Introduction

The resource of money is an important asset for nearly all organisations. The financial aims of private-sector organisations will include profit-making objectives. In contrast, for public-sector or charitable organisations the most efficient and effective use of often limited financial resources will be of primary importance.

Therefore the finance function of a business is concerned with the financing of the organisation, the use of finance by the organisation and the explanation of how the finance has been utilised to interested parties or stakeholders. The detailed activities undertaken by the finance function will vary in nature from the operational to the strategic and will be influenced by changes in the organisation's external environment. Therefore the finance function is concerned with financial management, management accounting and financial reporting. These areas of activity are looked at later in this chapter.

## External influences on the finance function

### The single currency

The 1991 Maastricht Treaty set the timetable for the first round of monetary union to occur in Europe in 1999. The deadline for meeting the necessary economic targets to join the first wave of European monetary union, EMU, was the end of 1997. The key economic target that countries had to meet was the deficit:gdp (gross domestic product) ratio, which had to be 3 per cent or less. All member states except Greece met this target. However Denmark, the UK and Sweden along with Greece did not join the single European currency in the first wave in 1999.[1] In June 1998 Robin Cook, British Foreign Secretary, strongly indicated that Britain expected to join the single currency soon after the next general election in the UK in 2002, shortly after the euro coins and notes come into circulation.

The new single currency is called the euro and is managed by the European Central Bank, based in Frankfurt. The role of the European Central Bank is to maintain the euro as a stable currency, as it is hoped that this will stop price rises caused by devaluing currencies and produce a European economy that is built around low inflation and low interest rates. In the longer term it is expected that such a European economy will attract world trade and international investors keen to be less dependent on the American dollar.

The euro became legal currency for trade and financial markets on 1 January 1999, in participating countries. Coins and notes will be introduced to participating countries on 1 January 2002, allowing individuals to be paid in euros, spend euros in shops and pay bills in euros. The two-stage changeover is aimed at allowing organisations to familiarise themselves with the euro and how it will operate. It also gives time for notes to be printed and coins minted. The notes will be in denominations of 5, 10, 20, 50, 100, 200 and 500 euros, and the eight euro coins will range from 1 cent to 2 euros (200 cents) in value. The circulation of old national currencies and the new euros will occur in parallel for up to six months after 1 January 2002. During this time organisations will display the price of goods and services in both euros and a national currency, which along with charts and conversion tables will aid the individual consumer in converting from their old national currency to the new euro.[2]

The benefits of the euro for individual consumers include the greater ease with which the price of goods and services can be compared across participating EU member states. Purchasing goods or services from abroad via mail or the Internet will be less complex and the cost of foreign exchange and exchange rates when visiting different participating EU states will no longer exist.[3]

The benefits for organisations in participating countries include a decrease in the number of different currency bank accounts held, a rationalisation of cash operations and an overall saving in operating costs due to the removal of cash conversion costs and cross-border payments. In contrast, the disadvantages of a single currency arise from the intrinsic risk of the euro as a new currency with no past form and the associated hesitancy which will exist until exchange rates are finally established.[4]

## The minimum wage

The minimum wage was introduced in the UK in April 1999. The issues and debates surrounding the introduction of the minimum wage centred around the level at which it should be set and whether that level should vary for particular groups of people.

The old Wage Councils that set minimum wages in a number of low-paying industries, including for example retailing, catering and textiles, were abolished by the Conservative government. It argued that the minimum wage levels set by Wage Councils made employing staff too expensive and hence jobs were lost; this same argument was directed by the Conservatives at the idea of a national minimum wage. It has never been proved by the Conservatives that discontinuing the Wages Councils *created* jobs. Hence it would appear that their termination had no effect on employment, so the minimum wages set by the Wages Councils could have been an appropriate starting point in establishing a minimum wage level. The minimum wage rates at the time the Wages Councils were disbanded ranged from £2.72 per hour to £3.18 per hour, with an average of about £3 per hour. Revising these rates in line with average earnings growth would have produced a minimum wage rate of £3.85 in October 1998.[5]

Different interested parties have differing views on what level a minimum wage should have been set and what should influence its level.

### The TUC – representing the employees

The Trades Union Congress (TUC), which represents trade unions and employees, wanted the highest possible minimum wage and to get the best deal possible for workers. The TUC conference in 1997, voted in favour of a minimum wage of £4.26 per hour, the highest of all figures suggested. This figure was chosen because it was half male median hourly earnings. This may appear to be a simple and easy-to-apply method when deciding on the rate for a national minimum wage. However, there are different ways of calculating male median earnings. In the case of the TUC this was done by dividing weekly earnings, including overtime, by average hours excluding overtime. It is easy to see that if weekly earning including overtime were divided by the number of hours worked including overtime, then a lower hourly rate would have been arrived at.[6, 7]

#### The CBI – representing the employers

The Confederation of British Industry (CBI) expressed the view that a minimum wage level of around £3.50 could be acceptable. It views a high minimum wage rate as having the potential to harm businesses by increasing costs, leading to higher prices being charged and causing jobs to be terminated.

#### The academics – impartial

A group of academics (Richard Dickens, Stephen Machin and Alan Manning) advised the Low Pay Commission in October 1997 that a minimum wage of £3.85 would be the most appropriate, with lower rates for young people. A minimum wage at this level would have annoyed both the TUC and the CBI: the CBI wanted a lower minimum wage rate and the TUC a higher one. The figure of £3.85 per hour was arrived at on the basis of the average figure set by the different wage councils prior to their abolition of £3.07 per hour, which assuming average earnings growth would have stood at £3.85 in October 1998, when introduction of the minimum wage was expected. The academics recommended £2.50 per hour for 16–17-year-olds and £3 per hour for 18–20-year-olds. It was argued that the variation of minimum wage rates by age would have a noticeable effect on the pay packet of adults earning the minimum wage, without threatening the jobs of young people.[8, 9] The gap in earnings between young and old workers was significant, with average pay before introduction of the minimum wage being around £7.80 per hour, while teenagers averaged only £3.10 per hour. The introduction of a minimum wage at a level of £3.50 or more, without age differentiation, would have affected over half of teenagers in work in contrast to about 10 per cent of the whole population.[10]

In contrast, the claim for changing the minimum wage according to regions was nebulous, as regional wage differentials in the UK are not generally significant. London and the South East of England clearly have higher wage rates and Northern Ireland lower wage rates, but there is meagre regional variation in the rest of the UK. Therefore, as expected the minimum wage was introduced on a national and not a regional basis.[11]

It was estimated that the relative influence of a national minimum wage on the whole economy would be small even if it had been introduced at the high rate of £4.26 per hour. The effect on the wage bill would be smaller than the impact of the Equal Pay Act of 1970. However, the people affected by minimum wage legislation would be protected by law from the worst excesses and exploitation of the labour market and unfair employers.[12]

In the event, a minimum wage of £3.60 per hour for adults over 21 years old was announced in June 1998, along with a rate of £3.00 per hour for 18–21-year-olds, with no minimum for 16- and 17-year-olds. Both minimum wage rates came into operation in April 1999.

## Financial management

Financial management involves raising capital to finance the organisation's operations and ensuring that sufficient revenue is generated to cover the cost of any finance raised, e.g. interest payments. In the UK there are four main sources of funding for businesses: reserves, loans, share issues and the government.

### Reserves

Reserves are retained profits from previous years that have not been spent or distributed to shareholders. The use of reserves to fund expenditure means that the level of scrutiny to which the organisation's management is subjected is less exacting than if funds were raised from sources external to the company.

### Loans

Loans are either term loans or overdrafts and are commonly provided by banks. Term loans are long term and provide the borrowing organisation with capital in return for repayment of the capital with interest over a number of years, e.g. five or ten. In contrast, overdrafts are short-term loans for 12 months or less and will vary from day to day depending on the amount by which the organisation's current account is overdrawn. However, for many organisations an overdraft will be a 'permanent' source of temporary finance.

Loans are for a definite period and repayable with interest. The amount of interest payable depends on the interest rate charged, which is driven by the base rate of interest set by the Bank of England. The level of risk taken by the lender in lending money to the borrower will be influenced by the security offered against the loan and what the money is to be used for. All these will in turn also influence the interest rate charged on the loan. It is common for long-term loans to be secured against company assets. Securing the loan reduces the risk for lender and the interest rate for a secured loan will be lower than for an unsecured loan.

The interest payments come out of profits and cannot be reduced by the borrowing business if profits and trading conditions are unfavourable. This contrasts with dividend payments, which a company can alter depending on profits and trading conditions. A further characteristic of loans is that they do not carry ownership rights, which is in contrast to ordinary shares that do. If a company is unable to meet the loan and interest payments, then the bank or lender may decide to foreclose on the loan and appoint a receiver to take day-to-day control of the company. The receiver has to decide if the business is able to continue trading under its guidance and generate enough cash to pay the bank and other creditors, or if the business should be closed, the assets sold off and the cash generated used to pay the bank and other creditors.

In July 1991 Robert Maxwell used shares in the publicly quoted company Maxwell Communications Corporation (MCC) to secure loans to his own private companies of an estimated £300m. Earlier in the same year, Maxwell had used shares owned by the pension fund of Mirror Group Newspapers to secure loans to MMC. Much of this was irregular and illegal and on 31 October 1991 the value of shares in MMC started to plummet. Shortly after this, on 5 November, Maxwell died from drowning by going overboard from his yacht near the Canary Islands. This example raises two key issues to bear in mind when accepting assets as security for a loan. First, are the assets being offered as security those the borrower is legally allowed to offer? Maxwell was not legally in a position to offer the shares belonging to the Mirror Group Newspaper Pension Fund as security, but he did so and they were accepted as security for loans to MMC. Secondly, will the assets offered as security be of a value equal to that of the loan, should the lender have to foreclose on the loan and liquidate the security given? The value of MCC shares used as security for loans plummeted in October 1991, leaving the value of the loan security as minimal.

**Ordinary shares**

A company will issue ordinary shares to raise capital. Once issued shares can be traded on the stock market if the company is listed, where individuals or organisations may buy and sell them. Shareholders will usually fall into one of three categories: the individual investor, institutional shareholders and existing shareholders (rights issues are looked at in the next section). The value of a shareholder's investment in a company will rise or fall depending on how the company itself performs and on how the stock market as a whole performs.

SmithKline Beecham, the pharmaceutical company, saw the price of its shares rise from below 650p in April 1996 to nearly 950p in February 1997, slipping back to 720p in June 1998. The overall increase in the company's share price was in part due to the good performance of the company's products in the marketplace. In particular, SmithKline achieved a strong position in the market for oral health products, illustrated by the success of toothpaste brand Aquafresh. For further details see the introductory case study to this chapter.

Ordinary shares carry ownership rights entitling the holder of the shares to one vote per share owned at Annual General Meetings and Extraordinary General Meetings of the company and the right to receive a dividend if the company issues one. Dividends are a distribution of profits to the owners/shareholders of the company and can be raised or lowered as the company sees fit. They are usually expressed in the form of so many pence per share owned. For example, if a company announces a dividend of 17.5p per share, a shareholder owning 200 shares would receive a payment of £35. Most companies seek stability or a slight upward trend in their dividend payments with the aim of keeping shareholders satisfied.

Shareholders are either individual members of the public investing relatively small sums of money (thousands of pounds) in shares or corporate investors with large amounts of money to invest (millions of pounds) for the company or their clients. For the individual small investor, shares are a form of investment that is more risky than keeping savings in a building society or bank, but shares are highly likely to give a better return in the long term than a building society or bank.

In the 1980s individuals were greatly encouraged to become shareholders by the Conservative government of Margaret Thatcher. This was done by encouraging individuals to buy shares in the privatisation offers that occurred in the 1980s through to the early 1990s. For many individuals it was the first time they had purchased shares and the companies in which they acquired shares included British Telecom, British Gas, electricity distributors, power generators, British Airways and Railtrack. These flotations were aimed at individual and corporate investors. Corporate investors in the stock market include insurance companies and pension funds, which invest money to enable them to pay pensions and attractive bonuses on insurance policies to their clients.

Margaret Thatcher's view of Britain's citizens as shareholders continued in the 1990s with privatisations in the financial services industry. In 1997 the insurance company Norwich Union and the following building societies all floated on the stockmarket: Halifax, Alliance & Leicester, Woolwich and Northern Rock. These flotations meant that the account and policy holders of these companies received shares in them, hence releasing many millions into the pockets of the individuals who sold the shares they received. Other shareholders viewed the Norwich Union and the building societies as good investments and kept their shares.

**Rights issues**

A rights issue is similar to a share issue, but is for existing shareholders only. A rights issue is a way of raising capital that is viewed as less threatening than a share issue, as the shareholder breakdown remains the same. Hence there is no opportunity for another company to build up a stake in the business with the aim of mounting a hostile takeover bid. In a rights issue ordinary shareholders are offered new shares in proportion to those they already own. For example, if a shareholder owns 500 shares in a company and the rights issue is a one for ten offer, then the shareholder will be entitled to purchase one new share for every ten already owned, i.e. 50 new shares in total. Rights issue shares are usually offered at a price equal to or below the stock market price to make the investment appear attractive to the shareholders.

If shareholders take up the rights issue by purchasing the shares, then the shareholder breakdown remains the same. If the existing shareholders do not buy the new shares and the rights issues is a failure, then the shares are offered on the open market by the underwriters and the shareholder profile of the company may change, offering the opportunity for a hostile takeover bid. This is possible as when the shares are sold by the underwriters they are offered at price less than market value. Therefore a company with a failed rights issue provides the opportunity for a potentially powerful and unwanted shareholder to build up a large stake in it, hence the hostile takeover threat.

A failed rights issue may point to the possibility that the company is not doing well. Therefore investors should examine closely a company's reasons for undertaking a rights issue. A company using the money raised by a rights issue to expand the business, with the aim of seeing profits, share price and dividends rise in the future, will be a better investment than a company using the money raised via a rights issue to pay off old debts and avoid business failure.

In June 1991 Sainsbury's announced a £489m rights issue, the first in its 18 years as a quoted company. The aim of this rights issue was clear: the cash was for building 160 new and replacement supermarkets, 70 Homebase stores and 10 Savacentre stores. Sainsbury's rights issue was a one for ten issue at £3.12 and shares were trading at £3.52 on the day of the announcement. The preferred source of finance was a rights issue as the market was receptive to the idea and Sainsbury's share price strong.

### Government

The government in the UK has over the years invested in industry and commerce in a number of different ways. Depending on the government in power, state funding takes several forms. The British government has invested in prestige and successful projects such as Concorde, a joint project between Air France and British Airways in the 1960s and 1970s when both airlines were state owned.

An example of a failed government-funded project, from the 1970s–80s, is the transputer project. The transputer was a parallel processing microprocessor chip developed by Inmos, which was set up by state funding from the National Enterprise Board under a Labour government in 1978. Inmos was loss making and was privatised in the early 1980s by the Tory government, which sold it to Thorn-EMI. In 1989 Thorn-EMI sold the loss-making Inmos to the French-Italian company SGS Thompson.

More recent Government funding of industry includes the Labour government in November 1997 making a £200m investment in Rolls-Royce for the development and extension of the Trent aero engines for Boeing and Airbus aircraft. The government funded this project as in the following 20 years £50 billion of orders, 16 000 jobs and a 'commercial return' for the taxpayer would be secured.[13]

Historically the state has supported declining industries, usually with a view to preventing mass unemployment in economically deprived regions. Companies that received state funding in the 1970s included British Steel and British Leyland. In the 1980s the Tory government stopped state support and privatised many of the organisations that had received state funding. This allowed the companies such as British Steel and British Airways to control their own finances and raise money on the open money markets. The type of state funding of business and industry occurring in the 1980s under the Tory government was financial support for small businesses. At this time small businesses were viewed as the economic salvation of industry, given the decline and contraction of larger and more traditional heavy industries, such as steel and shipbuilding. The contraction and decline of heavy industry in the UK was due to competition from overseas companies and hence the need to be cost competitive, which led to cost cutting and redundancies.

Government funding for small businesses generally took the form of grants and loans of government money distributed via agencies. The rules and conditions governing the grants and loans depended on the type of business, industry or location to which the grant or loan was being made. Schemes for distributing government funding to business and industry included the following.

#### DTI Enterprise Initiative

Grants of money were given by the Enterprise Initiative to help small and medium-sized enterprises (SMEs) work on a product or project in one of a number of areas of activity. The grant of money was to be used by the business to employ a consultant in the following areas: marketing, business planning, design, information technology or production.

#### British Coal Enterprise

This scheme provided low-interest loans to businesses setting up or expanding in coalmining areas where many mining jobs had been lost due to pit closures. The aim of the scheme was to create jobs in ex-coalmining areas to replace some of the jobs lost. The scheme was open to ex-miners and non-miners.

#### Enterprise Allowance

This was one of the earliest schemes for providing funding to small businesses and offered a small weekly allowance to new small businesses and advice on setting up and running a small business.

### The use of capital raised and leasing assets

Organisations use the capital raised to acquire assets. Many organisations are more concerned with using assets than actually owning them and in this situation they will have to make the decision to lease rather than purchase assets. The leasing of assets has become more common during the last 20 years. Leasing is one way of acquiring assets without paying the full price. The most commonly leased assets are cars, plant and machinery, and information technology and office equipment. The types of lease used to rent out a particular asset may vary. The common types of leasing arrangement are an operating lease, a finance lease and a sale and leaseback.

An operating lease is a short-term contract in which the supplier of equipment, for example information technology hardware, makes the equipment available to a business. The business will enter into an operating lease and make a series of lease payments to the provider of the equipment. The tax advantages of leasing assets accrue immediately rather than over a number of years, as would be the case with depreciation allowances on purchased assets.

A finance lease is used when suppliers of an expensive item of capital equipment are paid directly in full for the item they have supplied to the company by the financing organisation, e.g. bank, finance house or merchant bank. Therefore the financing organisation provides the finance and the company makes leasing payments to the financing organisation.

A sale and leaseback is used when a company is concerned with using rather than owning an asset. The key principle here is that the company sells the asset to release the capital tied up in the asset and leases it back. Organisations such as pension funds may purchase a large, expensive fixed asset, like a head office building, and lease it back to a company. Hence the capital is released in the form of cash for the company to spend on other projects, without the upheaval of moving office.

## Management accounting

Management accounting generates information for managers to use in planning and decision making when the financial resources of the organisation have to be allocated. The information generated is largely quantitative in nature and is generated by application of management accounting techniques. The type of management accounting techniques examined in this chapter are costing, budgeting and investment appraisal.

## Costing

Costing involves looking at and defining the costs involved in producing a product or service. How costs are defined influences their use in the techniques of absorption and marginal costing, which are looked at later in this chapter. The cost of producing a product or service has three elements to it, the costs of materials, labour and overheads. These costs will be direct or indirect; and/or fixed or variable.

Material and labour costs can be either direct or indirect costs. Direct costs are the expenditure on elements that goes straight into producing the product or service. They include expenditure on raw materials and components and the wages of production staff or front-line service delivery staff. Indirect costs are often called overheads. Overheads or indirect costs are the expenses that do not contribute directly to the product or service being produced and therefore cannot be attributed to a particular job. Indirect costs or overheads include indirect labour costs, indirect material costs and indirect expenses. Office cleaners, catering staff and security staff are good examples of indirect labour costs for most organisations. Indirect material costs are the expenditure on cleaning materials, maintenance materials and subsidised food for employees. Indirect expenses are charges for items that have to be met, but that have no direct relationship to the cost of production. Indirect expenses include rent, heating and lighting bills and insurance.

Fixed costs do not change directly in relation to the level of activity or production, but have to be paid out, usually on a short-term basis such as monthly or quarterly. Fixed costs include many of the indirect costs just mentioned such as rent, insurance and maintenance contracts. Variable costs are sometimes called marginal costs. This is quite simply due to the marginal costing technique attributing variable costs only to the units of production/cost. Marginal costing is looked at later in this chapter. Hence variable costs are those that do vary in relation to the level of activity or production and include direct labour costs and direct materials costs. Prime costs are the sum of direct wages and direct materials (prime costs = direct wages + direct materials).

## Absorption costing

The use of absorption costing to provide quotes for jobs takes account of both fixed and variable costs. Absorption costing aims to ensure that all the overhead costs incurred by the business are covered by the revenues it receives. The information generated by absorption costing aims to provide information that can be used to give quotes for jobs. If an accurate quote for a job is to be provided, then a correct amount has to be included in the quote to cover the portion of overheads incurred in carrying out the job. This requires a decision to be made on what basis the organisation is going to allocate overheads. Is it going to allocate overheads on the basis of materials used in jobs, the amount paid in direct wages to complete the job or merely the number of units produced?

There are six methods of absorbing production overheads. These overhead absorption rates (OAR) are outlined in the worked example below. Comment on the calculations being carried out is also provided.

**Example question**

### Absorption costing

(a) Calculate six different overhead absorption rates for Job 99/2014 and indicate the total overhead cost incurred for each OAR.

(b) Using the most appropriate overhead absorption rate, calculate a quote for Job 99/2014.

**Production Department – Monthly Costs Report – March 1999**

| | |
|---|---|
| Total cost centre overheads (TCCO) | £60 000 |
| Number of cost units | 1600 units |
| Direct labour hours | 5000 hours |
| Machine hours | 4000 hours |
| Direct wages | £50 000 |
| Direct materials | £40 000 |

(Prime cost = Direct wages + Direct materials)

**Job Number 99/2014**

| | |
|---|---|
| Direct material cost | £5000 |
| Direct wages paid (£6.00 per hour) | £3750 |
| Time taken on machine | 500 hours |
| Number of units produced | 250 units |

***Workings, answers and comments – part (a)***

*1 Cost unit OAR*

$$\frac{\text{TCCO}}{\text{Number of cost units}} = \frac{£60\,000}{1600} = £37.50 \text{ overhead per unit}$$

The use of the cost unit OAR means that overheads are charged at a rate of £37.50 per unit supplied. Therefore the use of the cost unit OAR for Job 99/2014 would result in an overhead charge of £9375 (£37.50 × 250 units produced).

The cost unit OAR is appropriate only if all units of production are the same.

*2 Direct labour OAR*

$$\frac{\text{TCCO}}{\text{Number of direct labour hours}} = \frac{£60\,000}{5000}$$

$$= £12 \text{ overhead per direct labour hour (DLH)}$$

The use of the direct labour OAR means that overheads are charged at a rate of £12.00 per hour spent directly working on the job. The use of the direct labour OAR for Job 99/2014 is shown in the worked answer for Question 2. This method of absorbing or allocating overheads requires accurate records to be kept regarding the number of direct labour hours worked. However, this absorption rate is particularly suitable if the production department is labour intensive, as there will be a direct relationship between time spent on production and the overheads incurred.

*3 Machine hour OAR*

$$\frac{\text{TCCO}}{\text{Overhead per machine hour}} = \frac{£60\,000}{£15} = 4000 \text{ machine hours}$$

The use of the machine hour OAR means that overheads are charged at a rate of £15.00 per hour spent using machines to complete the job. Hence the use of the machine hour OAR for Job 99/2014 would result in an overhead charge of £7500 (£15 × 500 units produced). This method of absorbing or allocating overheads requires accurate records to be kept regarding the number of machine hours used. However, this absorption rate is particularly suitable if the production department is machine intensive, as there will be a direct relationship between machining time and the overheads incurred.

*4 Direct wage percentage OAR*

$$\frac{\text{TCCO}}{\text{Direct wages}} = \frac{£60\,000}{£50\,000} \times 100 = 120\%$$

The use of the direct wage percentage OAR will result in an overhead charge of £1.20 for every £1 of direct wages paid for completion of the job. Consequently, the use of the direct wage percentage OAR for Job 99/2014 results in an overhead charge of

£4500 (£3750 × 120 per cent). This overhead absorption rate is in many cases an appropriate OAR as it takes account of rates of pay and the number of hours worked, which usually relates directly to the time a unit takes to produce. However, if there is considerable variation in the rates of pay workers receive then this method of overhead absorption or allocation is much less suitable.

*5 Direct material cost percentage OAR*

$$\frac{\text{TCCO}}{\text{Direct material cost}} = \frac{£60\,000}{£40\,000} \times 100 = 150\%$$

The use of the direct material cost percentage OAR will result in an overhead charge of £1.50 for every £1 of direct material used in the job. The overheads for Job 99/2014 resulting from the use of the direct material cost percentage are £7500 (£5000 × 150 per cent). This overhead absorption rate does not take account of time, i.e. if two jobs take the same amount of time to complete, but one job uses more expensive materials, then the overhead absorption rates will differ.

*6 Prime cost percentage OAR*

$$\frac{\text{TCCO}}{\text{Prime costs}} = \frac{£60\,000}{£90\,000} \times 100 = 66.7\%$$

The use of the prime cost percentage OAR will result in an overhead charge of £0.67 for every £1 of prime cost for the job. Hence the use of the prime cost percentage OAR will give an overhead charge of £5836.25 (£8750 × 66.7 per cent). This overhead absorption rate combines the downside of both the direct materials OAR and direct wages OAR, without having any benefits of its own.

***Workings, answers and comments – part (b)***

The next step is to apply one of the overhead absorption rates to Job 99/2014 and calculate a quote or total cost for Job 99/2014. There is considerable variation in the overheads to be charged depending on the overhead absorption rate used, from £9375 with the cost unit OAR down to £4500 with the direct wage OAR. In theory, any of the six overhead absorption rates can be used to provide a quote for a job. However, the company must choose one overhead absorption rate to use in its production department and use it consistently. In most instances the overheads incurred relate to the time a job spends in production, so it is usually best to choose a rate that takes account of time, such as the direct labour hours OAR or the direct machine hours OAR.

The production department in the example has marginally more labour hours (5000 hours) than machine hours (4000 hours), hence the company incurs overhead predominantly to provide labour. Therefore use of the direct labour hours OAR would be most appropriate and give a total cost of £97 500, as shown in the following calculation.

**Job No 99/2014**

1 Number of hours worked
 = Direct wages paid/hourly rate
 = £3750/£6 per hour
 = 625 hours worked on Job 99/2014

2 Calculation of overheads
= Number of hours worked × Direct labour OAR
= 625 × £12 (from answer to Q1)
= £7500

3 Calculation of quote for job 99/2014

| | £ |
|---|---|
| Direct materials | 40 000 |
| Direct wages | 50 000 |
| Prime cost | 90 000 |
| Overheads | 7 500 |
| Total Cost | 97 500 |

**Marginal costing**

Marginal costing is the technique of charging variable costs to the cost or production units. This is a direct contrast with absorption costing where, in addition to variable costs, fixed costs or overheads are charged to cost units by use of one of the overhead absorption rates examined in the previous section.

A key figure in marginal costing is the contribution. This is the difference between sales and variable costs and is a contribution towards fixed costs. Therefore fixed costs for a period are written off against the contribution for that period to give the final profit or loss for the period. The worked example below demonstrates the calculation of profit and contribution. Contribution can be calculated for one unit or any chosen level of sales.

**Example question**

**Calculation of contribution**

A product sells for £75 and has variable costs of £45. During the period ending 31 March 1999 the product sold 3500 units. Fixed costs for the period were £30 000. Calculate the total profit, contribution per unit and contribution for sales of 3500 units.

***Working***

| | *per unit* | *3500 units* |
|---|---|---|
| Selling price | £75 | £262 500 |
| less Variable costs | £45 | £157 500 |
| Contribution | £30 | £105 000 |
| less Fixed cost | | £30 000 |
| Profit | | £75 000 |

***Answer***

Total profit is £75 000, contribution per unit is £30, and contribution for 3500 units is £105 000.

**Marginal costing and decision making**

Marginal costing can assist in decision making, particularly when deciding:

- which products to manufacture;
- if production of one or more products should be ceased;
- whether to accept a special order or contract;
- whether to make or buy a component.

Example question

### Which products to manufacture

A company has a choice of manufacturing two out of three products. Indicate which products should be manufactured when each of the following ranking methods is applied to the information given below.

1 Ranking by contribution.
2 Ranking by profit/volume ratio.
3 Ranking by total contribution.

***Information provided***

| | *Product A* | *Product B* | *Product C* |
|---|---|---|---|
| Selling price per unit | £35 | £60 | £45 |
| Variable costs per unit | | | |
| Materials | £10 | £24 | £18 |
| Labour | £6 | £14 | £8 |
| Overheads | £4 | £6 | £6 |
| Total cost | £20 | £44 | £32 |
| Contribution per unit | £15 | £16 | £13 |

**1 Ranking by contribution**

***Working and comment***

This method of deciding which products to manufacture assumes that there are no limits on production resources or sales that can be achieved. Therefore the amount of the unit contribution can be the decision criteria.

***Answer***

Products A and B have the highest unit contributions, £15 and £16 respectively, and would therefore be the preferred products for manufacture.

**2 Ranking by profit/volume ratio**

***Working and comment***

If the products have a maximum sales income that can be achieved from any of the three products, then the profit/volume ratio can be used to rank the products. If maximum sales of £100 000 could be achieved for each of the three products, then the calculation would be

$$\text{Profit/Volume ratio} = \frac{\text{Contribution}}{\text{Selling price per unit}} \times 100$$

$$\text{Product A} = \frac{£15}{£35} \times 100 = 42.9\%$$

$$\text{Product B} = \frac{£16}{£60} \times 100 = 26.7\%$$

$$\text{Product C} = \frac{£13}{£45} \times 100 = 28.9\%$$

With sales of £100 000 the contribution from Product A would be £42 900, 42.9 per cent of £100 000; for Product B the contribution is £26 700, 26.7 per cent of £100 000; and for Product C the contribution is £28 900, 28.9 per cent of £100 000.

***Answer***
In this instance the company should choose Products A and C as they make the biggest contributions to fixed costs.

**3 Ranking by total contribution**

***Working and comment***
If the maximum sales achievable for each product vary, then ranking should be by the total contribution that each product makes to fixed costs. If maximum sales achievable are 5000 units of Product A, or 2500 units of Product B or 3500 of Product C, the following ranking exercise helps establish which products should be produced.

Total contribution = sales units × contribution per unit

Product A – Total contribution = 5000 units × £15 = £75 000
Product B – Total contribution = 2500 units × £16 = £40 000
Product C – Total contribution = 3500 units × £13 = £45 500

***Answer***
Products A and C have the highest total contribution and should therefore be the products manufactured.

Example question

**Ceasing manufacture of a product**

A company produces a range of three products. The profit and loss account for the period ending 31 March 2001 shows that Product B has made a loss. The managing director suggests that Product B should be dropped as this will not influence sales of the other products. Produce a marginal cost statement for the period ending 31 March 2001 and advise the managing director as regards dropping Product B.

***Information provided***
*Profit and loss account, year ended 31 March 2000*

| | *Product A* | *Product B* | *Product C* |
|---|---|---|---|
| Sales | 250 000 | 200 000 | 140 000 |
| Direct materials | 62 500 | 110 000 | 32 500 |
| Direct labour | 55 000 | 60 000 | 40 000 |
| Variable overheads | 35 000 | 12 000 | 29 000 |
| Fixed overheads | 50 000 | 65 000 | 30 000 |
| Total costs | 202 500 | 247 000 | 131 500 |
| Profit/(Loss) | 47 500 | (47 000) | 8 500 |

***Working and comment – marginal cost statement***
*Marginal cost statement, year ended 31 March*

| | *Product A* | *Product B* | *Product C* | *Total* |
|---|---|---|---|---|
| Sales | 250 000 | 200 000 | 140 000 | 590 000 |
| Direct materials | 62 500 | 110 000 | 32 500 | 205 000 |
| Direct labour | 55 000 | 60 000 | 40 000 | 155 000 |
| Variable overheads | 35 000 | 12 000 | 29 000 | 76 000 |
| Total variable costs | 152 500 | 182 000 | 101 500 | 436 000 |
| Contribution | 97 500 | 18 000 | 38 500 | 154 000 |
| *Less* total fixed costs | | | | 145 000 |
| Profit | | | | 9 000 |

The marginal cost statement above shows that Product B makes a contribution of £18 000 to fixed costs. If the company ceased to manufacture Product B, then the contribution of £18 000 would be lost and the company would make a total loss of £9000, calculated in the following way: profit £9000 – Product B contribution £18 000 = –£9000 loss. The general rule is that it is usually expedient to continue with a product that makes a contribution to fixed costs.

***Answer***
Advise the managing director to continue manufacturing Product B.

Example question

**Acceptance of a special order or contract**

A company manufactures a product that has variable costs of £15 per unit and a selling price of £19.50. A regular customer asks for an additional 3000 units as well as its normal order, but wants to negotiate a special price of £18 per unit for the additional units. Should the company agree?

***Working and comment***
In this situation there are clearly issues of customer relationships and the behaviour of competitors, particularly if the additional order is not accepted. However, the financial viewpoint is straightforward. The variable costs of the product are £15, hence any selling price above £15 will give a contribution. Accordingly for financial reasons it is worth accepting the order at the reduced price of £18 per unit. In this case the general rule applies that any product giving a contribution is worth manufacturing.

In contrast it would not make financial sense to sell the product at a price less than variable costs, as this would result in a negative contribution. Similarly, if the company could supply the required additional 3000 units at £18, but to do so had to reduce its current sales at £19.50, accepting the additional order would reduce the total contribution and therefore it should not be accepted.

***Answer***
In this case accept the additional order at a price of £18.

Example question

**Making or buying a product**

Company CHC can make Component X itself and incur variable costs of £27 per unit or purchase the component from Company RBC for £30 per unit. Which is the best option for Company CHC?

***Working and comment***
If Company CHC has unused capacity, it makes sense for the component to be produced in-house as the variable cost of £27 per unit is lower than the buying in price of £30 per unit. It should be noted that fixed costs are omitted from the comparison as they will continue to be paid even when none of the factory facilities is in use. If the company does not have unused capacity and in-house manufacture of the component requires manufacture of another product to be stopped, then further analysis is required.

***Answer***
In this case Company CHC should manufacture the component in-house.

## Investment appraisal

Investment appraisal techniques are used by management to help in making decisions concerning investment in long-term projects and spending capital finance. The application of the investment appraisal methods discussed in this chapter provides quantitative data for use in this type of management decision making. Quantitative data can be useful and pertinent when decisions are being made concerning the investment of capital that an organisation has raised. Hence there is a clear overlap between investment appraisal, an area of management accounting, and the use to which the funds or capital raised are put (*see* section on financial management earlier in this chapter). Capital will be invested in major projects such as the purchase of a new equipment or machinery, the acquisition of another company or the development and launch of a new product or service.

Thorough and objective evaluation of an investment opportunity before capital is spent will include assessment of both quantitative and qualitative information. Assessment of any investment opportunity will also include estimating the risk involved in not making the investment and an evaluation of the risk if the project fails. The quantitative assessment of an investment opportunity can be carried out by using any of the investment appraisal methods discussed in this chapter: payback, accounting rate of return, net present value and internal rate of return. An investment represents the commitment of money now for gains or returns in the future and the quantitative data generated by investment appraisal will help answer the following questions concerning an investment:

1 Will the investment provide an adequate financial return?
2 Is the investment the best alternative from the options the company has available?
3 What is the cost if the project fails?

### Payback method

The payback method measures the length of time taken for the return on the investment exactly to equal the amount originally invested. Hence where two or more investment proposals are being considered, the one that recovers the original investment in the shortest time is the more acceptable.

**Example question**

**Payback method**

Consider each of the three investment opportunities outlined below. Apply the payback method and state which would be the preferred investment.

| | *Investment A* | *Investment B* | *Investment C* |
|---|---|---|---|
| Year 0 original investment | (£20 000) | (£30 000) | (£40 000) |
| Year 1 net cashflow | £12 000 | £8 000 | £16 000 |
| Year 2 net cashflow | £6 000 | £8 000 | £12 000 |
| Year 3 net cashflow | £6 000 | £6 000 | £10 000 |
| Year 4 net cashflow | £4 000 | £6 000 | £8 000 |
| Year 5 net cashflow | £4 000 | £6 000 | £8 000 |

**Example calculation – Payback method – Investment A**

*Investment A* = £20 000

| | *Cashflow* | *Cumulative cashflow* |
|---|---|---|
| Year 1 net cashflow | £12 000 | £12 000 |
| Year 2 net cashflow | £6 000 | £18 000 |
| Year 3 net cashflow | £6 000 | £24 000 |
| Year 4 net cashflow | £4 000 | £28 000 |
| Year 5 net cashflow | £4 000 | £32 000 |

1 The investment has been paid back by the end of Year 3.
2 But exactly when in Year 3 does payback occur?
3 At the end of Year 2 £18 000 has been paid back, leaving £2000 of the original £20 000 investment still to be paid back.
4 The total payback in Year 3 is £6000, therefore the calculation below shows that the £2000 is paid back in the first four months of Year 3.

$$\frac{\text{£2000 (to payback after Year 2)}}{\text{£6000 (Year 3 net cashflow)}} \times 12 \text{ (months in a year)} = \frac{2000}{6000} \times 12 = 4 \text{ months}$$

Therefore the total payback period is 2 years 4 months.

The payback period for Investments B and C can be calculated in the same way.

| | *Investment A* | *Investment B* | *Investment C* |
|---|---|---|---|
| Payback period | 2 years 4 months | 4 years 4 months | 3 years 3 months |

***Answer***

Hence the preferred investment is A, as this pays back in the shortest time.

**Advantages and disadvantages of the payback method**

The advantages of the payback method are that the payback period is simple to calculate and easy to understand. However, the payback method takes no account of profit, loss and depreciation from the sale of fixed assets, although it does recognise the uncertainty of the future. Therefore the payback method acknowledges that the sooner the investment is recovered, the smaller the risk involved and thus uses the earliest cashflows first. If the money is to be invested in a project that is subject to rapid technological change, then the project with the most rapid payback and turnaround will be the most favourable.

Further drawbacks of the payback method are that it ignores all cashflow after the payback period and the total life of the project. Other difficulties are that no allowances are made for the lower value of money paid back in the future. The value of money paid back today is worth more than money paid back in the future, i.e. £100 today is worth more than £100 in five years' time. An added problem with the payback method is that profits are disregarded and an investment with a shorter payback period may be selected even if it is less profitable overall than a project that takes longer to pay back. For example, an investment of £5000 will pay back quicker than a £50 000 investment, regardless of the fact that the £50 000 investment may be more profitable in the long run. Another disadvantage is that the payback method relies on estimates of net cash inflows and the timing of their receipt.

**Accounting rate of return**

The accounting rate of return (ARR) expresses the profit generated by an investment or project as a percentage of the capital invested.

$$\text{ARR} = \frac{\text{Profit}}{\text{Capital employed}} \times 100$$

In the ARR calculation the net profit before interest and taxation figure is used and averaged over the lifetime of the project. The capital employed can be either the initial capital employed or the average capital employed over the lifetime of the project. The average capital employed over the lifetime of the project takes into account the residual value of the project at the end of its working life, *see* method 2 below.

Therefore two methods for carrying out ARR calculations exist:

**Method 1**

$$\text{ARR} = \frac{\text{average net profit per annum}}{\text{initial capital employed}} \times 100$$

**Method 2**

$$\text{ARR} = \frac{\text{average net profit per annum}}{\text{average capital employed*}} \times 100$$

$$\text{*average capital employed} = \frac{\text{initial capital employed} + \text{residual value}}{2}$$

The worked example below uses method 2, average capital employed.

**Example question**

**Accounting rate of return**

Consider each of the three investment opportunities outlined below. Apply the accounting rate of return (ARR) method and state which would be the preferred investment.

| | *Investment A* | *Investment B* | *Investment C* |
|---|---|---|---|
| Year 0 original investment | (£20 000) | (£30 000) | (£40 000) |
| Year 1 net profit | £12 000 | £8 000 | £16 000 |
| Year 2 net profit | £6 000 | £8 000 | £12 000 |
| Year 3 net profit | £6 000 | £6 000 | £10 000 |
| Year 4 net profit | £4 000 | £6 000 | £8 000 |
| Year 5 net profit | £4 000 | £6 000 | £8 000 |

**Example calculation – accounting rate of return – method 2**

| | *Investment A* | *Investment B* | *Investment C* |
|---|---|---|---|
| Total net return | £32 000 | £34 000 | £54 000 |
| *less* original investment | £20 000 | £30 000 | £40 000 |
| Net profit | £12 000 | £4 000 | £14 000 |
| Years of life | 5 | 5 | 5 |
| Average profit per annum (profit/years of life) | £2400 | £800 | £2800 |
| Residual value | £5000 | £5000 | £7000 |
| Average capital employed | $\frac{£20\,000 + £5000}{2}$ | $\frac{£30\,000 + £5000}{2}$ | $\frac{£40\,000 + £7000}{2}$ |
| | = £12 500 | = £17 500 | = £23 500 |
| Average rate of return | $\frac{£2400}{£12\,500} \times 100$ | $\frac{£800}{£17\,500} \times 100$ | $\frac{£2800}{£23\,500} \times 100$ |
| | = 19.2% | = 4.6% | = 11.9% |

***Answer***

Hence the preferred investment is A, as this shows the highest return.

**Advantages and disadvantages of the accounting rate of return method**

The ARR encompasses the entire life of the project and all expected profits. Its other principal advantage is its comparative simplicity to calculate and understand. However, there are a number of disadvantages relating to this method of investment appraisal. The disadvantages include the net profit being defined in different ways, for example should net profit before or after depreciation on the investment be used? Other difficulties include the ARR using profits as a key factor in the calculation, whereas in investment decisions cashflow is the crucial factor. Also the ARR does not make allowances for the different value of money over time, compared with discounted cashflow methods (see the next method of investment appraisal to be looked at, net present value). High returns in the early years of an investment have a greater net worth and are easier to predict, but these factors are not taken account in ARR calculations. Interpretation of the ARR results can be ambiguous as often there will be no indication of what an acceptable ARR would be for a specific project, and furthermore different ARRs will be acceptable in different situations.

Finally, the use of residual investment values can substantially affect the ARR calculated. The use of residual values and an average capital employed can make a notable difference in the ARR values for a project. Remember that residual values are difficult to estimate and the difference they can make is illustrated below. The higher the residual value, the lower the ARR.

**Example**

Average net profit per annum = £100 000
Initial capital employed = £300 000
Residual values = £10 000 and £50 000

1 Residual value = £10 000

$$\text{Average capital employed} = \frac{£300\,000 + £10\,000}{2} = £155\,000$$

$$\text{ARR} = \frac{£100\,000}{£155\,000} \times 100$$

ARR = 64.5%

2 Residual value = £50 000

$$\text{Average capital employed} = \frac{£300\,000 + £50\,000}{2} = £175\,000$$

$$\text{ARR} = \frac{£100\,000}{£175\,000} \times 100$$

ARR = 57.1%

Despite the disadvantages, the ARR method of investment appraisal is suitable if the project is short term and accurate estimates can be made for any residual values to be included in the calculations.

**Net present value**

The net present value (NPV) method of investment appraisal makes allowances for money received in the future being worth less than if they were received today. Therefore the NPV method converts future net cashflows into present-day values by discounting the value of money that is expected to be received in the future.

If the discounted net cashflows exceed the original investment then the project could go ahead. When choosing between two or more projects, the project with the highest positive NPV exceeds its original investment by the greatest amount and is usually the preferred project. If the discounted net cashflows are less than the original investment, then the investment should not be allowed to proceed as money will be lost. An acceptable interest rate or rate of return has to be decided on and the discounting factors to be used in the NPV calculation need to be read from a discount table.

In selecting a discounting rate the following need to be considered:

- the rate of interest that the company could obtain if the money were invested outside the business;
- the cost of capital required to make the investment;
- the rate of return (internal) that the company expects to gain on investments.

If the company is to be profitable, then in the long run the rate of return (internal) needs to exceed the external rate that it can earn investing outside the business. The cashflows are multiplied by the discount cashflow value from the discount tables to give the discounted value of future net cashflows. The discounted net cashflows are added up and the original investment subtracted – if the total NPV is positive, then the project is acceptable.

Example question

### Net present value method

Consider each of the three investment opportunities outlined below. Apply the NPV method of investment appraisal and state which would be the preferred investment. The cost of capital is 8 per cent.

| | *Investment A* | *Investment B* | *Investment C* |
|---|---|---|---|
| Year 0 original investment | (£20 000) | (£30 000) | (£40 000) |
| Year 1 net cashflows | £10 000 | £6 000 | £12 000 |
| Year 2 net cashflows | £8 000 | £6 000 | £12 000 |
| Year 3 net cashflows | £6 000 | £6 000 | £12 000 |
| Year 4 net cashflows | £4 000 | £8 000 | £12 000 |
| Year 5 net cashflows | £4 000 | £12 000 | £12 000 |

***Calculation – net present value***

| | *Investment A* | | | *Investment B* | | | *Investment C* | | |
|---|---|---|---|---|---|---|---|---|---|
| | *Cashflow* | *DCF 8%* | *Net Cashflow* | *Cashflow* | *DCF 8%* | *Net Cashflow* | *Cashflow* | *DCF 8%* | *Net Cashflow* |
| Yr 0 | (20 000) | 1.000 | (20 000) | (30 000) | 1.000 | (30 000) | (40 000) | 1.000 | (40 000) |
| Yr 1 | 10 000 | .926 | 9 260 | 6 000 | .926 | 5 556 | 12 000 | .926 | 11 112 |
| Yr 2 | 8 000 | .858 | 6 864 | 6 000 | .858 | 5 148 | 12 000 | .858 | 10 296 |
| Yr 3 | 6 000 | .794 | 4 764 | 6 000 | .794 | 4 764 | 12 000 | .794 | 9 528 |
| Yr 4 | 4 000 | .735 | 2 940 | 8 000 | .735 | 5 880 | 12 000 | .735 | 8 820 |
| Yr 5 | 4 000 | .681 | 2 724 | 12 000 | .681 | 8 172 | 12 000 | .681 | 8 172 |
| | | | +6 652 | | | –480 | | | +7 928 |

***Answer***
Hence the preferred investment is C, as this has the highest positive NPV.

### Advantages and disadvantages of the net present value method

The main advantage of the NPV method is that cashflows and hence the relevance of liquidity are both taken account of in the calculations. The other fundamental advantage is that it takes account of the time value of money, unlike the payback and ARR methods. The comparison between projects' NPV values is straightforward and it is easy to judge which is the most profitable project. The disadvantages of the NPV method lie in the difficulty of accurately estimating the initial cost of the project, cash in-flows and the time periods in which cash in-flows will occur.

## Internal rate of return

The internal rate of return (IRR) method of investment appraisal is similar to the net present value (NPV) method. The key difference between NPV and IRR is that IRR seeks a discount rate at which the net cash in-flows, when discounted, exactly equal the amount originally invested, i.e. NPV = 0. The initial pieces of information that have to be sought for an IRR calculation are a discount rate giving a positive return or positive NPV, and a discount rate giving a negative NPV. There is no easy way of ascertaining which discount rates will give a positive or negative NPV, except by trial and error coupled with careful judgement.

The IRR is determined by interpolation between the discount rate giving a positive NPV and that giving a negative NPV; *see* guidelines below. The interpolation between the two rates is done by use of the formula shown below. This is not an exact method and the closer together the two discount rates used in the calculation, the more accurate the answer will be.

The formula for calculating the IRR is as follows;

$$\text{IRR} = \text{Positive rate} + \frac{(\text{Positive NPV} \times \text{Difference between discount rates})}{(\text{Range of NPV values})}$$

#### Guidelines for choosing discounting cashflow rates for IRR calculations

1 You have to find two DCF rates:
 (a) one giving a positive NPV value;
 (b) one giving a negative NPV value.

2 Which DCF rates do you choose?
 (a) Start with a rate of around 10 per cent. If this gives you a *total negative NPV*, then you need to choose a rate that is less than 10 per cent to get a *total positive NPV*, say 5 per cent.
 (b) If the first DCF rate you choose, say 10 per cent, gives you a *positive total NPV*, then the second rate will have to be greater than 10 per cent: you could try 15 per cent.
 (c) The two DCF rates used do not want to be more than 5 per cent apart, although they can of course be less than 5 per cent apart.

3 An alternative to starting with, say, 10 per cent is to start with the DCF rate you may have used in a previous NPV calculation on the same data. To do this the NPV given must be fairly close to zero.

Example question

#### Internal rate of return method

The company considering each of the three investment opportunities shown below requires a return of at least 10 per cent on all the investments it makes. Using the IRR method of investment appraisal, state which of the three investment opportunities the company should seriously consider.

| | *Investment A* | *Investment B* | *Investment C* |
|---|---|---|---|
| Year 0 original investment | (£20 000) | (£30 000) | (£40 000) |
| Year 1 net cashflow | £10 000 | £6 000 | £12 000 |
| Year 2 net cashflow | £8 000 | £6 000 | £12 000 |
| Year 3 net cashflow | £6 000 | £6 000 | £12 000 |
| Year 4 net cashflow | £4 000 | £8 000 | £12 000 |
| Year 5 net cashflow | £4 000 | £12 000 | £12 000 |

***Example calculation – internal rate of return – investment A***

| | *Cashflow* | *DCF 22%* | *Net Cashflow* | *Cashflow* | *DCF 23%* | *Net Cashflow* |
|---|---|---|---|---|---|---|
| Year 0 | (20 000) | 1.000 | (20 000) | (20 000) | 1.000 | (20 000) |
| Year 1 | 10 000 | 0.820 | 8 200 | 10 000 | 0.813 | 8 130 |
| Year 2 | 8 000 | 0.672 | 5 376 | 8 000 | 0.661 | 5 288 |
| Year 3 | 6 000 | 0.551 | 3 306 | 6 000 | 0.537 | 3 222 |
| Year 4 | 4 000 | 0.451 | 1 804 | 4 000 | 0.437 | 1 748 |
| Year 5 | 4 000 | 0.370 | 1 480 | 4 000 | 0.355 | 1 420 |
| | | | +166 | | | –192 |

Internal rate of return = 22% + (166/358 × 1) = 22.464%

***Example calculation – internal rate of return – investment B***

| | *Cashflow* | *DCF 7%* | *Net Cashflow* | *Cashflow* | *DCF 8%* | *Net Cashflow* |
|---|---|---|---|---|---|---|
| Year 0 | (30 000) | 1.000 | (30 000) | (30 000) | 1.000 | (30 000) |
| Year 1 | 6 000 | 0.935 | 5 610 | 6 000 | 0.926 | 5 556 |
| Year 2 | 6 000 | 0.873 | 5 238 | 6 000 | 0.857 | 5 142 |
| Year 3 | 6 000 | 0.816 | 4 896 | 6 000 | 0.794 | 4 764 |
| Year 4 | 8 000 | 0.763 | 6 104 | 8 000 | 0.735 | 5 880 |
| Year 5 | 12 000 | 0.713 | 8 556 | 12 000 | 0.681 | 8 172 |
| | | | +404 | | | –486 |

Internal rate of return = 7% + (404/890 × 1) = 7.454%

***Example calculation – internal rate of return – investment C***

| | *Cashflow* | *DCF 16%* | *Net Cashflow* | *Cashflow* | *DCF 15%* | *Net Cashflow* |
|---|---|---|---|---|---|---|
| Year 0 | (40 000) | 1.000 | (40 000) | (40 000) | 1.000 | (40 000) |
| Year 1 | 12 000 | 0.862 | 10 344 | 12 000 | 0.870 | 10 440 |
| Year 2 | 12 000 | 0.743 | 8 916 | 12 000 | 0.756 | 9 072 |
| Year 3 | 12 000 | 0.641 | 7 692 | 12 000 | 0.658 | 7 896 |
| Year 4 | 12 000 | 0.552 | 6 624 | 12 000 | 0.572 | 6 864 |
| Year 5 | 12 000 | 0.476 | 5 712 | 12 000 | 0.497 | 5 964 |
| | | | –712 | | | +236 |

Internal rate of return = 15% + (236/948 × 1) = 15.249%

***Answer***

The investments that should be seriously considered are investments A and C, as both have an IRR of greater than 10 per cent.

### Advantages and disadvantages of the internal rate of return method

The two clear advantages of the IRR method are that emphasis is placed on liquidity in the calculation and it results in a clear percentage return required on investment. The IRR is a measure of the intensity of capital use and also gives a return for risks. In the worked example for Investment A, if the cost of capital is 15 per cent and the IRR is 22.464 per cent, therefore the return for the risk is 7.464 per cent, the difference between the two figures (22.464 – 15). In general IRR is a more difficult method to apply than NPV. In most cases IRR and NPV will give the same answer as to acceptance or rejection of an investment, but may vary in ranking, thus possibly leading to different choices.

The disadvantages of the IRR method are that the reasoning behind the calculation is not easy to understand and it is difficult to determine two interest rates to interpolate between. Careful judgement should be exercised when making decisions based on IRR calculations. For example, an IRR of 25 per cent may appear attractive, but if the original investment is only £500, then an IRR of 15 per cent on an original investment of £10 000 would be more sensible.

## Financial reporting

In the UK companies are obliged by law to produce an annual report and accounts that relate to their financial and business performance. A company's annual report and accounts have to be audited and a copy filed with the Registrar of Companies at Companies House in London. The Companies Acts of 1981, 1985 and 1989 specify the layout and format of modern-day published accounts. The published accounts have to contain certain pieces of information to satisfy the legal requirements. The types of information required include information on the directors and their report; financial statements; information to clarify the details of the report and accounts; and an auditor's report.

The section of the published accounts giving information about the directors is often at the front. In the case of many large well-known companies, e.g. Marks & Spencer or British Aerospace, a photograph of each director is accompanied by a vignette on their career to date and their role on the board. These directors have to produce a report to be included in the published accounts. The report must provide fair comment on the company's performance over the financial year and important events that have occurred since the year end. Details of transactions involving its own shares by the company must also be included in the directors' report. Comment must also be included on probable future developments and on any research and development that the company is undertaking. In the introductory case study reporting on the publication of SmithKline Beecham's 1996 results, the company's share price clearly rose in late 1996 and early 1997. This linked directly with the company reporting promising results on the research and development of new drugs for illnesses such as Alzheimer's disease and migraines.

The financial statements that need to be included in published accounts are the profit and loss account, the balance sheet and a cashflow statement. The profit and loss account and the balance sheet are summary statements. The profit and loss account provides a summary of the company's income or sales revenue and expenditure, leaving a profit or loss on the bottom line. This is complemented by the balance sheet, which summarises the company's financial position at the end of the financial year, showing its assets and liabilities. In contrast, the cashflow statement shows in detail how the business has financed its operations. The details included in a cashflow statement are the particulars of how money has been raised – shares, loans or profits – and how the money has been spent – to acquire fixed assets such as buildings and machinery, or to buy current assets such as stock, raw materials and components.

The other sections that need to appear in published accounts are a statement of accounting policies, notes to accounts and statistical information. Notes to accounts provide a large amount of detail relating to the accounts and activities of the company. Some of the details included in the notes are how the operating profit and turnover figures are calculated; how individual asset and liability figures are calculated; loans and interest payments; tax details; paid and proposed dividends; and changes to accounting policies. A complete list can be found in Dyson listed in further reading for this chapter. The statistical information section of the published accounts may compare the current year's financial performance with results from previous years. This can be done by use of the financial ratios for the business over two or more years; *see* Michael Brett's book.[14]

A necessary item in any annual report and accounts is the auditor's report. The auditor should be independent of the directors of the company and a member of one of the chartered professional accounting bodies approved to perform audits. The auditor's report is addressed to the shareholders and should ideally state that the report and accounts provide a true and fair view of the company's activities for the financial year examined. To do this the auditor needs to satisfy him/herself that the reported assets exist and have been correctly valued in the accounts. The auditor also needs to ensure that the disclosure of liabilities is complete and thorough. The auditor may also inspect data and documents to confirm that entries in the books are genuine, as well as checking the accuracy of the books. Finally, the auditor should check that all benefits are accounted for and have been collected by the proper recipients.

The annual report and accounts, which give a true and fair view of a company's activities for the financial year examined, will provide significant information on the company and its activities and will therefore be of interest to a variety of different players who want to know about the business. This will allow concerned parties or financial stakeholders to make informed judgements regarding their role in relation to the company.

## Financial stakeholders

### The state

The financial information provided will determine the level and amount of taxation that a company will pay to the state. The current level of corporation tax is 30 per cent in the UK. The Labour government elected in May 1997 promised that corporation tax will not rise above 30 per cent for the remainder of its parliament.

### Current and potential investors

The information provided in the annual report and accounts will allow current and potential investors to make informed judgements about future investments and current investments. For example, in February 1997 SmithKline Beecham's shareholders would have very likely decided to hold on to their shares in the company, as dividends rose by 25 per cent and the company was showing significant growth in a number of areas. For further details see the introductory case study at the start of this chapter.

### Employees

Companies that choose to involve employees in the running of the business may see the disclosure of financial information as an important element of the employees' participation. This is especially so if a profit-sharing or an employee share ownership scheme operates.

### Creditors and banks

These interested parties are concerned with the company's liquidity and need to assess the risk involved in offering credit or loans. The information disclosed in the annual report and accounts will be useful in assessing this risk.

### Customers and debtors

Customers and debtors are stakeholders who purchase the company's products and services and may find the information disclosed in the annual report and accounts useful in assessing the risk of placing long-term or large orders.

### Competitors

Competitors as stakeholders typically find the information provided by the annual report and accounts a practical yardstick against which to measure their own performance. Insights into which directions competitors could be heading may be offered in their annual reports and accounts.

## Conclusion

This chapter examined two of the key external environmental influences on any organisation and its finance department in the UK at the start of the twenty-first century, namely the single European currency and the minimum wage. The bulk of the chapter then examined the main activities of a finance department and the types of decisions that have to be made concerning money and financial stakeholders.

ETHICAL ISSUES CASE STUDY 8.2 FT

# Battle over late payers

**By Tim Burt**

Lawrence Chapman is irritated by late payment of commercial debt. A partner at Countryside Art, a small Lincolnshire textiles company, he thinks many customers simply use late payment as a form of free credit.

'We feel we are being used – it starves us of the cash we need to grow,' says Mr Chapman, whose company employs 16 people and has annual sales of about £500 000 a year.

Mr Chapman is one of a growing number of entrepreneurs who have called for the imposition of a statutory right to interest on overdue bills, and he recently lent his support to a campaign by Britain's opposition Labour party to introduce the necessary legislation.

The proposal, however, has divided the small business community. Some business leaders back Labour's call for statutory interest, while a large group, including the main employers organisations, have sided with the Conservatives in opposing it.

Although most EU countries already enforce statutory interest on late payment, the UK government has vowed to fight any attempt to introduce it, and claims that it has proved ineffective where it has been applied.

In the run up to the general election, the issue has become the focal point of the political battle for the small business vote. The Conservatives, self-proclaimed champions of small- and medium-sized enterprise, claim Labour is failing to listen to business and accuses it of opting for 'clumsy, heavy-handed legislation'.

Labour and the Liberal Democrats say the government has fostered a business culture that encourages companies to ignore bills and allows small firms to be held to ransom by larger ones. Both opposition parties favour interest on late debts and have criticised the Tories for not following the example of countries, such as Sweden and Germany, in using it to help improve payment schedules.

Their stand, nevertheless, has been criticised by employers organisations such as the Confederation of British Industry and the Institute of Directors, which have also enlisted the backing of Federation of Small Businesses – representing 96 000 companies – and the Small Business Bureau.

This informal coalition claims that statutory interest is too blunt an instrument to change the UK culture of late payment. They argue that it would simply legitimise late payment, with companies treating it as a form of overdraft.

Mr Jon Ainger, senior policy adviser on small- and medium-sized enterprises at the CBI, says a more subtle approach is needed to persuade UK companies to change ingrained habits. 'Most other European countries have a statutory right to interest but it does not seem to have had much effect,' he adds. 'Italy, Spain and Cyprus have introduced it and remain among the worse payers.'

His view is echoed by Mr Stephen Alambritis, spokesman for the Federation of Small Business, who believes that even if interest charges were introduced many companies would not collect it for fear of souring relations with customers, often bigger businesses.

He wants a much broader approach to the problem and applauds the government's recent move requiring companies to publish their payment records in their annual reports. The federation plans to use that information to draw up a blacklist of late payers, partly to 'shame' them into more prompt payments but also to give businesses a chance to avoid those with a dubious record.

To be fair, the opposition parties have advocated such measures for some time. And, like the government, they have backed the CBI's prompt payment code and the introduction last year of BS7890, a British standard for late payers. Although only voluntary, under the standard fines of up to £5000 can be imposed on signatory companies that fail to meet its terms.

But that is too little, too late, for the Forum of Private Business, which represents 25 000 companies. It claims the CBI and IoD are ignoring the wishes of their members in opposing statutory interest. Mr Stan Mendham, its chief executive, says voluntary codes and publication of payment records are a step in the right direction but do not go far enough. 'Late payment costs people their existence in the end and we believe most businesses would welcome [the imposition of] statutory interest,' he says.

While the political parties and rival business organisations slug it out, the debate over statutory interest threatens to overshadow the real problem: that in the UK, most large businesses believe lengthy payment schedules are simply a fact of business life.

If you accept that argument, then it is up to companies to work around it to secure the best terms to protect their cash flow. That can mean pricing contracts to cover late payments, and offering discounts for early or prompt receipts.

Credit insurance brokers claim companies can minimise the risk of late payment by taking out insurance cover as part of their credit risk management. Alternatively, they can use factoring businesses to collect debts or chase debtors.

That all comes at a cost, of course. And for many small businesses credit insurance and factoring is an expensive way of retrieving money owed. 'Factoring is not an option because the rates are simply not competitive for a company of our size,' says Steven Morrell of SGM Management & Design.

His architectural project management business, employing six people with a turnover of £250 000, tries to agree fixed payment terms before beginning work on a contract.

'We need to move to a situation where companies pay bills when they fall due,' he adds. 'But it is very difficult to see that happening – despite the new measures – when the philosophy in this country is that a bill is worth chasing even when it is 30 days overdue.'

*Source*: *Financial Times*, 7 March 1997. Reprinted with permission.

**Questions for ethical issues case study 8.2**

1 Consider the issues discussed in the case study and state whether you agree or disagree with the idea of legislation introducing a statutory right for businesses to impose interest on overdue bills.

2 (a) If you agree with the introduction of the legislation mentioned in Question 1, then suggest a minimum of five rules and regulations that you would like to see the new legislation introduce.

*or*

(b) If you do not agree with the introduction of the legislation mentioned in Question 1, then explain why and list a minimum of five alternative actions that businesses and industry could take to tackle the problem of late payers.

## EXIT CASE STUDY 8.3

# Ingenuity fails to bridge Tube's finance gap

**By Charles Batchelor**

An auction of London Underground station signs; sponsorship of the famous Tube map by a telephone directory service; a £400m private finance deal to fund the renewal of the dilapidated Northern Line . . .

Managers could hardly be accused of ignoring the market place in their search for more funds for the capital's publicly-owned underground railway network. But in spite of all their efforts – and the growing contribution from the surplus they make on operations – the Tube remains dependent on government support for the bulk of its funds.

The announcement last week of plans to postpone a range of investment projects, from new escalators to stable embankments, was the result of the government's decision to remove nearly £700m from the Underground's budget over the next three years. The basic grant has been cut by £380m while cost overruns on the Jubilee Line extension, previously met by the government, will require £296m which otherwise would have been spent elsewhere.

The Tube has three main sources of funding:

- The government's core grant which goes on general maintenance and small-scale improvements. After peaking at £680m, core funding fell steadily and is expected to amount to £358m this year, Tube managers told MPs last week. It will fall further over the next two years, though transport officials insisted yesterday that sufficient funds would be available to maintain the network before privatisation. The 1996–97 allocation is roughly in line with the £350m which the Underground estimates it needs to maintain the system in a 'steady state' – providing a decent standard of service though without room for any improvements. But the steady state figure will rise to £380m when the Jubilee Line extension opens in March 1998 and there is a backlog of £1.2bn in essential investments to make up.
- Ring-fenced funding which goes to particular large-scale projects. Most of these funds have gone to the extension in recent years but small sums have also gone in work on stations which will link to the high-speed Channel tunnel rail link and on the now suspended CrossRail project – an underground line for main line trains between Paddington and Liverpool Street. Ring-fenced funds have risen steadily in recent years as work on the extension has increased and are forecast to reach £547m this year, compared with £487m in 1995–96.
- Revenues from ticket sales are expected to reach £815m, up slightly on the £808m in 1995–96. With the exception of slight decline in the two years following the economic downturn in the late 1980s, Tube revenues have risen steadily.

A move to appoint managers for specific lines in 1988 and the introduction of competitive tendering by outside contractors in the late 1980s have tightened management control and boosted recent operating surpluses. From an operating loss in 1992–93, the figures have improved to an expected £210m this year.

These traditional sources of funding are being supplemented by a range of new devices although they have yet to make an impact on overall financing levels.

As well as the Northern Line, information technology systems have been upgraded under a £13m private finance initiative project while the Tube hopes shortly to decide whether or not to go ahead with three more schemes, covering automated ticketing, power generation and supply and an integrated radio network.

The government calculates that PFI will contribute nearly £400m over the next three years but Mr Peter Ford, London Transport chairman, believes it must be seen as an add-on which cannot be applied to all areas of Tube operations.

**Exit case study 8.3** ***continued***

'The PFI has a life of its own. You cannot fund all the projects you need and you don't know if you will succeed in getting the money until the last minute,' he said.

Ambitious plans to seek sponsorship for stations have yet to produce results although a £2.5m contribution from the Corporation of London towards a £43m upgrade of Bank station was rewarded with the installation of four pairs of cast bronze dragons, the corporation's motif.

Property investments make a small contribution to the Tube's coffers – these amounted to £43m in 1995–96 – and some commentators view retail and property development as an underexploited opportunity.

The Tube is working on big property developments at Victoria and White City but persuading travellers to stop and shop on a crowded underground station is not managers' first priority.

'There is a direct conflict between the Underground's operational requirements and the retailers' needs,' says Mr Andrew Shackel of LT Property. 'You have to accept that retail comes second to the core business of moving people around the system.'

The Tube was regarded as having been more tightly managed and of having a clearer view of its finances than was the case at British Rail, according to one former Treasury official.

This has not prevented the government seeking further efficiencies in the private sector although, like the former British Rail train operations, continued subsidy in some form will still be needed.

*Source*: *Financial Times*, 26 February 1997. Reprinted with permission.

**Questions for exit case study 8.3**

**1** (a) Identify the financial players in London Underground and rank them according to the financial contribution that each makes to London Underground.

**1** (b) In your opinion, will the financial contributions made by each of the financial players increase or decrease in the current economic and political climate?

**2** Identify the conflicts that may occur between the different financial players in London Underground and suggest how they may be resolved.

## Short-answer questions

**1** Explain the role of the finance function.

**2** Define financial accounting.

**3** When did European monetary union occur?

**4** When was the national minimum wage introduced in the UK and what level was it set at initially?

**5** Name four sources of finance for a company.

**6** State the key difference between a share issue and a rights issue.

**7** What is a sale and leaseback and when do companies use one?

**8** Define management accounting.

**9** What is the key difference between absorption and marginal costing?

**10** What is the payback method?

**11** State the key advantages of the NPV method of investment appraisal.

**12** Define financial reporting.

**13** Why is an auditors' report necessary in published accounts?

**14** Name the three financial statements that must be included in published accounts.

**15** What use would a competitor have for a company's published accounts?

## Assignment questions

1 Choose a private-sector organisation. Research and assess the impact of the euro on the organisation's activities. Present your findings in a 2000-word report.

2 Choose a charity or public-sector organisation. Research and assess the impact of the minimum wage on the organisation's activities. Present your findings in a 2000-word report.

## References


1 Helm, T (1998) 'Rome and Bonn clear the EMU hurdle', *Daily Telegraph*, 28 February.
2 http://europa.eu.int/euro/en/talk/talk04.asp?nav=en.
3 http://europa.eu.int/euro/en/talk/talk02.asp?nav=en.
4 Ibid.
5 Manning, A (1997) 'If it's good enough for everyone else, it's good enough for us', *Independent on Sunday*, 11 May.
6 Ibid.
7 Atkinson, M (1997) 'Experts press case for £3.85 minimum wage', *Observer*, 26 October.
8 Manning, op. cit.
9 Atkinson, op. cit.
10 Manning, op. cit.
11 Ibid.
12 Ibid.
13 *Daily Telegraph*, 14 November 1997.
14 Brett, M (1995) *How to Read the Financial Pages*, 4th edn, London: Century.


## Further reading


Brett, M (1995) *How to Read the Financial Pages*, 4th edn, London: Century.

Byers, S, Groth, J C, Richards, J C and Wiley, M K (1997) 'Capital investment analysis for managers', *Management Decision*, 35 (3&4).

Chadwick, L (1996) *The Essence of Financial Accounting*, 2nd edn, Harlow: Prentice Hall.

Chadwick, L (1997) *The Essence of Management Accounting*, 2nd edn, Harlow: Prentice Hall.

Dyson, J R (1997) *Accounting for Non-Accounting Students*, 4th edn, London: Financial Times Pitman Publishing.

Garrett, A (1999) 'Ready steady euro', *Management Today*, January.

Hannagan, T (1998) *Management: Concepts and Practices*, 2nd edn, Chapter 16, London: Financial Times Pitman Publishing.

Reid, W and Myddelton, D R (1996) *The Meaning of Company Accounts*, 6th edn, Gower.

Taggart, J and Taggart, J (1999) 'International competitiveness and the single currency', *Business Strategy Review*, 10 (2), Summer.

Vernon, M (1998) 'Dangers of the short term view (of the euro)', *Financial Times*, 5 November.

Wood, F (1999) *Business Accounting: Volume 2*, 8th edn, London: Financial Times Pitman Publishing.

Wood, F (1999) *Business Accounting: Volume 1*, 8th edn, London: Financial Times Pitman Publishing.

# 9 Human resource management

## Organisational context model

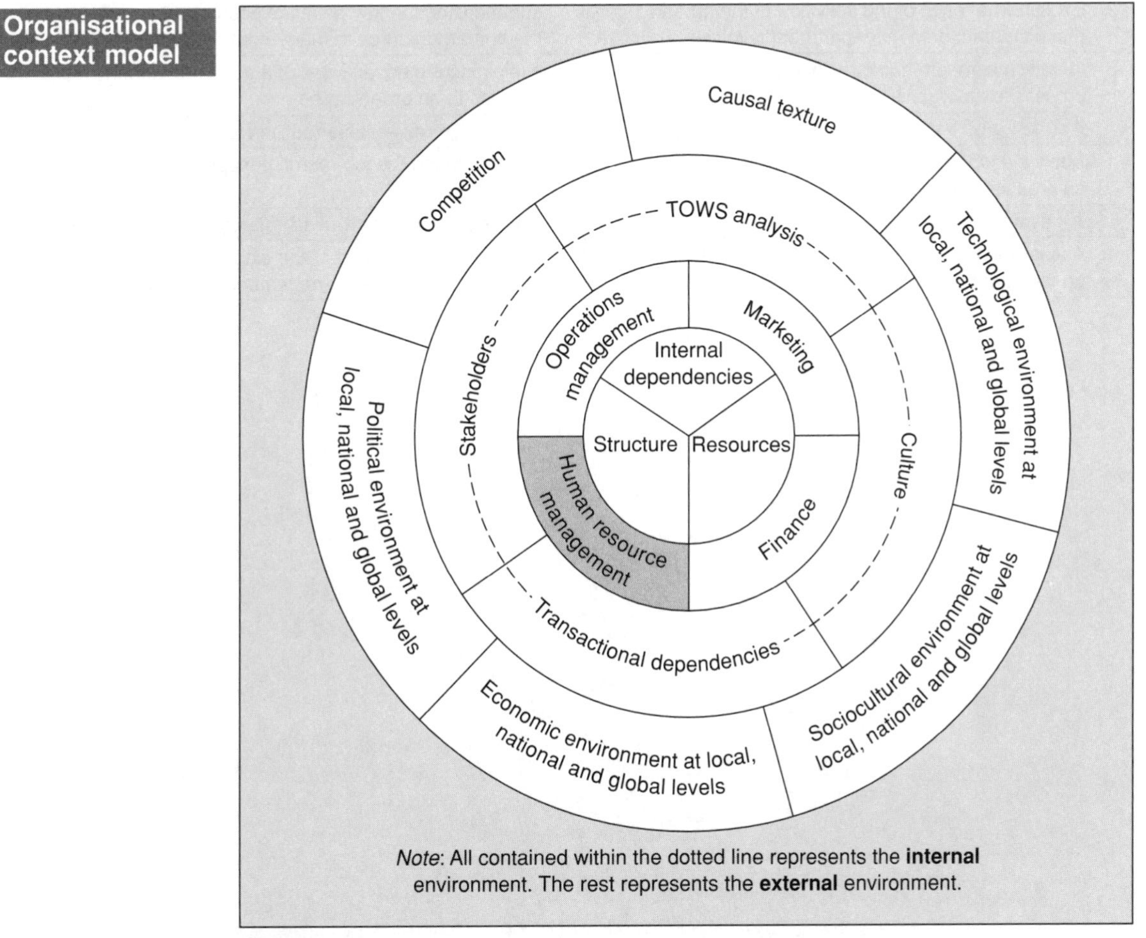


*Note*: All contained within the dotted line represents the **internal** environment. The rest represents the **external** environment.

**Exhibit 9.1**

## Learning outcomes

While reading this chapter and engaging in the activities, bear in mind that there are certain specific outcomes you should be aiming to achieve as set out in Exhibit 9.2 (overleaf).

**Exhibit 9.2 Learning outcomes**

| | Knowledge | Check you have achieved this by |
|---|---|---|
| 1 | Define human resource management | stating in one sentence what human resource management is |
| 2 | List the activities of the human resource management department | naming the tasks that the human resource management department performs |
| | **Comprehension** | **Check you have achieved this by** |
| 1 | Understand the organisational context of human resource management | identifying the role of the human resource management department in the organisation |
| 2 | Explain what each of the activities of the human resource management department involves | summarising the key points of each activity carried out by the human resource management department |
| 3 | Identify the importance of a sound job recruitment process to an organisation | explaining the advantages of a good recruitment process to an organisation |
| | **Application** | **Check you have achieved this by** |
| 1 | Apply the job recruitment process to a vacancy in an organisation with which you are familiar | demonstrating the job recruitment process using a job that you know |
| | **Analysis** | **Check you have achieved this by** |
| 1 | Assess the impact of events in the external environment on the human resource management department | anticipating the impact of the external environment on the human resource management department |

ENTRY CASE STUDY 9.1

FT

# Alternative to poaching

## How companies can collaborate to beat skills shortages

**By Vanessa Houlder**

Competition in the Scottish electronics industry for skilled technicians is famously intense and in the early 1990s recruitment problems were getting out of hand. 'We were basically poaching from each other and pushing salaries up. It was becoming a vicious circle,' says Morag McKelvie, a personnel manager at NEC Semiconductors.

The idea then that managers from competing companies would start working together to solve the skills shortage would have seemed incredible, she says. Russell Pryde, manager of Lothian and Edinburgh Enterprise (LEEL), one of Scotland's local enterprise companies, the equivalent of the Training and Enterprise Councils in England and Wales, agrees. 'I don't think they would have sat in the same room together.'

But in a striking turnround, a number of competitors have embarked on an unprecedented degree of co-operation on training over the past two years. Yesterday, their innovative approach was recognised when, together with a local college and LEEL, they received a special training award from the Institute of Personnel and Development at the National Training Awards, organised by the Department for Education and Employment.

The initiative dates back to mid-1993, when LEEL made a proposal to the main electronics companies in its area – NEC Semiconductors, Seagate Microelectrics and Motorola. It volunteered to put up £100 000 for a course to be run by West Lothian College to train people in the specific skills needed to maintain equipment – the main skills shortage faced by the companies. It envisaged that the students would be drawn from declining industries such as shipbuilding, who needed to be retrained.

The companies would choose the candidates who were awarded places on the 40-week course; employ them for a six-week placement during the course and have the first chance to recruit them at the end of the course. If they took on one of the students, they would pay LEEL £3000 to reimburse it for the cost of that individual's training.

Pryde admits that the scheme 'seemed hare-brained at first'. The perception was that foreign-owned companies investing in Scotland were used to being wooed with generous grants and would baulk at being asked to pay training costs in full.

In fact, however, the companies felt that the £3000 charge was good value, compared with the fees charged by recruitment agencies. They were enthusiastic about the chance to choose the students selected for the course. They also welcomed the opportunity to design the course. The skills required by the equipment technicians drew on both electronics and mechanics.

The first course began in November 1994. Since then, four have run, training about 50 technicians who ranged in age from early 20s to early 40s. About three-quarters of the total found jobs with sponsoring companies; a few others have found work with suppliers or other companies in the electronics industry.

On its own, the Lothian scheme can only address a small part of the skills shortages in British electronics. The UK's success in attracting semiconductor-related investment will mean much more training is needed to fill labour market gaps.

But LEEL participants believe that their course is a valuable model for others. 'It is a prime example of a training and enterprise council and a college reacting to industry's needs,' says McKelvie.

*Source*: *Financial Times*, 13 February 1997. Reprinted with permission.

## Introduction

Public- and private-sector organisations all employ people to enable customer needs to be met. Hence this chapter aims to explain the human resource management function of the organisational context model, Exhibit 9.1. This chapter focuses on human resource management at the national level, as global human resource management is covered in international business books, for example Daniels and Radebaugh.[1]

This chapter examines:

- the role of the human resource management function in organisations;
- the key legislation covering the recruitment and employment of human resources;
- demographics and the human resource management function;
- the flexible firm model;
- the employee recruitment process.

## Human resource management

Human resource management is an integrative general management activity that involves examining the organisation's demand for human resources with particular skills and abilities. This includes the recruitment of new staff along with the training and development of existing staff. Alternatively, if the organisation is over-staffed the issues of redundancy, retirement and non-replacement of staff who leave will all have to be considered.

Therefore human resource management considers the whole picture of staffing in the organisation and takes a strategic and operational view of human resource requirements. The operational role of human resource specialists includes them in offering guidance and support to enable other managers in the organisation to handle their role as managers of human resources. This has occurred because the activity of human resource management has become more decentralised and there is a growing need for a more flexible workforce. A flexible workforce and more decentralised human resource activities allow the organisation to respond more easily to influences and changes in the external environment.

In summary, the activity of human resource management includes recruitment and selection of appropriate staff, training and development of staff and the management of the employment relationship, which includes contracts, collective bargaining, reward systems and employee involvement.

### Human resource management and personnel management

Personnel management is a centralised and administrative function which focuses on the workforce. The activities covered by the personnel function include: recruitment; training; pay administration; dealing with employees work-related needs and problems; explanation of management expectations of the workforce; justification of management decisions; and negotiation regarding adaptation of management expectations and decisions if they are likely to produce a hostile reaction from the workforce.

In contrast HRM focuses on management needs for human resources to be utilised by the organisation. The main priority is planning, monitoring and adjusting the number and nature of the organisation's human resources. The issues relating to employees' work-related needs and management expectations of employees are discussed and dealt with as a management activity, with much

**Exhibit 9.3 Employment legislation**

| | |
|---|---|
| **Employment legislation** | • Industrial Relations Act 1971<br>• Trade Union and Labour Relations (Consolidation) Act 1992<br>• Trade Union Reform and Employment Rights Act 1993<br>• Employment Rights Act 1996<br>• Employment Rights (Dispute Resolution) Act 1998<br>• Employment Relations Act 1999 |
| **Health and safety at work legislation** | • Factories Act 1961<br>• Offices, Shops and Railway Premises Act 1963<br>• Fire Precautions Act 1971<br>• Health & Safety at Work Act 1974 |
| **Equalising employment legislation** | • Equal Pay Act 1970<br>• Sex Discrimination Acts 1975 and 1986<br>• Race Relations Act 1976<br>• Disability Discrimination Act 1995 |

less direct workforce involvement than in personnel management. The argument for this is that employees' interests are best served by effective overall management which takes account of their skills, abilities and career development needs.

In conclusion, personnel management is directed very much towards the personnel or workforce and the personnel department managing their work-related needs. In comparison, the HR department considers the strategic contribution that employees make to the organisation. Issues concerning the number of employees, the skills required and the cost to the organisation of employees with the required skills, are of prime importance to the HR department.

### External influences on human resource management

The human resource management function is influenced by the external environment. The main external influences currently include legislation, demographics and the move to a more adaptable workforce. Legislation that affects and influences the human resource management function falls into three categories, shown in Exhibit 9.3. Demographics and the flexible workforce are discussed later in this chapter. Sociocultural, economic and technological influences were discussed in Chapter 2.

## Employment legislation

The Contract of Employment Act 1963 specified, for the first time, that employees should receive the main terms and conditions of their employment in writing. This Act covered the formation, changes to and the ending of the contract of employment. The Industrial Relations Act 1971 allowed employees, again for the first time, to take their employer to an employment tribunal for unfair dismissal. Both Acts were assimilated in the Employment Protection (Consolidation) Act 1978. This Act put into place the condition that an employee must have been in full-time employment for two years or more with an employer before being able to go to an employment tribunal with a case of unfair dismissal.

The Conservatives were in power for a continuous period from 1979 to 1997. During the 1980s Margaret Thatcher and her government embarked on a methodical approach to legislation that reduced the power and rights of both the individual employee and the trade unions. The creation of an economic system that was increasingly free market was one of the aims of the Thatcher government. Consequently, it held the belief that the rights of the individual employee and unions, as established by the 1970s legislation, were obstructive to the establishment of a free-enterprise culture and job creation in the UK economy. Hence a significant amount of employment legislation was passed in the 1980s and some of the key acts are the Employment Acts of 1980, 1982 and 1988 and the Trade Union Act 1984. These four pieces of legislation all diminished the rights of individual employees and unions and expanded the legal regulation of industrial action and trade union activity. This meant there were changes to ballot requirements for calling a strike and nearly all types of secondary industrial action were outlawed. Secondary industrial action occurs when employees in one workplace take industrial action in support of employees in another workplace who are in dispute with their employer.

In the 1990s, it was clear that the Tory governments of Margaret Thatcher and later of John Major had won their battle with the trade unions. The Employment Act 1990 allowed those individuals refused employment due to their lack of membership of a trade union and hence of a closed shop, to take the union involved to an employment tribunal, further curtailing the power of the trade unions. This Act also made all forms of secondary industrial action unlawful. The Trade Union and Labour Relations (Consolidation) Act 1992 and the Employment Rights Act 1996 (as amended by the Trade Union Reform and Employment Rights Act 1993 and the Employment Relations Act 1999) consolidated and replaced previous employment legislation.

## Health and safety at work legislation

There are four key Health and Safety at Work Acts, outlined below. These deal with the health and safety of employees at work and were all passed between 1961 and 1974. These four key acts were updated in the 1990s by EU Directives. These Directives include regulations covering the provision of rest and no-smoking areas; minimum standards for safe use of machines and equipment; provision and use of protective equipment; lifting heavy loads and provision of lifting equipment; provision of free eye tests, glasses, regular breaks and training for employees using visual display monitors; and increasing health safeguards for employees working with carcinogenic substances. Other health and safety regulations include the Control of Substances Hazardous to Health Regulations 1988, introduced under the 1974 Health and Safety at Work Act. The regulations place a requirement on employers to pay close attention to the manner and the extent to which substances hazardous to health are handled, used and controlled.

### Factories Act 1961 and Offices, Shops and Railway Premises Act 1963

The Factories Act 1961 lays down minimum standards in factories regarding cleanliness, workspace, temperature, ventilation, lighting, toilet facilities, clothing, accommodation and first aid facilities. The Offices, Shops and Railway Premises Act 1963 extends the same general cover as outlined in the 1961 Act to other work premises. However, there are some small differences in the 1963 Act covering minimum workspace provision and temperature.

### Fire Precautions Act 1971

Workplace premises require a fire certificate and this Act allows the fire authorities to impose conditions on the certificate holder, including closure of a building if it does not comply with the conditions. These conditions can cover the means of escape and exit from the premises; instruction and training of employees on how to react in the event of a fire; and limits on the number of people on the premises at any one time.

### Health and Safety at Work Act 1974

The Health and Safety at Work Act 1974 is a comprehensive piece of legislation covering people at work and those who may be at risk from the activities of those engaged in work. It created the Health and Safety Commission, whose six to nine members are selected by the Secretary of State to represent employers, employees and local authorities. The Health and Safety Commission has to ensure that the Act is correctly implemented and hence one of its roles is to provide advice to local authorities, employers and others to enable them to understand their responsibilities under the Act. The Health and Safety Commission also has the power to carry out investigations and enquiries if it appears that the Act is not being adhered to in a workplace and the Commission may also issue codes of best practice and regulation.

The Health and Safety Commission and the Secretary of State select three people to form the Health and Safety Executive. This assists in the enforcement of the Health and Safety at Work Act 1974 by undertaking the daily administrative affairs relating to its implementation.

## Equalising employment legislation

### Equal Pay Act 1970

The Equal Pay Act 1970 came into force on 29 December 1975 and promotes equal pay for men and women. It stipulates that a man and woman should receive equal pay if either of the following types of circumstances applies:

- a man and a woman doing like work, for example a male shop assistant and a female shop assistant, would clearly both be entitled to the same pay;
- a woman can show that she is carrying out work rated as equivalent to that of a man, under the organisation's job evaluation scheme. The woman may

have a clerical post in the organisation and the man a technical job; if both jobs score the same points under the job evaluation scheme, then the woman can claim pay equal to that of the man.

Proceedings against the British government by the European Commission resulted in the Equal Pay (Amendment) Regulation 1983, which came into force in 1984. This makes it possible to claim equal pay for work that is considered to be of equal value to that done by a member of the opposite sex. Equal value considers the skills, effort and decision-making responsibility involved in carrying out the job and requires no job evaluation scheme to be in operation.

In July 1995 the Conservative government's policy of compulsory competitive tendering was significantly undermined when the Law Lords ruled in favour of 1300 school meals staff employed by North Yorkshire Council. The council had sacked the school meals staff, then re-employed them and reduced their conditions of service by reducing pay from £3.40 to £3.00 per hour, cutting holiday entitlement and abolishing the sick pay scheme. The council carried out these cuts in order to defeat an outside tender for the school meals service. It did so regardless of a 1988 job evaluation scheme that found the work of the school meals staff to be of equal value to that of road sweepers, gardeners and refuse collectors, who were predominantly male. This meant that the school meals staff were discriminated against under the equal pay legislation and entitled to more than £2m in lost pay and damages.[2]

The 1995 judgment by the Law Lords set a precedent and in July 1996 nearly 2000 school meals staff employed by the old Cleveland County Council won more than £1m in lost wages and damages at an industrial tribunal. The compensation was for a pay cut introduced to aid in the success of Cleveland County Council in compulsory competitive tendering against private firms.[3]

## Race and sex discrimination

There are two types of race and sex discrimination, direct and indirect. Direct discrimination occurs when workers of a particular sex, race or ethnic origin are treated less favourably than other workers, for example specifying a female secretary in a job advert when the job could be done equally well by a man. Indirect discrimination arises when a requirement is applied equally to all workers, but it is more difficult for one group of workers by virtue of their sex, race or ethnic origin to comply. For instance, a job advert requiring perfect written English for a manual labourer would discriminate against people of race, ethnic origin or nationality where English is not a first language.

### Sex Discrimination Act 1975

The Sex Discrimination Act 1975 came into force at the same time as the Equal Pay Act 1970, at the end of 1975. This allowed organisations five years to achieve parity of pay between the sexes. The Sex Discrimination Act 1975 promotes the equal treatment of men and women in the areas of recruitment, training, promotion, benefits and dismissal. It also established the Equal Opportunities Commission, whose role is to eliminate discrimination on grounds of sex or marital status; promote equal opportunities between men and women; and monitor implementation of the Equal Pay Act 1970 and the Sex Discrimination Act 1975.

If a genuine occupational qualification applies to the post, then discrimination is allowed. Vacancies in the field of modelling or acting have a genuine occupational qualification, for example a woman to model female clothes.

### Race Relations Act 1976

The Race Relations Act 1976 is similar to the Sex Discrimination Act 1975. It makes it unlawful to discriminate against someone on grounds of their race, colour, nationality or ethnic origin and established the Commission for Racial Equality.

The exceptions to the Race Relations Act are similar to those for the Sex Discrimination Act 1975 and include the areas of entertainment, acting, artistic or photographic modelling, specialised restaurants and community or social workers providing personal services to members of particular racial group. However, discriminatory treatment in terms and conditions of employment is not allowed.

## Examples of race and sex discrimination

The existence of race and sex discrimination legislation does not guarantee a complete lack of discrimination in the workplace and it is the role of the Commission for Racial Equality and Equal Opportunities Commission to monitor and ensure effective implementation of the legislation.

In recent years clear examples of race discrimination have been found in the armed forces and the legal profession. For example, in December 1996 the Government Legal Service, which employs solicitors and barristers to work in government departments, was found guilty of breaking the laws relating to both race and sex discrimination with regard to its recruitment procedures. It was taken to an industrial tribunal by a black woman, Chineme Nwoke, who was supported by the Commission for Racial Equality.[4] When she was interviewed for a job by the Government Legal Service she was given a very low grade, which effectively precluded her from applying to the organisation again. She won her case and was awarded £2000 for injury to feelings and a very unusual exemplary payment of £1000 for 'aggravated damages'. In its ruling the tribunal reported that while Chineme Nwoke did not do sufficiently well to be recommended for appointment, her application was not sufficiently unsatisfactory to permit her exclusion from future shortlisting for interview. The tribunal also pointed out in its judgment that during the recruitment process for the Government Legal Service:

- all white candidates with local government experience received a C grade or above, while no applicant from the ethnic minorities with a corresponding background achieved such a rating;
- eight white candidates with a 2.2 degree or lower received a C grade, but no one from the ethnic minorities with the same qualification received such a grade;
- 2.4 per cent of black and Asian candidates gained a C grade or above, compared with 7.6 per cent of white men and 11.2 per cent of white women. Despite women gaining higher grades, they were less likely to be recommended for a vacancy and if recruited were paid less than men;
- 50 per cent of black and Asian shortlisted candidates were finally appointed, compared with 76.2 per cent of white men and 63 per cent of white women.[5]

Discrimination in the armed forces was highlighted in March 1997 when an independent report commissioned by the Ministry of Defence was published.[6] This report detailed widespread incidents of racial harassment in the armed forces. The existence of racial harassment and the low number of ethnic minorities in the armed forces, 1.4 per cent of personnel, led the Ministry of Defence to aim to improve its recruitment, training and appraisal systems to remove any racial discrimination and bias. This also includes the need to train and educate existing staff in the seriousness of racial inequality and the handling of racial issues. The report also featured the example of the Navy, where good statements and policies on equal opportunities exist, but implementation is unplanned, uncoordinated and unmonitored.

#### Disability Discrimination Act 1995

The Disability Discrimination Act 1995 defines disability as:

> a physical or mental impairment which has a substantial and long term adverse effect on a person's ability to carry out normal day to day activities. People who have had a disability, but no longer have one, are covered by the Act.[7]

Discrimination happens when a disabled person is dealt with less favourably than an able-bodied person due to the disabled person's disability and this cannot be justified. Employers are not allowed to justify unfavourable treatment of a disabled person if adjustment could be made to overcome the reasons for unfavourable treatment. For example, an employer could not refuse to promote a disabled person because the equipment they had to use could not be accessed, if by rearranging the equipment the disabled person could do the job.

Organisations employing more than 20 people must not discriminate against disabled employees in the areas of recruitment, working conditions or employee benefits – bonuses, promotion, training and dismissal. This covers permanent members of staff, temporary workers and contract workers. However, it does not cover the armed services, police officers, active members of a fire brigade and prison officers, or people working on ships, aircraft or hovercrafts.

The advertising of jobs must not indicate that a person might not get a job because of their disability or because the employer is unwilling to make adjustments. Selection process must not favour people who have no disability, offer jobs on less favourable terms to a disabled person, or refuse employment because of disability.

## Demographics and human resource management

Demography is the statistical study of changes in the nature of a population. The USA, western Europe and Japan all experienced a growth in their working-age population from 1950–80. In contrast, all expect to experience a decline in their working-age population between 2010 and 2050. In 1990 a quarter of the working-age population in OECD countries was less than 25 years old and about a third was 45–64 years old. It is anticipated that by 2040 less than a fifth of the working-age population will be under 25 years old and over two-fifths will be above 45 years old. This is significant as younger employees are viewed as more adaptable, with better and more up-to-date training. Moreover, not all people of working age choose to work: some will be in full-time higher or further education and others, mostly women, may choose to stay at home to care for young children or sick relatives.[8]

From the mid-1970s through to the mid-1990s early retirement was an accepted method of reducing the size of the workforce.[9] Older employees are seen as being more expensive and less flexible. The notion of early retirement has been particularly prevalent in the public sector. The combination of early retirement and reduced numbers of young people in the population means that there is a likelihood of labour shortages in western Europe, North America and Japan in the early part of the twenty-first century, from 2000 to 2020. The likely response of organisations to this is the active recruitment of older staff to do part-time and temporary jobs traditionally done by younger people.

In the UK the supermarket chain Tesco and the DIY chain B&Q are examples of companies whose employment policies include the active recruitment of older staff. In the future, retaining and retraining older employees will be a more appealing option for organisations as the competition for skilled younger employees will be increasingly fierce.[10, 11] The advantages of an older workforce include the tendency of older people to change jobs and move around the country less frequently than younger people, hence contributing to lower labour turnover.

To attract and retain the required level of skilled labour, lifetime or long-term employment contracts may have to be offered. This will contrast strongly with the period from the 1980s to the early part of the twenty-first century when flexible labour contracts[12] and workforces fashioned on the flexible firm model[13] were more common.

## The flexible workforce and human resource management

In the late twentieth century, the advent of the flexible workforce owed much to the popularity of downsizing in organisations and the general shift in the economy from manufacturing to service provision. A manufacturer can either employ extra staff to meet peak demand or stock goods produced in a quiet time to meet demand in peak times. However, a service provider is unable to stock products to meet peaks in demand and therefore meets that demand by employing extra part-time or temporary staff. This is less expensive than employing a larger number of permanent full-time staff, as during troughs in demand the extra full-time permanent staff would be inactive. Therefore downsizing occurs when organisations delayer, removing layers of employees, so that they become structurally flatter, or when organisations reduce the number of core employees and recruit more part-time or subcontracted employees as and when required. The concept of the flexible workforce means organisations operating flexible working patterns to cope with peaks and troughs in demand by redeploying staff across different activities and locations or employing and laying off staff as cheaply as possible.

Classic examples of flexibility in the labour market include retailers taking on extra staff over Christmas; frozen vegetable companies, like Bird's Eye, taking on extra staff during summer months when the vegetables are harvested; restaurants or pubs taking on extra staff on Friday and Saturday evenings when they are busiest; and universities taking on part-time lecturers during an academic year when student over-recruitment has occurred.

The idea of an organisation and its flexible workforce is best summed up by the flexible firm model.[14] This divides employees into three categories: core, peripheral and external. The core permanent employees have highly skilled jobs, with relatively good job security and career prospects. These employees are expected to be flexible in terms of their role in the organisation and working location.

The first peripheral group are those employees with full-time jobs, not careers. In this group labour turnover is higher than for core employees. Employees in this group often require more vocational skills than core employees. They include supervisors, secretaries, assemblers and administrators.

The second peripheral group provides the major component of flexibility in the workforce and includes employees on short-term contracts, part-time employees, job-sharing employees and subsidised trainees. Employees on short-term contracts will hold full-time skilled jobs, for example a software engineer may be employed on a short-term contract to work full time on developing a piece of software for an organisation. In contrast, part-time employees work on a permanent basis for the organisation, but only for a fraction of the hours of a full-time employee. For example, a part-time employee may only work mornings or certain days of the week. Job sharing occurs when two people split a job, the pay and rewards between them. This type of arrangement is common among women who wish to work full time but not for the full working week. Subsidised trainees work for the organisation while learning a trade and/or gaining qualifications. A portion of the costs are covered by government funding. A good example of a subsidised training initiative is the Modern Apprenticeships Scheme. Launched in September 1995, a modern apprenticeship takes three years to complete and involves the apprentice studying for a National Vocational Qualification in an area of craft, technical or junior management skills. The government covers around 25 per cent of the cost to the organisation of employing and training the apprentice.[15]

The final category of staff in the flexible firm model is external staff. They can be brought in quickly to meet increased demand and include self-employed consultants, subcontractors and temporary agency staff. The latter may include secretarial staff, administrators and supervisors. The common subcontracted activities are those that are non-core and can be done more cheaply and satisfactorily by contracted staff, such as cleaning, catering, provision of IT support and running the payroll. In contrast, self-employed consultants are used on a project-by-project basis, particularly when the necessary skills are not available in-house.

Organisations have been able to develop and create further flexibility in the workforce by the use of technology. The arrival of faxes, mobile phones, laptop computers and modems has resulted in employees being able to work anywhere, in hotels, on trains, planes or at home. The ultimate use of technology is in the activity of teleworking or telecommuting. Telecommuters work at home, with technology that enables them to receive and send work and messages to their employing organisation. The potential advantages of telecommuting are that it saves commuting time, especially in the South East of England where traffic congestion is common; environmental benefits, such as reducing the 19.8bn gallons of exhaust fumes produced per day by cars in the UK; saving money, as estimates suggest that homeworkers spend less money on working clothes and there is no cost of commuting, resulting in a real pay rate 50 per cent higher than if commuting; and savings to companies estimated at between £1500 and £3000 per employee per year.

The potential disadvantages of telecommuting include homeworkers feeling isolated, lonely, overworked and neglected by their work colleagues and the organisation. The danger exists that home and the office are the same place and work takes over the home environment, hence home is no longer an escape from work. The telecommuter also loses the companionship and social side of office life, so augmenting feelings of isolation, loneliness and neglect. The potential benefit of being able to work at any time of the day or night may not materialise for telecommuters as their clients may only work from nine to five. Those involved with high-volume, low-margin work may be reluctant to take time off as they fear loss of business and earnings, leading to feelings of overwork.

The company DEC has developed a home-working scheme that involves employees working from home but includes regular contact with other telecommuters via technology and occasionally in person at an office or telecentre, into which telecommuters may call to meet other telecommuters. This provides a social focus for office life and helps to combat feelings of isolation, loneliness and neglect among teleworkers.[16]

## The employee recruitment process

The recruitment of employees to an organisation is crucial if the organisation is to acquire, retain and maintain the skills and abilities to provide customers with the products and services required in an efficient and effective manner. This also contributes directly to helping the organisation achieve its strategic goals. The development and use of a suitable recruitment procedure allow appropriate applicants to be matched to suitable posts. This should result in employees remaining with the organisation and making efficient and effective contributions to its goals.

The lack of a suitable employee recruitment process or poor implementation of the process may result in initial low recruitment costs. However, a lack of forethought and planning is false economy, resulting in unsuitable employees being recruited. The appointment of over- or under-skilled and qualified employees leads to tasks and activities not being done effectively and employees leaving their jobs relatively soon after taking up employment. This rapid turnover of employees results in the recruitment process being repeated to find a replacement employee and incurring the associated costs. Costs are incurred in the areas of defining the job and the type of applicant, reading and shortlisting completed applications, and interviewing shortlisted applicants. There will also be the costs of any advertising undertaken and employment agencies used.

Rapid turnover of employees and associated high recruitment costs can result from a shortage of appropriately skilled labour. The electronics companies in the entry case study for this chapter experienced a high turnover of skilled technicians due to a shortage of people with suitable skills in the labour market. They bore the costs normally associated with high labour turnover and paying increasing salaries to attract the required skilled technicians from competitor companies. These skilled technicians normally remained with their new employer for a relatively short period, before moving on to another company offering higher pay and more perks. The electronics companies confronted the shortage of skilled technicians by collaborating with each other and with a local college to fund and develop a training course that produces technicians equipped with the required skills and abilities. The manner in which they sought to deal with the skills shortages is an excellent example of organisations managing, influencing and changing a key aspect of their external environment.

The shortage of skills can be addressed on a number of fronts: the recruitment of employees; the use of training and development; the redeployment of existing staff; and work patterns and practices.

The recruitment of employees is one area to consider and build on in attempt to ease a skills shortage. The advertising of vacancies can be boosted by widening the geographic area of search and increasing the range of advertising media employed. In addition, the advertising effort could be directed towards non-traditional groups of potential employees, for example the nursing profession targeting men; the engineering or construction professions targeting women; or, as B&Q has done, targeting people over fifty to do jobs traditionally done by younger people. Other tactics involve requiring successful applicants recruited in one part of the country to relocate to the organisation's premises in another area where a skills shortage exists. The direct approach of offering higher pay and more perks is always available, but expensive. However, the danger of this lies in the vicious circle that can be created when organisations poach skilled staff from one another and drive pay ever upwards, as in the entry case study. The lowering of entry requirements, such as qualifications and relevant work experience, is another means of reducing a shortage of skills. For example, the Army has on occasion lowered the standard of physical fitness required of recruits and extended the basic training to allow new recruits to attain an acceptable level of physical fitness. Alternatively, for some organisations the use of consultants, subcontractors and agencies could be initiated or extended to reduce a skills shortage.

Another way of handling a shortage of skills is to train existing employees up to the required skills level. This will help diminish the skills shortage, although training and acquiring skills can take time and this is not a short-term method. However, if the training is ongoing and developmental, it may well help in the retention and increased motivation of existing employees. The hazard of providing appropriate and good training is that well-trained and qualified employees leave for a better job elsewhere, with more pay and perks. One way of stemming a loss of recently skilled employees is to offer opportunities and promotion within the organisation.

An alternative to closing the skills gap by increasing the skills available in the organisation is to reduce the need for skills by altering work practices to accommodate existing skill levels. This includes reducing output, improving productivity, overtime working and adjusting shift patterns, which is particularly practical if it leads to increased productivity and reduced overtime. The skills gap could also be reduced by restructuring and reorganising the workforce and their jobs, which may entail multi-skilling, which in turn may necessitate further training.

Another option for dealing with a lack of skills is to find staff with the required skills from among those already in the organisation and second or promote them into the positions where the skills shortages exist for a temporary or permanent period.

The four stages of the employee recruitment process are looked at in theory (*see* Exhibit 9.4) and in relation to a city-centre café/bar recruiting part-time staff.

**Exhibit 9.4 Employee recruitment process**

| | |
|---|---|
| Stage 1 | **Assessment of the job**<br>• Job analysis<br>• Job description |
| Stage 2 | **Assessment of the type of applicant required**<br>• Person specification<br>• Key results areas |
| Stage 3 | **Attracting applicants**<br>• Placement of the advertisement<br>• The advertisement |
| Stage 4 | **Assessing applicants**<br>• Assessment of application forms<br>• Assessment of applicants |

## Stage 1 – Assessment of the job

### Job analysis

This is the first step in the employee recruitment process and involves gaining the correct information relating to the vacant job, as this will allow an accurate job description and person specification to be drawn up. Job analysis normally starts when the current job holder hands in their notice. Information has to be collected in order that the job description for the vacant job is clear and up to date. This is crucial, as the job description and the person specification drawn up from it will be central to the employee recruitment process.

The information collected should cover the areas of the job and its position in the organisation; the tasks and activities the job involves; the responsibilities of the job; and the conditions under which the job is carried out. The current job description is a good initial source of information. Further and more up-to-date information can be gathered by interviewing the current job holder before they leave the organisation and key personnel who surround the vacant job. These key personnel include the manager or supervisor of the current job holder and the latter's peers and subordinates. The utilisation of these sources of information gives a good indication of the tasks that the job involves and the key reporting and working relationships that the job holder must maintain. This information, along with consideration of the organisation's current and future plans and strategy, permits an up-to-date and accurate job description to be compiled.

### Job description

The job description defines and outlines the job and covers the areas of the job shown in Exhibit 9.5. In the sections of the job description and personnel specification, all areas are examined and related to the part-time job of a server in a city-centre café-bar, the type of part-time job a full-time student might seek to supplement their grant and student loan.

#### Identification of the job

This section of the job description sets the job in its organisational context by stating the job title and the location of the job in the organisation. The

**Exhibit 9.5**
**Job description**

| | |
|---|---|
| 1 | Identification of the job |
| 2 | Summary of the job |
| 3 | Content of the job |
| 4 | Working conditions |
| 5 | Performance standards |

location is defined by stating which section, department or division the job is situated in and where it will be based. This is important if the organisation has several sites in different geographic locations. The role of the job should also be put in context by outlining which staff the new appointee will be responsible to and how many staff the appointee will be responsible for. Staff with whom the new appointee will be expected to work and liaise, both inside and outside the department, are also to be indicated. The city-centre café-bar seeking to recruit part-time staff must identify which position part-time staff are going to occupy. Is the vacancy for a part-time relief manager, part-time bar/serving staff or part-time kitchen staff?

### Summary of the job

This segment of the job description examines why the job is done. This is accomplished by stating the overall objectives and purpose of the job, as well as comparing and contrasting it with other jobs in the organisation that are close or similar to it. The resulting differences between the vacant job and any similar jobs must be clearly demonstrated. Finally, the overall objectives and purpose of the job are linked with the overall objectives of the department and organisation. The café-bar recruiting part-time bar/serving staff may summarise such a job as 'welcoming customers and ensuring they receive friendly and accurate service'.

### Content of the job

This portion of the job description depicts the content of the job in detail and indicates what is done and how it is done. In writing this, the objectives of the job and why the job is done are also considered closely (*see* previous section).

It is necessary to specify the tasks and activities constituting the job. The main tasks and activities are listed first, followed by those that are secondary. This allows outsiders to the organisation, such as potential applicants, to understand what is involved for the job holder. The people and equipment available and required for carrying out the job tasks and activities are also indicated in this portion of the job description. Issues to consider here are:

- Whether the individual does the job single handed, as part of a team, or perhaps the job requires a mixture of individual and teamwork (the café-bar server is required to work as part of a team).

- Whether the job requires physical strength, intelligence, application of individual judgement or a combination of these traits (the café-bar server needs to be of average intelligence).
- Whether the organisation provides resources and equipment to enable the job holder to carry out the required tasks and activities (the café-bar server is provided with a uniform of black trousers and purple shirt, which must be kept clean and always be worn on duty).

Resources provided by the organisation could include access to office equipment such as computers, photocopiers, telephones, faxes and e-mail. Other resources made available to employees to enable them to perform their jobs may include portable personal computers, mobile phones, corporate uniforms, protective clothing, company cars, hand tools, specialised technical equipment and interest-free loans for season tickets.

If the job is for part-time bar/serving staff, the tasks and activities involved would include showing customers to their table, taking orders for food and drinks, serving food and drinks, laying tables, clearing tables, collecting payment, ensuring the customer is happy with the food and drink they have ordered and serving behind the bar. Such part-time bar/serving staff would be responsible to the manager and have to work with the kitchen staff in ensuring that correct orders for food are taken, passed on to kitchen staff and served to the right customer.

#### Working conditions

This category of the job description looks at the working environment and circumstances of the job. The physical working environment should be reliably portrayed, e.g. a noisy factory, a clean, hygienic factory or an open-plan office. The hours of work and circumstances in which the job is carried out should also be accurately portrayed, particularly if they are unusual, for example the job may require the job holder to work shifts or travel away from home regularly. The server in a café-bar faces the working conditions of a trendy city-centre café-bar catering for all types of customers. The job is part time, the hours being 12–5pm Wednesday afternoon and Thursday and Friday evenings, which are very busy, from 6–11.30pm.

#### Performance standards

This part of the job description gives an indication of the normal level of performance or productivity expected from the job holder. This can be expressed in terms of number of hours of work and in terms of meeting the objectives of the job as laid out in the summary of the job. Examples of performance standards for different jobs, including a part-time server in a city-centre café-bar, are shown in Exhibit 9.6.

The job description serves as the basis for the next logical stage in the job recruitment process, the production of a person specification. A person specification is a series of criteria outlining the ideal person for the job. All applicants for the job will be assessed against these criteria.

**Exhibit 9.6 Performance standards for different jobs**

| | |
|---|---|
| **University lecturer in marketing** | • 450 hours' class contact per year<br>• Organise and run successful open days to attract potential applicants to the business school<br>• Four publications per academic year |
| **Shop assistant in newsagents** | • 37.5 hours' work per week<br>• Organise correct delivery of morning papers<br>• Pleasant appearance and personality |
| **Computer salesperson** | • Generate £15 000 sales per month<br>• Provide relevant customer demonstrations on request<br>• Clean driving licence |
| **Part-time server in café-bar** | • 15 hours per week, 1–6pm on Wednesday, 6–11pm on Thursday and Friday<br>• Clean, neat and tidy appearance<br>• Provide friendly and accurate service to customers |

## Stage 2 – Assessment of type of applicant required

### Person specification

The person specification is derived from the job description by translating the job activities into the specific skills and abilities required to perform the job effectively. Hence the employee recruitment process needs to ensure that the organisation fills vacancies by attracting, recruiting and retaining ideal candidates, who possess the skills and abilities required.

The person specification can be drawn up by using a predetermined framework. Frequently used frameworks include Rodger's seven-point plan and Fraser's five-fold grading system. Both these frameworks are used to draw up criteria that can be employed to assess applicants' suitability for employment; *see* Exhibits 9.7 and 9.8 respectively. The criteria drawn up should be identified as either essential or desirable. The essential criteria are those that applicants must meet if they are to be considered competent to carry out the job. However,

**Exhibit 9.7 Rodger's seven-point plan**

| | |
|---|---|
| 1 | Physical make-up |
| 2 | Attainment |
| 3 | General intelligence |
| 4 | Special aptitudes |
| 5 | Interests |
| 6 | Disposition |
| 7 | Circumstances |

*Source*: Torrington, D and Hill, L (1995) *Personnel Management: HRM in Action* (3rd edn), Harlow: Prentice Hall.

**Exhibit 9.8 Fraser's five-fold grading**

| | |
|---|---|
| 1 | Impact on others |
| 2 | Qualifications or acquired knowledge |
| 3 | Innate abilities |
| 4 | Motivation |
| 5 | Adjustment or emotional balance |

in addition to meeting essential criteria, applicants are usually expected to meet some, but not all, desirable criteria.

The benefits which accrue from the use of such criteria are that the areas that are important to the recruitment decision are clearly defined before the recruitment process takes place, hence ensuring that they are covered; that the areas to be covered in the interview are divided into focused chunks that can form the basis for the main part of the interview; and that there is more consistency in the recruitment process and thus a greater likelihood that all candidates will be treated and assessed fairly.

## Rodger's seven-point plan

### Physical make-up

The physical requirements for effective performance in a job may cover the areas of general health, physical fitness, appearance, manner and voice. For example, the emergency services and armed forces require higher than average standards of physical fitness, as do the jobs of physical education teacher and aerobics instructor. Positions in organisations that involve employees in making an impact on others, clients or members of the public usually require them to be of smart appearance and possessing a pleasant disposition. Alternatively, if the job is in a national call centre and involves frequent contact with clients or customers over the telephone, then a pleasant voice will be important. In the case of the student seeking a job as a server in a café-bar, a neat, clean and tidy appearance will be important.

### Attainment

This deals with the level of education and experience required for the job to be carried out successfully. The level of education is assessed by considering the qualifications that an applicant has gained. Many jobs will have minimum qualification expectations of applicants. The qualifications sought for particular jobs may include some of the following: minimum number of GCSEs (or equivalents), including English and Mathematics; A-levels (or equivalents); a degree; a postgraduate qualification; professional, technical or vocational qualifications. For example, the post of marketing assistant in a publishing company specialising in scientific books and journals may require someone with a first degree in science, such as chemistry, biology or physics, and a postgraduate marketing qualification.

The experience required for a job relates to the previous type of employment that suitable applicants are likely to have held. Senior vacancies in most fields of employment will require significant relevant work experience in a similar or related job, whereas posts for new graduates require much less relevant work experience, with the experience limited to that which a new graduate could have acquired via work placement or holiday employment. For example, the post of European sales manager for a chemical company may require a chemistry graduate with a proven track record in European sales and at least eight years' sales experience. In contrast, a vacancy for a graduate trainee systems analyst in one of the main high-street banks may specify a computing or mathematics graduate with six or twelve months' commercial experience in systems analysis in a banking environment, gained via a placement or holiday employment.

The student seeking a job in a café-bar as a server is likely to be considered for the job as they will have a good number of GCSEs and A-levels (or equivalent), which are the qualifications to get into university or college. The manager of the café-bar may also look for part-time staff who have previous experience in a similar type of job. This may include working in a bar or restaurant, or a job where the applicant has to deal with members of the public, e.g. a Saturday job in a shop.

### General intelligence

Applicants with an appropriate level of general intelligence will be required if a job is to be undertaken properly, hence the relevant level of general intelligence needs to be looked for in applicants. Jobs requiring complicated work patterns and activities to be picked up quickly will demand a different level of general intelligence when compared with jobs that are repetitive and very routine. Therefore students seeking a job as a server in a café-bar are likely to be considered because they are expected to possess a good level of general intelligence, which is demonstrated by the fact that they are at university.

### Special aptitudes

Special aptitudes are knowledge and skills that are vital for effective performance in the job. Ideally, applicants who already possess or have the propensity to acquire the necessary special aptitudes should be sought. The acquisition of essential knowledge and skills could simply require an applicant to be prepared to adapt or update existing skills and knowledge. The types of knowledge and skills covered by this section of Rodger's seven-point plan are presentation skills, interpersonal skills, telephone skills, numeracy skills, report writing skills, information technology skills and knowledge, and specific job-related knowledge.

The manager of the café-bar may seek part-time employees who already possess some relevant skills. These skills could include waiting on tables; changing a barrel or optic; and working a computerised till.

### Interests

The issue being examined in this section of Rodger's seven-point plan is the requirement of the applicant to have any out-of-work activities that support the application for employment. For example, a person playing in a weekend football team could be viewed as more predisposed to being an active team participant who gets on well with colleagues at work.

The manager of the café-bar may look for students who have experience of teamwork from either a sporting or social activity or as part of their studies at university or college. Students undertaking a business studies or catering course at university or college may use this fact to support their application to work part-time in the café-bar.

#### Disposition

The personal characteristics relevant to the job are considered in this section. If the job requires the successful applicant to work as a member of a team, then the ability to work well with others will be important. Alternatively, if the job is a supervisory or managerial position, then leadership and responsibility will be integral. Other personal characteristics include the ability to cope with pressure and meet tight deadlines and the competence to work on one's own initiative.

The part-time server in the café-bar should be able to get on with customers, work as part of a team and cope with the pressure of a very full bar on a Thursday and Friday evening. Wednesday afternoons are less busy and involve less bar work, but more waiting on tables for people ordering lunch, teas and coffees.

#### Circumstances

In this section the applicant's ability to conform to the circumstances in which the job has to be undertaken are considered. The circumstances may include shift work, weekend work, working away from home or being on call during evenings, weekends and public holidays. Examples include bus and train drivers working shifts, security guards working during the night and hospital consultants being on call at weekends.

The part-time server in the café-bar has to be able to work Wednesday afternoon 12–5pm and Thursday and Friday evenings 6–11.30pm.

### Key results areas

The person specification defines the ideal candidate for the job, but it is becoming more common also to define the key results areas expected of the ideal candidate. Therefore the key results areas are assembled at the same time as the person specification. They declare the important results expected of the job holder. Key results are the outputs and outcomes produced by the job holder and are assessed by use of explicit success criteria. Success criteria express the expected outcomes and outputs in terms of quality, quantity, cost and time. The use of key results areas provides clear goals for the job holder, with a strong emphasis on outputs and expected results, hence providing a clear basis for appraisal of the job holder.

## Stage 3 – Attracting applicants

**Placement of the advertisement**

Organisations have to be able to attract potential employees with the required qualifications, work experience, aptitudes and disposition for the vacancy being filled. The success of the organisation in attracting appropriate potential employees depends in part on its selection and use of suitable advertising media and third-party recruitment bodies.

### Direct advertising

The interests of the organisation lie in placing advertisements in publications that will circulate among the largest number of suitable potential applicants. The type of publication used will depend on the vacancy being filled. A vacancy for a shop assistant or office cleaner is best advertised in the local press, as these are jobs for which people are unlikely to move into the locality from a long way away. In contrast, suitably qualified applicants for the vacancy of a marketing manager for a multinational company or a university professor are more likely to be attracted from a larger geographic area and therefore such vacancies are advertised at least in the national press or even in international or overseas press.

The local press in the UK varies depending on the geographic area. For instance, the area of Sheffield and South Yorkshire is served by several local papers including the *Sheffield Star*, the *Yorkshire Post* and the *Barnsley Chronicle*. Large cities also have local papers, many of which are published as evening papers, including the *Evening Times* in Glasgow and the *Evening Standard* in London. The café-bar seeking serving staff would be most likely to use a local paper.

The UK has a number of national broadsheet newspapers published either daily during the week or on a Sunday. Sunday broadsheet newspapers include the *Independent on Sunday*, the *Observer*, the *Sunday Telegraph* and the *Sunday Times*. These Sunday broadsheets advertise senior jobs in a variety of fields. Daily broadsheet newspapers include the *Times*, the *Financial Times*, the *Independent*, the *Daily Telegraph* and the *Guardian*.

A number of these daily broadsheets devote a particular day of the week to certain types of jobs, for example legal jobs are advertised in the *Times* on a Tuesday and the *Guardian* advertises jobs in education on the same day. This approach to advertising vacancies means that a sizable selection of jobs in a particular field are advertised together on a specific day. The advantage is that individuals seeking employment in a particular field can be guaranteed a good variety of vacancies in that field if they purchase the newspaper on the relevant day. The advantage to the advertising organisation of such an approach is that their vacancy is advertised on a day when a considerable number of individuals seeking vacancies in a particular field are likely to purchase the newspaper.

The alternative to advertising in a broadsheet newspaper is to advertise in the industry or technical press, as nearly all professions have at least one such publication. Examples include the *New Scientist* for research scientists and technicians; the *Engineer* for different type of engineering jobs, civil and mechanical; and the *Bookseller* for vacancies in publishing. The key advantages in using the industry or technical press are similar to those for advertising in daily broadsheet newspapers on an appropriate day. In addition, however, interested applicants are perhaps more likely already to be employed in the industry. The industry or technical press is appropriate for unusual or very specific technical jobs, which are not normally advertised extensively in the local or national press. Many have Internet sites listing vacancies.

Other methods of advertising vacancies are equally valid and may be used as an alternative to or in addition to press advertising. Advertising within the organisation itself is relevant if it is large enough that staff in one part of the organisation are unlikely to hear about a vacancy in another part by word of mouth. Such internal advertising also opens up opportunities for skills mixing and development for existing staff. The cost of advertising within the organisation is much less than press advertising, although it reaches a much smaller audience. However, recruiting an employee who already works for the organisation can have advantages in that the applicant will already know about the organisation and its business. Internal advertising is usually through a vacancies bulletin circulated throughout the organisation to employees or via a vacancies noticeboard. The latter can be accessible to staff only or to staff and the general public. For example, vacancy noticeboards are sometimes located inside supermarkets close to the main public entrance and are used to advertise jobs such as a cashier, on a full- or part-time basis.

**Third-party advertising**

An alternative to the direct advertising of vacancies is to use a third-party organisation to help attract applicants. Third-party organisations bring together applicants seeking work and organisations offering the type of employment sought. Therefore they often focus on particular types of employment or employees.

#### Commercial employment agencies

There are many commercial employment agencies, which either specialise in one particular type of employment, for example accountants or HGV drivers, or operate in a range of fields. The Advance employment agency covers a range of occupations, including office support staff; engineering and technical staff; sales and marketing staff; and hotel and catering staff. Payment is made to place a vacancy with a commercial employment agency and in return the agency advertises the vacancy to its clients, monitors responses to the vacancy and performs the preliminary interviews.

#### University employment agency

Universities may set up their own employment agency dealing in part-time jobs for students. This type of agency brings together students seeking part-time work and local employers seeking part-time staff. The university agency can refuse to 'advertise' jobs that are likely to result in students being exploited, by for example offering less than the minimum wage.

### Management selection consultancy

Management selection consultancies advertise on behalf of organisations seeking to fill managerial posts. They advertise in the press, sometimes not revealing the name of the organisation on whose behalf they are acting. The management selection consultancy will carry out the first interviews and produce a shortlist for the client company to assess further. Alternatively, the consultancy may act on behalf of a client and approach an identified person to assess their interest in the vacancy, which is commonly known as headhunting.

### Schools, colleges and universities

An organisation specifically seeking a school, college or university leaver may use the university, school or college careers service. Schools, colleges and universities employ careers advisers and run careers libraries where brochures, application forms and careers literature are kept for the use of those students about to leave. Organisations supply schools, colleges and universities with brochures, applications forms, industry literature and posters free of charge. If the organisation frequently employs school, college or university leavers, then developing and maintaining a close relationship with schools, colleges or universities may be an appropriate strategy. The schools, colleges or universities may also supply the opportunity for organisations to carry out initial interviews on their premises, hence allowing the visiting organisation to see a number of applicants in one visit.

### Job centres

Government job centres are free to advertise in and provide help with shortlisting. The use of job centres is practical when there is a considerable pool of available candidates.

## The advertisement

The cost of advertising vacancies in the press or via a third party such as a recruitment agency varies according to the publication or third party used; length and size of the advert; and duration for which the advert is on display. The successful advertising of jobs requires the use of appropriate publications or third parties. This is important, as suitable applicants have to see the advertisement to be attracted to the vacancy. Failure to attract suitable applicants means that the organisation will have squandered time and money on advertising in the wrong place and on dealing with unsuitable applicants.

The information in job advertisements needs to be clear and reliable and to tell potential applicants, in brief terms, what the organisation does, what the job involves and the key requirements looked for in applicants. This allows suitable applicants to assess if the vacancy is relevant to them and to make an informed judgement about whether to apply or not. It is also usual to indicate the expected salary and how potential applicants should apply for the vacancy. A suggested outline advertisement is shown in Exhibit 9.9.

**Exhibit 9.9 Suggested outline of job vacancy advertisement**

| | |
|---|---|
| **Brief details** | • Name of the organisation<br>• Line of business – main products/services |
| **Job and duties** | • Job title and main tasks – summarised from job description |
| **Key requirements of successful applicant** | • Key qualifications, work experience, skills and circumstances – summarised from personnel specification |
| **Salary** | • State the salary or salary scale for the job<br>• Tell applicants how to apply, e.g. CV and covering letter or application form |

## Stage 4 – Assessing applicants

The assessment of applicants occurs in two steps. The first is an assessment of the applications received, with some applicants being rejected at this juncture. The second step is a further assessment of selected applicants by interviews and aptitude tests, before finally choosing the successful applicant.

**Assessment of application forms**

The initial sorting of application forms can be a difficult and tedious task, particularly if there has been a large response to the advertising. However, the initial sorting of application forms and shortlisting of candidates for interview needs to be methodically and rationally undertaken. The staff doing this initial sorting process need to be involved in the recruitment process and fully understand the relevance and use of job descriptions and personnel specifications, if the sorting process is to be fair and accurate. A thorough shortlisting process aids in ensuring the accuracy and fairness of the recruitment process and lessens the likelihood of inappropriate applicants being called for interview. Therefore this initial shortlisting process makes use of the criteria from the person specification and applicants meeting the essential and some of the desirable criteria are those who will be called for a first interview. Applicants not meeting the essential criteria in the person specification are rejected.

**Assessment of applicants**

The next step is to assess the shortlisted applicants by interview. There are variations in the interview process: with some vacancies only one interview is required and for other jobs the interview process is in two stages. It is unusual for a candidate to experience more than two interviews for a vacancy. Equally, the number of interviewers can also vary. There are generally two views governing the number of interviewers. One view is that an effective interview and discussion can only take place on a one-to-one basis, so candidates meet one interviewer. The other view is that the interview process should be more open, so the interview is carried out and the appointment decision made by a panel of interviewers.

**Exhibit 9.10 Interviewing applicants**

| | |
|---|---|
| **One-interview vacancies** | • Interview<br>• Select successful applicant<br>• Check references<br>• Confirm appointment |
| **Two-interview vacancies** | • Interview<br>• Select successful applicants for second interviews<br>• Aptitude tests and activities<br>• Second interview<br>• Select successful applicant<br>• Check references<br>• Confirm appointment |

The sequence of events for one- and two-interview vacancies are shown in Exhibit 9.10. In two-interview vacancies aptitude tests and activities are usually taken after the applicant has been selected for a second interview, but prior to that interview being conducted. References should be checked after the selection of the preferred candidate and prior to the confirmation of appointment, as references are used to confirm the interviewers' opinion of the successful applicant.

### Interviewing

The job interview is a two-way communication process. It is an opportunity for the interviewer to get to know the candidate and further assess them against the criteria in the person specification. It is also an opportunity for the interviewee to gain a greater insight into the organisation and the work being offered. The interviewer should commence the interview by putting the candidate at ease and outlining the structure of the interview. A possible interview structure is shown in Exhibit 9.11.

**Exhibit 9.11 Interview structure**

| | |
|---|---|
| 1 | Scene setting |
| 2 | Application form and person specification |
| 3 | Candidate's questions |
| 4 | Close the interview |

### Scene setting

The physical scene or location needs to be appropriate, usually a quiet room away from the disruption of telephones, noise and interruptions. The interviewer should set the scene at the start of the interview and recap on what the job involves and the type of person sought. Information from the job description and person specification can be used to do this.

### Application form and person specification

The main body of the interview is structured around the completed application form and the person specification. This allows the information provided on the application form to be tied in with the relevant section of the person specification. For example, qualifications and work experience listed on the application form will link to the attainment section of the person specification. This allows the interviewer to ask the candidate questions about the information provided on their application, while also making sure that the candidate is assessed against the criteria in the person specification.

The questions used in interviews to assess a candidate need to be open ended, as this allows the candidate's reasons for wanting the job to be examined. Open-ended questions give applicants the opportunity to explain the knowledge, skills and experience they have gained and justify how these equip them as suitable for the vacancy. If the candidate provides incomplete answers to the questions, then the interviewer should ask follow-up questions and probe for the information required to assess the candidate against the person specification.

#### Candidate's questions

As previously stated, the interview is an opportunity for the interviewee to gain a greater insight into the organisation and the work being offered. Although having to ask questions may be seen as a nightmare for the nervous applicant, the well-prepared candidate will have researched the company and have a few intelligent questions to ask. Alternatively, from the interviewer's point of view it is always appropriate to give applicants the opportunity to seek any further information they require, as even the best recruitment process may have forgotten something.

#### Close the interview

The candidate should be thanked for attending the interview and told how to claim any interview expenses to which they are entitled. Candidates also need to be advised of when they will know the outcome of the interview.

#### Informing candidates

All candidates who were interviewed need to be informed of the outcome of their interview. The unsuccessful candidates need to be told that they have not succeeded and, if appropriate, why they have not been successful. Equally, the successful candidate needs to be told that they have gained the job. This should be confirmed in writing, along with information on the salary and arrangements for the candidate to start work. A contract of employment should be drawn up and issued to the successful candidate.

## Conclusion

This chapter defines human resource management, examines the key external environmental influences on a human resource management department, including legislation, and finally covers the recruitment process in detail.

ETHICAL ISSUES CASE STUDY 9.2

FT

# There's no accounting for magic

**By Richard Donkin**

There may be those who care little for sport. Some may even profess lack of interest in football. But few who saw Chris Waddle's spectacular side-footed chip over the goal-keeper's head from 40 yards, which helped secure victory for Bradford City against Everton in Saturday's FA cup tie, would deny they had witnessed something special.

How can you place a value on such goals? How can you place a value on Chris Waddle? Such questions must tax the minds of football managers, particularly since the 36-year-old Waddle, approaching the end of his playing career, was given a free transfer to one of the English league's less fashionable clubs.

Beyond football, the value of employees is beginning to attract increasing interest from business leaders, some of whom are showing signs of frustration at the failure of the accountancy profession to devise a satisfactory way of assessing the worth of human talent.

This is apparent in a survey of directors among 120 of the UK's top service sector companies carried out for Theodore Goddard, the London law firm.

Some two-thirds of the directors harboured frustrations, they said, because they believed that accountants placed more value on tangible assets, such as property and equipment, than on staff.

But recently the accountancy profession has been spending more time debating goodwill accounting, and if the proposals outlined in FRED (Financial Report Exposure Draft) 12 are adopted by the UK's Accounting Standards Board, its nature will change. One effect would be that the value of intangible assets – like footballers – with a limited useful life would be reduced year by year over their estimated life.

But this is a highly contentious issue – made more uncertain by the Bosman ruling in the European Court which allows football players to move between EU member states without a transfer fee once their contracts have expired.

The biggest difficulty in valuing company employees is that, unlike fixed assets, they are not owned by the business. They can and do walk out of the door.

Karen Moloney, of Moloney and Gealey, a human resource consultancy, says evaluations of football players could hold the key to finding a process for assessing the human or intellectual capital of a company.

Defining intellectual capital, she argues, is important if employees, the so-called human resources, are to be seen as an investment rather than as a cost.

Kate Olley, a human resources consultant at Arthur Andersen, says that if companies could work out the real costs of their employees they could measure the expected return and, where necessary, identify potential cost savings.

Some of these issues are beginning to attract the attention of the personnel profession as it attempts to quantify its contribution to business development.

It may, for example, be incumbent upon personnel to point out the need to reorganise a company's pay and management structure to give better rewards and recognition to technical staff whose value, under a flatter employment structure, has risen beyond that of many of the managers.

This would only be possible, however, if the business had an effective system of valuing the contribution of individual employees.

But most existing evaluations, including employee appraisals, are extremely subjective. Returning to the football field, David Myddelton, professor of finance and accounting at Cranfield School of Management, likes to cite the French national squad's rejection of Eric Cantona, the Manchester United midfield player, because the manager does not regard him as a team player.

Yet Cantona, under skilful club management, has shown that he is one of the finest players gracing the English football league. Such observations make Myddelton doubtful that any accounting system could deal with the vagaries of human ability.

It may indeed be a red herring to look towards accountancy for a solution. That, at any rate, is the belief of Skandia, the Swedish insurance and financial services company.

For the past five years Skandia has included a supplement to its annual accounts highlighting value creation through human development.

Leif Edvinsson, director of intellectual capital at Skandia, explains it as the difference between harvesting – the job of the chief financial officer who focuses on the profit and loss account – and that of looking after the roots of the organisation, the feeding and nurturing necessary to maintain healthy future growth.

The Skandia approach embraces a more holistic view of a company within society that recognizes there is a relationship between profit, sustainability, renewal and employment.

Accountancy, argues Edvinsson, tends to focus on the concerns of the stock market that concentrate on financial values, such as earnings per share – another view of harvesting.

Skandia is seeking to free up what it calls its structural capital – what is left at the office when people go home.

In some ways this concept is similar to that employed by Ricardo Semler at Semco, the Brazilian industrial products manufacturer. Semler has become less concerned with who the company employs than how the fixed assets are best exploited for profit.

**Ethical issues case study 9.2 *continued***

Skandia has developed a series of alliances, collaborative ventures and partnerships that generate profit from professionals not directly employed by the company. In Skandia AFS, its assurance and financial services business, the ratio of direct Skandia employees to those employed in the network of alliances is 1:30.

The company has developed an accounting-style format for displaying its intellectual capital, a table it calls the Skandia Navigator. This includes such measures as a satisfied customer index, an empowerment index, and training expenses per employee. But the company also attempts to set out the processes involved in developing employees.

Such processes are beginning to spin off into the overall business. This concept of renewal is driving negotiations with the Swedish government for an insurance-led scheme that will allow Swedish employees to take time off work to learn new skills or renew their expertise.

The scheme, called competency insurance, is based on the premise that employees will need to spend between 20 and 30 per cent of their time reinvesting in training and learning if they are to keep up with changes in the international marketplace.

Employees would invest 5 per cent of their salaries in an insurance policy which would fund leave of absence to undergo re-training. The company is hoping to make the insurance payments tax deductable.

The perception of the Theodore Goddard survey that, to quote the report, 'bean-counters value bricks and machines more than people' may give accountants some food for thought. In the meantime they will receive few complaints from Bradford City where they will be talking about that goal for years.

*Source*: *Financial Times*, 29 January 1997. Reprinted with permission.

**Questions for ethical issues case study 9.2**

1 Do you agree that accountants should place 'more value on tangible assets, such as property and equipment, than on staff'? Explain your answer.

2 The accounting function is ill equipped to value the human resources of an organisation. Do you agree or disagree? Write 100 words defending or criticising the statement.

3 Summarise the benefits and costs to the employees and the organisation of investing in training, as occurs in Skandia ASF. In your opinion, do the costs outweigh the benefits to the employees and the organisation? Explain your answer.

## EXIT CASE STUDY 9.3

FT

# Flexible friends

**By Vanessa Houlder**

The increasing emphasis on part-time and temporary work has changed the job market beyond recognition. But while this trend has improved the productivity and flexibility of many businesses, it has raised difficult issues for managers.

How do companies find high-calibre staff prepared to work unconventional hours, often for relatively low pay? And how do they combat the widespread perception that these workers, who often have an important role in dealing with customers, are poorly trained and undervalued?

Recent research by the Roffey Park Management Institute in Sussex concluded that businesses that are successful at managing part-time and temporary employees have put more effort into meeting their needs.

Burton Group, which replaced 1000 full-time staff with 3000 part-time jobs in 1993, has responded to recruitment problems by allowing store managers to write contracts to suit individual employees. At one Topshop outlet a job is shared by a student who works in the vacation and a woman who works during the school term.

Asda, the retail group where 80 per cent of the workforce work part-time, has also felt the need to be more flexible towards part-time workers. Consultations showed that one reason for staff turnover reaching 30 per cent a year was that many wanted a longer working week.

'Employees who work very short hours, below the National Insurance threshold, are very cheap employees,'

**Exit case study 9.2 *continued***

says David Smith, employee relations manager of Asda. 'But we found people were leaving. So we are going against the trend and offering longer hours.' With the help of longer contracts and other measures, such as improved maternity leave, career breaks and the ability to swap shifts, staff turnover fell by 2 percentage points.

The issues concerning flexible workers are nothing new for Oxfordshire County Council, where more than two-thirds of the 16 000-strong workforce work part-time. That partly reflects the need to cover round-the-clock services, such as in residential homes and fire fighting, but offering flexible working is also a way to attract professional employees, such as legal staff, who might be paid more in the private sector. Valued staff may want time off for childcare, further education and other part-time jobs or may want to continue part-time after taking early retirement.

But much of the debate about flexible working concerns the other end of the spectrum, the poorly-paid and lowly-valued employees who do not qualify for National Insurance, for employment protection or for statutory benefits. Attention has particularly focused on zero-hour contracts, introduced by companies such as Burger King, in which individuals are only paid when dealing with customers. More than half the part-timers interviewed in a 1995 TUC study said employers regarded them as 'second-class staff'.

The tendency to treat flexible workers as poor relations is creating a dilemma for employers, according to Christina Evans, a research associate at Roffey Park Management Institute. 'In many organisations, it is those who work "flexibly" who have the most responsibility for customer service, at peak trading periods. In other words it is the "flexible" employees who are the ones who have the most impact on sales,' she says.

She says better use of flexible workforces, would bring benefits including 'better customer service, lower staff turnover and a more motivated workforce'.

Her research uncovered many positive aspects of flexible working. Companies often commented that part-time workers were more disciplined in their time management and so more productive.

Nonetheless, the research highlighted a number of barriers to making the best use of flexible employees. Managers are often unenthusiastic about supervising them since it complicates scheduling and rota arrangements.

The assumption that part-time employees are less career-minded can be self-fulfilling. Unless companies examine their promotion policies, there may be unintended barriers to the promotion of part-time workers. For example, at First Direct, the telephone banking service, the Roffey Park study noted an assumption that an employee could not be rated 'very good' unless they worked a full shift and were exposed to all the trading activities.

Training is a particularly vexed issue for flexible workers. Most companies would agree that part-time staff need at least as much training as full-time staff, yet few provide that. The training of 80 per cent of those in full-time employment is funded by their employer, compared with 36 per cent of those working part-time, according to the Labour Market Quarterly Report.

The widespread practice of offering flexible workers worse pay, training and conditions than the full-time workforce could backfire, says the Institute of Management. It warns that the current attitude towards flexible workers could 'lead to the development of two-tier workforces – with all of the difficulties inherent in managing them.'

More effort is needed to integrate flexible employees with the company's core workforce: 'What is needed is a new approach to managing flexible workers, which acknowledges the needs of this group of employees to be valued, included and invested in,' she says.

*Source*: *Financial Times*, 29 January 1997. Reprinted with permission.

**Questions for exit case study 9.3**

1 List the advantages and disadvantages to an organisation of employing part-time and temporary workers.

2 What is needed is a new approach to managing flexible workers, which acknowledges the needs of this group of employees to be valued, included and invested in.

You are an HR manager for a national cinema chain and have the task of drawing up a plan to ensure that flexible staff will be 'valued, included and invested in'. Identify the positions in which your flexible staff operate and draw up the required plan. How do you measure your plan's success and ensure that it works? (Your annual bonus depends on the success of your plan!)

## Short-answer questions

1 Define human resource management.

2 Explain the difference between human resource management and personnel management.

3 Name the three categories of legislation that most directly influence the human resource management function.

4 Why in most industrialised countries are population demographics likely to influence the human resource management function in the future?

5 Explain why the move to a more flexible labour force has occurred.

6 State the main advantages of telecommuting.

7 State the main disadvantages of telecommuting.

8 Indicate two ways in which an organisation may manage a skills shortage in a particular area.

9 List the four stages of the employee recruitment process.

10 Explain the purpose of the job description.

11 Explain the purpose of the person specification.

12 Name the points in Rodger's seven-point plan for the person specification.

13 Give two examples of where each of the following can take place: third-party advertising and direct advertising.

14 Indicate a suitable structure for a job interview.

## Assignment questions

1 Summarise and illustrate how an organisation with a workforce that is too small and lacking in the required skills and abilities could remedy the situation.

2 Name and explain all the elements of the job recruitment process and discuss which of these elements you consider to be key in terms of ensuring equality of opportunity.

3 Produce a job description for a university lecturer in marketing. Use the job description you have produced to draw up a personnel specification. Indicate where the job could be advertised and state what information you would include in any advertisement used.

## References

1 Daniels, J D and Radebaugh, L (1997) *International Business*, 8th edn, Reading, MA: Addison Wesley Longman.

2 Clement, B (1995) 'Dinner ladies' equal pay win undermines competition law', *Independent*, 7 July.

3 Clement, B (1996) 'Dinner ladies awarded £1m over council's unfair pay cut', *Independent*, 30 July.
4 Clement, B (1996) 'Legal service found guilty of race bias', *Independent*, 7 December.
5 Ibid.
6 Gary, B (1997) 'Armed forces under attack for racism', *Financial Times*, 21 March.
7 The Minister for Disabled People (1996) *The Disability Discrimination Act – Employment*, DL 70, October.
8 Johnson, P (1990) 'Our ageing population – the implications for business and government', *Long Range Planning* 23 (2), April.
9 Ibid.
10 Ibid.
11 *Independent on Sunday*, 29 March 1992.
12 Johnson, op. cit.
13 Atkinson, J (1984) 'Manpower strategies for flexible organisations', *Personnel Management*, August.
14 Ibid.
15 Bolger, A (1997) 'Thoroughly modern training', *Financial Times*, 24 March.
16 Penman, D (1994) 'No workplace like home', *Independent*, 6 June.

## Further reading

The following books and articles are concerned with human resource management:

Apgar, M (1998) 'The alternative workplace: changing where and how people work', *Harvard Business Review*, May/June.

Bratton, J and Gold, J (1994) *Human Resource Management: Theory and Practice*, Basingstoke: Macmillan.

Clement, B (1999) 'Nice suit. Is that the union rep?', *Management Today*, March.

Cowling, A and James, P (1994) *The Essence of Industrial Relations and Personnel Management*, Hemel Hempstead: Prentice Hall.

Curtis, T (1994) *Business and Marketing for Engineers and Scientists*, Chapter 4, Maidenhead: McGraw-Hill.

Grundy, T (1997) 'Human resource management – a strategic approach', *Long Range Planning*, 30 (4), August.

Hannagan, T (1998) *Management Concepts and Practices*, 2nd edn, Chapter 11, London: Financial Times Pitman Publishing.

Jebb, F (1998) 'Flex appeal', *Management Today*, July.

Merrick, N (1999) 'The key shift', *People Management*, 11 March.

Mitchell, A (1998) 'New model unions', *Management Today*, July.

McKenna, E and Beech, N (1995) *The Essence of Human Resource Management*, Chapter 8, Harlow: Prentice Hall.

Thompson, J L (1997) *Strategic Management: Awareness and Change*, 3rd edn, Chapter 11, London: International Thomson Business Press.

Torrington, D and Hall, L (1995) *Personnel Management: HRM in Action*, 3rd edn, Hemel Hempstead: Prentice Hall.

Ulrich, D (1998) 'A new mandate for human resources', *Harvard Business Review*, January/February.

The following books and articles address demographics and the flexible workforce:

Braid, M (1995) 'Tomorrow belongs to them', *Independent*, 2 October.

Brown, M (1997) 'Design for working', *Management Today*, March.

Gwyther, M (1992) 'Britain bracing for the age bomb', *Independent on Sunday*, 29 March.
Handy, C (1995) *The Age of Unreason*, Arrow.
McIntyre Brown, A (1997) 'The pay band wagon', *Management Today*, August.
Nicholson-Lord, D (1995) '"Greys" take over from the young as big spenders', *Independent*, 27 January.
Smith, D (1997) 'Job insecurity and other myths', *Management Today*, May.
Smith, D (1998) 'Skills shortage – we're learning (at last)', *Management Today*, July.
Worthington, I and Britton, C (2000) *The Business Environment*, 3rd edn, Chapter 5, Harlow: Financial Times Prentice Hall.

The following articles consider the recruitment process:

Gabb, A (1997) 'University challenge', *Management Today*, December.
Lucas, E (1999) 'Virtual realities of candidate selection', *Professional Manager*, March.
Merriden, T (1997) 'Vacancies in the skills department', *Management Today*, May.
Wheatley, M (1997) 'Open all hours', *Management Today*, September.

# 10 TOWS analysis

## Organisational context model

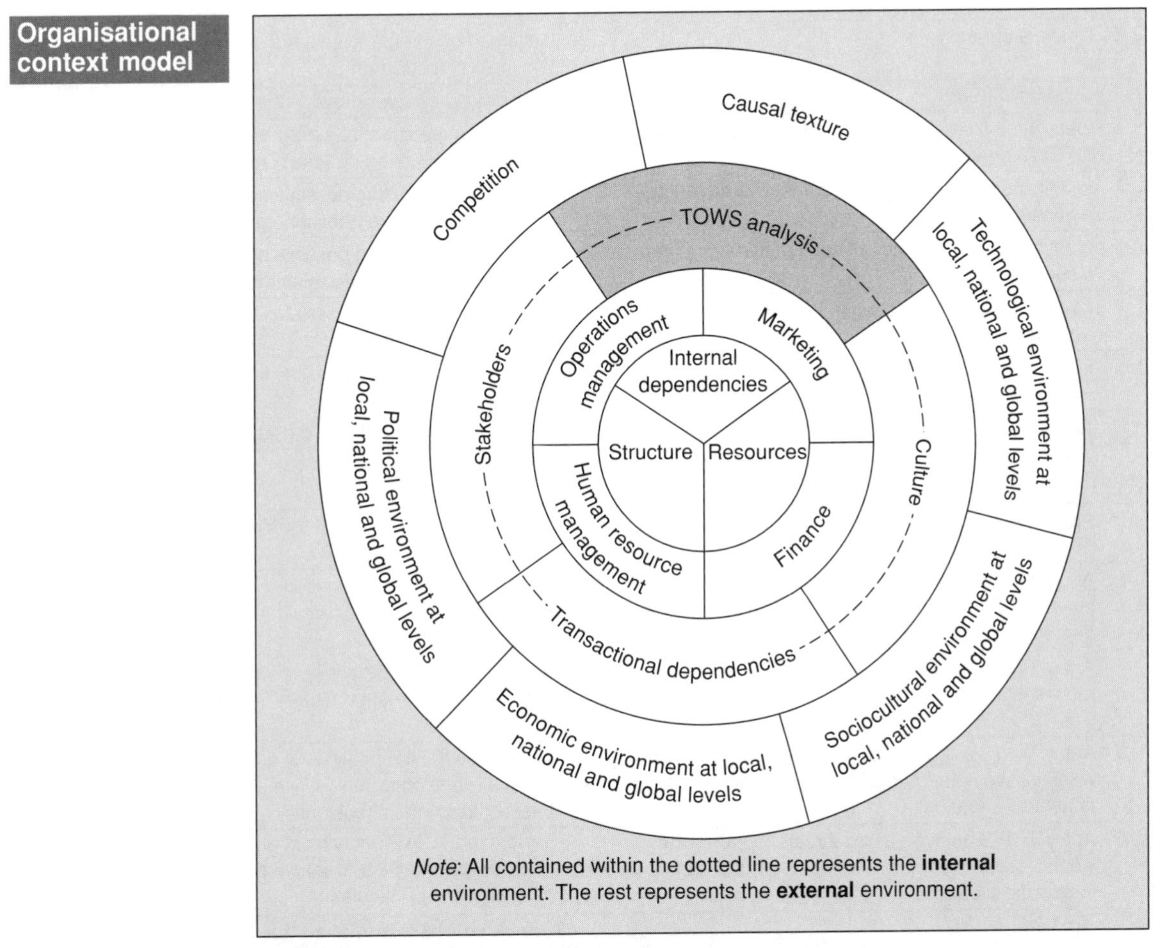


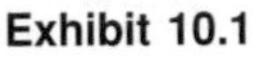

**Exhibit 10.1**

## Learning outcomes

While reading this chapter and engaging in the activities, bear in mind that there are certain specific outcomes you should be aiming to achieve as set out in Exhibit 10.2 (overleaf).

**Exhibit 10.2 Learning outcomes**

| | Knowledge | Check you have achieved this by |
|---|---|---|
| 1 | Define TOWS analysis | stating in one sentence the purpose of a TOWS analysis |
| 2 | Define a threat | stating in one sentence what a threat is and giving an example of a threat |
| 3 | Define an opportunity | stating in one sentence what an opportunity is and giving an example of an opportunity |
| 4 | Define a weakness | stating in one sentence what a weakness is and giving an example of a weakness |
| 5 | Define a strength | stating in one sentence what a strength is and giving an example of a strength |
| | **Comprehension** | **Check you have achieved this by** |
| 1 | Identify the issues that should be discussed as part of a TOWS analysis | discussing the questions and issues that should be covered in summing up any TOWS analysis |
| 2 | Identify possible sources of competitive advantage for a general situation | listing and discussing possible sources of competitive advantage for a hypothetical organisation |
| 3 | Identify possible sources of competitive advantage for a specific situation | listing and discussing possible sources of competitive advantage for a real organisation |
| 4 | Identify possible key success factors for a general situation | listing and discussing possible key success factors for a hypothetical organisation |
| 5 | Identify possible key success factors for a specific situation | listing and discussing possible key success factors for a real organisation |
| | **Application** | **Check you have achieved this by** |
| 1 | Apply basic TOWS analysis to general situations | performing a basic TOWS for a hypothetical organisation |
| 2 | Apply basic TOWS analysis to specific situations | performing a basic TOWS for a real organisation |
| 3 | Apply TOWS analysis using external environmental factors, strengths and weaknesses to general situations | performing TOWS analysis of a hypothetical organisation using external environmental factors, strengths and weaknesses |
| 4 | Apply TOWS analysis using external environmental factors, strengths and weaknesses to specific situations | performing TOWS analysis of a real organisation using external environmental factors, strengths and weaknesses |
| 5 | Apply TOWS analysis using threats, opportunities, weaknesses, strengths and the four functions of business to general situations | performing TOWS analysis of a hypothetical organisation using threats, opportunities, weaknesses, strengths and the four functions of business |
| 6 | Apply TOWS analysis using threats, opportunities, weaknesses, strengths and the four functions of business to specific situations | performing TOWS analysis of a real organisation using threats, opportunities, weaknesses, strengths and the four functions of business |
| | **Analysis** | **Check you have achieved this by** |
| 1 | Choose information and data sources that best assist managers in organisations with scanning and understanding the internal and external environment | comparing information and data sources, evaluating the most suitable media and sources of information, from both within and outside the organisation |

ENTRY CASE STUDY 10.1 FT

# Business keeps rolling

**By Nicholas Lander**

Rolling out is the phrase employed at present by restaurateurs keen to exploit the growing interest in eating out.

Crucially, 'rolling out' involves finding a central – but not prime – location to attract a seven-day-a-week clientele; creating a friendly, unobtrusive interior design with a bar; and writing a medium-priced menu – say around £20 for three courses – that has a catch-all feel.

Once you decide to 'roll out' a restaurant, you can roll on. Oliver Peyton opened his micro-brewery and restaurant, Mash & Air, in Manchester last December and will open another in Great Portland Street, London W1, early next year.

Jean-Christophe Novelli opened his first restaurant in Clerkenwell, east London, and then moved west to open Novelli W8 in Notting Hill. He is branching out on a global scale next month when he opens a Maison Novelli in Constantia, South Africa.

Antony Worrall Thompson recently opened Woz, in Ladbroke Grove (0208–968 2200), where a five-course fixed dinner costs £22.50. Once the restaurant has been fine-tuned he hopes to open other sites around London.

However, in spite of a buoyant market, success is not guaranteed. When Cindy Pawlcyn and her partners in the successful Real Restaurants Group decided to move their style of California cooking to the midwest, they encountered insurmountable obstacles. 'We could not find the right ingredients, the right cooks or the right management,' Pawlcyn confessed.

Such is the confidence in London at present that two restaurant groups have opened prototype restaurants with a view to 'rolling them out' around the UK. The first, MPW, bears the initials of Marco Pierre White and is part of his joint venture company with Granada; the second, Zinc, is the mid-price offering from the Conran restaurant group.

The service at MPW left a lot to be desired. My colleague, a former chef, had already described it as 'minus two' by the time we had ordered. From the door we had spotted two waiters eating hurriedly behind the bar; the receptionist failed to explain that different menus were in operation – one on the terrace, the other in the main restaurant – and we had to ask for bread.

The cause, if not the explanation, was that MPW had served 140 customers during lunch. White has infused the chef, talented, peripatetic Gary Hollihead, and the kitchen staff with honourable intentions.

The menu offers a wide range of predominantly French classic dishes (one of which, pajarski of salmon, an up-market fishcake, my colleague had not seen on a menu for at least 10 years), at reasonable prices. But the delivery at MPW must get better.

There are 16 starters ranging from Escargot Chablis to a galantine of duck with Cumberland sauce; 12 fish dishes, including the rarely seen trout *au bleu* (sic) and a fresh shrimp omelette; and six meat main courses.

Our two starters, risotto of girolles and hot asparagus with a gleaming mousseline sauce, were well executed and proved far more enjoyable than the smoked haddock Monte Carlo and calves liver *à l'anglaise* which followed – after a considerable longeur.

The pleasure of sitting outside Zinc on almost pedestrianised Heddon Street on a bright sunny day proved transitory. We had just ordered our drinks as a truck crammed with scaffolding arrived to our left, while to our right came the joyful noise of hammers, saws and drills working on what will soon be a theme pub.

Unlike MPW's menu, Zinc's is written in designer-faded typewriter script, contains no surprises and is lacking in imagination. It is standard fare at standard prices. Starters include smoked salmon (£5), crab cake (£4.75) and *terrine de campagne* (£3.75). Middle courses include Caesar salad (£5.95), steak sandwich (£7.50) or a club sandwich at £6.95. Main courses range from fish and chips (£7.50) to rib of beef (£11.50) and sausage and mash (£6.50).

Zinc also serves its own version of a lobster club sandwich (£9.95), an idea created in New York by Anne Rosenzweig, chef/owner of Arcadia, where it became a great favourite of Sotheby's chairman, Alfred Taubman.

*Source*: *Financial Times*, 25 October 1997. Reprinted with permission.

## Introduction

A fundamental overview of the essential issues faced by an organisation can be provided by a thoroughly carried out TOWS analysis. Such an analysis is concerned with identifying and evaluating:

- the influence of the external environment (threats and opportunities) on an organisation;
- the ability of an organisation to use its strengths and weaknesses to operate successfully in its external environment.

TOWS analysis first examines the threats and opportunities that arise from the external environment of an organisation, which is followed by an assessment of the internal weaknesses and strengths that the organisation possesses. The identification of external influences on an organisation is aided by the use of PEST analysis, LoNGPEST analysis, competitor analysis and market analysis; refer to Chapters 1, 2, 3 and 6 respectively. Once the external influences on an organisation have been identified, they can then be judged to be either a threat or opportunity. The internal environment of an organisation is considered in a similar manner by examining strengths and weaknesses. The following sources of strengths and weaknesses are assessed: resources, structure, culture and the tasks carried out by the four functions of business (marketing, operations management, finance and HRM); refer to Chapters 6, 7, 8 and 9 respectively.

## Threats

Threats have the potential to damage an organisation's performance in the marketplace or external environment. Threats often arise from competitors or factors that are outside the control of the organisation. The competitor that cuts prices by 50 per cent today and brings out a new generation of technologically advanced products poses a clear threat to all other organisations operating in the same industry. Threats may also arise from changes in legislation or taxation relating to the industry in which an organisation operates. For example, the imposition of value added tax (VAT) on newspapers and books or children's clothes would affect both the manufacturers of such products and the amount of business done by retailers selling such products to members of the public.

Clearly, threats from the external environment may endanger an organisation. However, threats may also jeopardise good opportunities of which the organisation expects to take advantage. In 1997 the American company WorldCom tendered £19bn for MCI, another American company. This offer outbid BT's proposal of £11bn for MCI by over 70 per cent. The WorldCom offer became a realised threat that snatched an opportunity from under BT's nose and left a gap in the latter's strategy to secure a global partner.

## Opportunities

Opportunities are openings or chances in the external environment or marketplace that an organisation may pursue to obtain benefits. The identification of a new geographic market in North America for a firm's products is an opportunity. Such opportunities can be exploited by manufacturing the product in the firm's home country and exporting the product to North America, or by forming a strategic alliance with a local US company and having the benefits of the greater understanding of the local and national external environment offered by that partner in the strategic alliance. This type of arrangement will also need to confer benefits on the alliance partner, otherwise it is unlikely to be successful in the long term. The third alternative is to manufacture the product locally, which is perhaps the most time and resource consuming of the three options as it involves setting up from scratch in a foreign country. This will be more difficult than operating in a home environment or with advice from an organisation for which the foreign market is a home environment. In 1997 before WorldCom outbid BT, the identification of MCI as a global strategic partner was an opportunity for BT.

### Key success factors

The degree to which an organisation is successful depends on its ability to meet its key success factors (KSF). Key success factors are what an organisation must do well and better than its competitors if it is to succeed. They can arise from a number of different sources.

First, key success factors may be established by the industry in which an organisation operates. For example, in the clothes mail-order business, being able to provide speedy delivery to the customer's home is a key success factor for all companies in the industry. Secondly, key success factors may be determined by the organisation itself. When the Midland Bank initially set up telephone banking with its subsidiary First Direct, the key success factors were to provide an accurate banking service that required a minimum amount of paperwork; could be accessed 24 hours a day; and did away with the need for high-street branches. Other banks have adopted these key success factors by providing their own telephone banking services and will continue to try to meet these key success factors by providing banking services in the home via the Internet, private dial-up services or TV-based services.

Finally, key success factors may be signified by customers indicating that they require products with particular features or services. One example is the demand from consumers for high-quality take-away coffee from coffee bars and sandwich shops. The demand from consumers is not only for high-quality coffee, but also for a range of different types of coffee, from the familiar cappuccino and espresso to the more exotic sounding cafe latte, mocha and arabica. Therefore a key success factor for coffee and sandwich bars is the provision of a wider range of high-quality coffees.

## Weaknesses

A weakness is defined as lacking a competence, resource or attribute that an organisation needs to perform better than its competitors in the external environment. A company producing tableware for the domestic and catering markets will rely in part on styling and designs to make products appealing to customers. If it relies on the designs that have always been used or occasionally on shopfloor staff coming up with new patterns, it is likely to lack competence in design, a key success factor. The lack of a key resource, such as a new piece of technologically advanced equipment, is also a weakness, particularly if your competitors do have access to that equipment.

## Strengths

A strength is a competence, valuable resource or attribute that an organisation uses to exploit opportunities in the external environment or to help it counter threats from the external environment. Strengths could include a resource such as a well-motivated and skilled workforce, with low turnover; or an attribute such as a strongly established brand image or reputation. Examples include Cadbury's Dairy Milk brand and Marks & Spencer's reputation for good quality.

### Stakeholders and key success factors

Customers are stakeholders in an organisation and fulfilling key success factors involves the organisation in meeting the needs and expectations of its customers and other stakeholders; *see* Chapter 11. For example, a key success factor may be a good relationship with a reliable supplier. This will be especially true if the supplier is the only supplier or one of very few supplying a key component or part. Meeting the key success factors will require the organisation to meet the supplier's expectations, which will include regular orders of a certain minimum size, with little room for negotiation on price if the supplier is powerful.

In seeking to satisfy stakeholders, especially customers, while at the same time outperforming competitors, organisations should seek to:

- fulfil the key success factors for the industry or market;
- develop competencies that provide competitive advantage (*see* discussion on competitive advantage and premium prices);
- utilise competencies to meet the requirements of specific customers and aim to charge a premium price.

### Competitive advantage and premium prices

Competitive advantage arises from the unique features or 'extras' that a product or service possesses and for which customers are prepared to pay a higher or premium price. For example, some dry cleaners offer a standard service and a gold service. The standard service includes dry cleaning the item of clothing, pressing it by machine and returning the item to its owner on a wire hanger with a polythene cover over it. In contrast, the gold service includes dry cleaning the garment, hand pressing and finishing it, before it is returned to its owner on a more robust plastic hanger and in a more substantial plastic holder. There will be a small group of customers who will be prepared to pay a higher or premium price for the extras that the gold service provides. Being able to offer the gold service will provide the dry cleaner with a competitive advantage over nearby dry cleaners who do not offer this service.

## Different types of TOWS analysis

The most basic TOWS analysis will examine how threats and opportunities can be dealt with while allowing the organisation to utilise its strengths and weaknesses to meet its key success factors. Merely producing long lists of threats, opportunities, weaknesses and strengths indicates a lack of thought and seriousness as regards the organisation. Lists should be brief and specific, indicating the key and important issues. The threats, opportunities, weaknesses and strengths should be judged and assessed in relative terms. For example, there is little merit in expressing the view that a particular strength or opportunity is 'good'. Its worth should be expressed relative to how a competitor fares with the same strength or opportunity. Organisations need to aim to be better than competitors when it comes to possessing strengths and exploiting opportunities. The same holds true for weaknesses and threats: organisations need to seek to minimise the effect of these to a greater extent than their competitors.

A basic TOWS analysis should discuss, illustrate and debate the threats, opportunities, weaknesses and strengths identified and how the organisation may build on the strengths, exploit the opportunities and minimise the weaknesses and threats to a greater extent than its competitors. This will include an assessment of where the company is at the current time and where it wishes to be at some point in the future. The organisation needs to decide how far away that future is; it will vary from a few months to many years depending on the organisation, the nature of its business and its current situation.

## Guidelines for carrying out TOWS analysis

1 Identify *key* threats, opportunities, weaknesses and strengths. Do not produce a long list.
2 Once the key threats, opportunities, weaknesses and strengths have been identified, clear discussion and debate concerning them is required. The appropriate discussion and debate may be generated by considering the questions outlined below.

**Threats**

Do threats need managing more urgently than the opportunities pursued? Which threats need to be dealt with immediately and in the short term? Which threats are issues for the organisation to consider when undertaking longer-term planning? How can critical threats be offset or turned into new opportunities?

**Opportunities**

Identify new markets and market segments that might be suitable given the organisation's existing strengths and competencies. Identify changes that are occurring to existing customers and within existing markets. Consider using strategies of market penetration and market development to take advantage of any opportunities arising from existing and changing markets; *see* Chapter 6. Identify changes that need to be made to products and services. Consider strategies of product development and diversification to take advantage of any opportunities arising from changes to existing products; *see* Chapter 6.

**Weaknesses**

Decide if remedying weaknesses is more urgent than building on strengths to exploit opportunities. Does ignoring important weaknesses make the organisation vulnerable to threats, which could result in its going out of business or being taken over? How can critical weaknesses be offset or converted into strengths?

**Strengths**

Decide if the organisation has the appropriate strengths on which to build and exploit its opportunities. How can it best exploit its strengths in relation to the opportunities available to it? Which strengths should the organisation seek to develop for the future?

## Advanced TOWS analysis

A more advanced TOWS analysis involves developing a two-dimensional matrix. There are several forms that a two-dimensional matrix can take for TOWS analysis, and two are considered here.

**TOWS analysis using external environmental factors, strengths and weaknesses**[1]

First, key issues or factors from the organisation's external environment are identified. Again, this list should not be lengthy, but should concentrate on the key issues in the external environment (*see* PEST analysis, LoNGPEST analysis, market analysis and competitor analysis in Chapters 1, 2, 3 and 6). A list containing a maximum of six to eight issues should in most situations be long enough. The same process needs to be undertaken for the organisation's strengths and weaknesses. These are then assessed and scored with + or – signs depending on how they are affected by the external factors or issues. This type of TOWS analysis is shown in Exhibit 10.3 for a new university in the UK at the end of the twentieth century.

### TOWS analysis – an example – the University Business School

The University Business School (UBS) is located in a new university in the centre of a large city in the middle of England. The UBS has a full-time academic staff of around 100, located in six subject groups: marketing, strategy, human resource management, computing and management sciences, international business, and public policy. All academic staff are qualified to at least master's degree level, with 46 staff also holding PhDs in a subject area relevant to their current job. A further 12 members of staff are currently undertaking PhDs by part-time study for which UBS pays the fees. UBS also employs around 30 part-time teaching staff, around 20 of whom are postgraduate students at UBS and 10 have taken early retirement from jobs in industry.

UBS has a portfolio of postgraduate taught programmes including a Master of Business Administration (MBA) and specialist master's courses in marketing, international marketing, international business, human resource management,

**Exhibit 10.3 TOWS analysis for UBS using external environmental factors, strengths and weaknesses**

| **External factors** ➡ | Competitive market | Attractive to school leavers | Demographic trends | Introduction of tuition fees for f/t students | Reduced government funding per student | Abolition of grants for f/t students | Introduction of means-tested loans | Total + | Total – |
|---|---|---|---|---|---|---|---|---|---|
| **Main strengths** ⬇ | | | | | | | | | |
| Capacity for developing vocational courses | ++ | ++ | + | + | –– | + | + | 8 | 2 |
| Good links with local FE collage | + | 0 | + | + | – | + | + | 5 | 1 |
| Strong part-time courses | + | 0 | + | + | + | + | + | 6 | 0 |
| Research rating | ++ | 0 | + | 0 | –– | 0 | 0 | 3 | 2 |
| Well-qualified academic staff | ++ | ++ | ++ | ++ | –– | + | + | 10 | 2 |
| **Main weaknesses** ⬇ | | | | | | | | | |
| Lack of staff accountability | –– | – | 0 | – | –– | – | – | 0 | 8 |
| High level of debt | – | 0 | 0 | – | –– | 0 | 0 | 0 | 4 |
| Inadequate student support systems | –– | –– | – | – | –– | – | – | 0 | 10 |
| Total of + from strengths and weaknesses | 8 | 4 | 6 | 5 | 1 | 4 | 4 | | |
| Total of – from strengths and weaknesses | 5 | 3 | 1 | 3 | 13 | 2 | 2 | | |

management sciences and public-sector management. The undergraduate provision mirrors this with degree courses in BA Business Studies (BABS), BA Industrial Marketing (BAIM), BA Human Resource Management (BAHRM), BA Public Policy (BAPP), BA International Business (BAIB) and BSc Management Sciences (BSMS). UBS also offers two BTEC courses, the Higher National Diploma in Business Studies (HND, full time) and Higher National Certificate in Business Studies (HNC, part time). The BABS, BAIM and BAHRM degree programmes are also available on a part-time basis.

UBS has a very strong reputation locally and regionally as a provider of top-quality business and management education and many part-time postgraduate students will travel to UBS rather than attend the university in their home city. The average distance travelled by a part-time postgraduate student from outside the home city of UBS is a return trip of 63 miles. The part-time undergraduate courses attract students from a slightly more contained geographic area, with the average distance travelled by a part-time undergraduate from outside the home city of UBS being 48 miles.

A TOWS analysis of UBS is shown in Exhibit 10.3. If such an analysis is done carefully, it is possible to identify that the important strengths of the business school are its ability to develop vocational courses and its retention and development of a highly qualified academic staff. Therefore the business school's core competencies lie in course development and its HR policy. It is also very clear that the biggest weaknesses and areas for development are student support systems and academic staff accountability. This may require the business school to develop a more extensive system of course leaders and personal tutoring, along with liaising more closely with the business school's administrative staff who are responsible for communication with the students regarding for example, exam dates and results. This is currently a lengthy and time-consuming process that means students have to wait up to eight weeks after examination boards have met to receive written confirmation of exam results. The issue of staff accountability can only be addressed via the successful implementation of a suitable appraisal system, which is likely to require a change in culture among the academic staff concerned.

It is also clear that the biggest threat faced by the business school is the decreasing amount of government funding that each student attracts and there is little to counter this in what UBS is currently doing. The next main threat is to the business school's competitive position, with other universities and colleges possibly able to offer similar courses and perhaps lower fees. However, this threat is fairly well countered by the fact that the business school has vocational courses, which are well taught and are also available as on a part-time and full-time basis.

### TOWS analysis using threats, opportunities, strengths, weaknesses and the four functions of business

The second method for undertaking a matrix-type TOWS analysis is to consider the threats, opportunities, weaknesses and strengths in relation to the four key functions of business: marketing, operations management, finance and human resource management. This type of TOWS analysis is shown in Exhibit 10.4, which is an analysis of the restaurant MPW, named with the initials of the chef Marco Pierre White, who has opened the restaurant with his joint-venture partner Granada. The information for this TOWS analysis is from the entry case study for this chapter.

It can be seen from Exhibit 10.4 that the chief threats arise from other restaurants, particularly developing chains that may move into many of the locations in which MPW may be interested. Therefore MPW is going to have to be quick off the mark to obtain desirable locations into which to expand. MPW is also going to have to seek to retain good staff and prevent them being poached by other expanding chains. It is also going to have to ensure a good and stable relationship with its joint-venture partner the Granada group.

The key opportunity is expansion beyond London, via a strategy of market penetration and market development, which will involve offering the same type of reasonably priced food in cities around the UK.

The principal weaknesses arise from staff working flat out during peak periods and not being able to offer a high quality of service in the quieter times that directly follow those peak periods. This is very likely to influence some customer's opinion of the restaurant.

**Exhibit 10.4 TOWS analysis for MPW restaurant using threats, opportunities, weaknesses and strengths and the four functions**

| | Marketing | Operations | Human resources | Finance |
|---|---|---|---|---|
| **Threats** | • Other restaurants, particularly those that have a celebrity chef, like Anthony Worrall Thompson or Gary Rhodes | • The possible reducing number of locations available to expand into, particularly in London. This will be affected by the increasing number of similar restaurant chains seeking to expand in the same way and possibly drive up the cost of suitable locations that do exist | • Other restaurants opening in London and seeking to expand may seek to poach staff, particularly a talented chef who is a good cook and able to put together an imaginative menu | • If MPW is not very successful, Granada may choose to withdraw from the joint venture and cease to provide any expected financial resources |
| **Opportunities** | • Expansion around the UK, aim to have an MPW in every major city | • None mentioned in the case study | • None mentioned in the case study | • None mentioned in the case study |
| **Weaknesses** | • None mentioned in the case study | • Poor service from both waiting staff and the receptionist<br>• Some dishes served in the restaurant are not enjoyed by the diners | • Staff overworked during busy periods and offering poor service outside peak periods | • None mentioned in the case study |
| **Strengths** | • Marco Pierre White, MPW name<br>• Imaginative menu | • Some dishes served in the restaurant are greatly enjoyed by the diners | • Good chef, responsible for cooking and maybe putting together imaginative menu | • The MPW restaurant is the result of a joint venture with Granada, which will be an asset particularly if a chain of MPW restaurants is developed throughout the UK, as Granada has experience in managing chains of catering outlets<br>• Dishes on the menu are reasonably priced |

The fundamental strengths of MPW undoubtedly rest with both its name and the imaginative and high-quality food offered at reasonable prices. These strengths of imaginative and high-quality food at reasonable prices are key factors for success in the restaurant business. Therefore these strengths, supported by a sound joint venture with Granada, put MPW in a convincing position to pursue an expansion or 'rollout' of MPW restaurants.

## Conclusion

A TOWS analysis of an organisation by any of the methods outlined in this chapter provides a chance to draw together the key issues affecting the organisation, giving an overview of the organisation in both its internal and external environments. An assessment of the overall position of an organisation and a focus on the key issues provided by the TOWS analysis offer useful starting points when developing plans and strategies for the longer-term future.

ETHICAL ISSUES CASE STUDY 10.2 FT

# Business of travel

**By Amon Cohen**

Bob Taylor has got to know a good number of hotel chefs since becoming paralysed in a motor accident 21 years ago. As a wheelchair user, he finds that the front doors of many hotels are inaccessible to him so he enters via the kitchen service areas. 'Going in with the food and booze, I often get to see what is on the menu before anyone else,' says Mr Taylor.

Following his accident, Mr Taylor entered a 'new world of utter discrimination' whenever he tried to travel on business. As managing director of Birmingham airport until three years ago, he was at least in a position to rectify matters on his home patch. However, the overwhelming view among disabled travellers is that although facilities are improving in both the air and hotel industries, progress has been painfully slow.

Most disabled travellers have a horror story to tell. Martin Fortune, an aid worker in developing countries, had actually crossed the threshold of an Air India aircraft at London Heathrow when he ran into trouble.

Mr Fortune had told the airline that he was disabled and had obtained a signed form from his doctor confirming his fitness to travel. Nevertheless, as soon as an Air India cabin crew member saw that he was unaccompanied he was removed from the aircraft. Only hostile media coverage initiated by a journalist friend and a telephone call from his MP put Mr Fortune on a flight the following day.

The return journey was not uneventful, either. Although Mr Fortune made it back to the UK, his wheelchair did not – at least, not for another week.

Anne Begg, MP for Aberdeen South, also encounters discrimination as the user of a wheelchair. The Aberdeen to London sleeper train has no facilities for the disabled, nor a seating compartment where she could remain in her wheelchair. Consequently, she commutes from Scotland with British Airways, which looks after her well.

'Once they start to recognise you, it works well. If you shift airlines it can be awkward,' says Miss Begg.

That happened during the strike by BA staff a few months ago. She opted to fly instead with Air UK, which presented her with a form asking if her smell, appearance or behaviour were likely to cause offence to other persons.

'When I asked why I should fill this form in, they told me it was because I was ill. I said I wasn't ill; it's just that my legs don't work,' says Miss Begg. 'Like anyone else, I would not be travelling a lot if I were not healthy.' Air UK has since apologised and amended its form, but according to Bob Taylor, who is now chairman of a travel pressure group, the Tourism Advisory Forum, disabled business travellers are generally treated better than that.

'All UK airports are now well-equipped and so are UK airlines,' he says. Honourable mention should also go to the facilities offered by Eurostar, both on its trains and at its terminals.

Disabled travellers can expect high-quality airline and hotel facilities in Australia and – above all – in North America, thanks to a combination of progressive attitudes, more space and newer hotels.

There is also some improvement in accommodation east of the Atlantic. Forte Posthouses in the UK and the Irish Republic provide typical examples of the new thinking: many of the group's premises have specially converted rooms.

The main difference in these is that many of the features are nearer the ground, such as bed, light switches, coffee-maker and mini-bar. The same is true for the wardrobe, which is doorless. There is a sliding door to the bathroom, where the toilet, sink and bath have all been adapted, and there are two alarm cords. Furthermore, there is a connecting door to the adjacent bedroom for escorted travellers.

None of the facilities would make the rooms uncomfortable for an able-bodied person, and Forte says it does

**Ethical issues case study 10.2** ***continued***

not sell them only to the disabled. Prices are the same as for any room.

Two hotel chains which are particularly recommended for consistently good disabled facilities are Novotel (used regularly by both Mr Taylor and Miss Begg) and Travelodge. Holiday Inn and Sheraton are also said to be improving fast, while another progressive chain is Thistle Hotels.

The five adapted rooms at Thistle's Mount Royal property in central London include features such as remote-controlled door openers and – for the deaf – special telephones and a vibrating pillow alarm. There is a low-level desk in reception for checking in wheelchair users.

Robert Peel, Thistle's chief executive, appreciated the difficulties disabled guests encountered when he spent a day visiting his hotels in a wheelchair. 'I found that the wheelchair narrowed my options to the parts of the hotel where I knew I would not have problems,' he said.

Mr Peel has set a strategy of converting one in 50 of Thistle's entire stock of 13 450 bedrooms to disabled use over the next five years.

At the same time, Thistle is training its staff to ensure attitudes match infra-structure. 'Disabled guests don't want charity – they just need the sensitivity that things need to be easier to negotiate,' says Mr Peel.

*Source*: *Financial Times*, 20 November 1997. Reprinted with permission.

**Questions for ethical issues case study 10.2**

1. Perform a basic TOWS analysis of the strategy that Thistle Hotels are pursuing to make their facilities accessible to disabled travellers.
2. Identify the key success factors that a hotel will have to fulfil if a strategy to make facilities accessible to disabled travellers is to succeed.

## EXIT CASE STUDY 10.3

FT

# Bingo operators try to weigh up the numbers game

**By Scheherazade Daneshkhu**

Is bingo's number up? The recent rush by some of the largest operators to seek buyers for their bingo businesses appears to suggest so.

Bass yesterday blamed increased supply, falling demand and 'intense competition' for the poor performance of its Gala clubs. Yesterday it followed Rank and First Leisure in writing down the value of its bingo clubs.

Yet the industry is arguably better placed for recovery than at any time since the National Lottery's launch in 1994. Advertising restrictions were removed in April and other measures, such as increasing the size of prize money, are in the pipeline.

Last week, Vardon, the smallest of the four main operators, sold its 19 bingo clubs for £30.5m to a management buy-out team backed by HEV, the venture capital group.

Eyes are down at First Leisure, the third largest operator, to turn around the lossmaking business. Like Bass, the second largest, it is in talks to sell its bingo interests. That would leave Rank, the largest force, as the only quoted company with significant interests in the £1.3bn market, which attracts about 2m players a week.

Competition from the Lottery has hit the industry hard. Admissions are only recently recovering from a 15 per cent decline. But in the race to regain market share, mainly by building new clubs with parking and other improved facilities, infighting has broken out.

First Leisure has emerged with the most bruises, the only one of the four largest participants with losses in its bingo division. The business, which accounts for 15 per cent of group turnover, lost £1.3m at the operating level in the six months to April 30, despite a £61m development programme.

Profitability has also suffered at Bass and Rank, where bingo accounts for a much smaller proportion of the business. Gala's operating profits have almost halved over three years from £40m in the year to September 30 1994 to £24m this year.

Rank, which has invested more than £200m in its Mecca bingo division over the past four years, is beginning to experience recovery. It does not highlight bingo numbers but analysts estimate that bingo operating profits will rise from £34m last year to £37m this year – still well below £53m in 1994.

Yet First Leisure had been quick off the mark in building purpose-built clubs to appeal to younger and new players and to shake off the image of bingo as a game played by elderly women in smoky, converted cinemas. The age of the average bingo player is 48, down from 50 in 1991.

Fourteen of First Leisure's 20 clubs are new style and 13 of Vardon's 19 clubs were also purpose-built.

**Exit case study 10.3 *continued***

'Vardon and First Leisure had identified bingo as a solid business but one in which there had been no new thinking,' said Bruce Jones, leisure analyst at Merrill Lynch. 'Rank and Bass initially did little. Then the Lottery hit and Rank reacted by making a major investment in the business.'

More than a third of Rank's and Bass's estates are now new style. Each owns about 130 clubs.

Michael Grade, the executive chairman of First Leisure, said the group had underestimated the reaction from competitors, leading to head-to-head competition over new clubs in some areas. Prize money was also not as high as that offered in some competitors' clubs and 'although we succeeded in attracting new members to the clubs initially, the business evaporated once free prizes were gone', he said in June.

Others, such as Gala, have tried to attract new players with cabarets and shows. However, Andrew Hunter, leisure analyst at ABN AMRO Hoare Govett, believes the lure of bingo still resides in offering large prizes – currently capped at £250 000. 'Various efforts have been made to attract new players but ultimately they tend to go where the big prize money is.'

Peter Taylor, European Investment director at HEV, believes good returns are to be had in bingo, particularly by opening clubs in towns too small to interest Rank and Bass, where he says return on investment can be as high as 30 per cent.

Vardon says its decision to leave bingo is linked to its desire to accelerate development in the fast-growing health and fitness industry. It points to the rising profitability of its bingo division as evidence that it was not squeezed out by the larger players. Bingo operating profits almost doubled last year to £3.2m from £1.6m in 1995 on turnover up 23 per cent to £21.8m.

But the Lottery has taken its toll and bingo deregulation has been slow. 'What had been a stable business became unstable,' said Steven Palmer, Vardon's finance director. The group's new bingo clubs 'struggled' to make investment returns of 20 per cent, unlike health and fitness which he says is producing higher returns.

Despite leaving bingo, he believes it has a future. 'People like gambling and they will get bored with the Lottery. After all, bingo lasts all night whereas the Lottery is over in a few seconds.'

*Source*: *Financial Times*, 4 December 1997. Reprinted with permission.

**Question for exit case study 10.3**

1 Using the information from the case study, perform an advanced TOWS analysis for a bingo operator. The TOWS analysis should use external environmental factors, strengths and weaknesses.

2 Identify the key success factors that a bingo operator will have to possess if it is to surpass past profits. Identify competitors in the industry and suggest how competitive advantage could be achieved.

## Short-answer questions

1 Describe the key concerns of a TOWS analysis.

2 In the context of this chapter, define a threat.

3 In the context of this chapter, define an opportunity.

4 In the context of this chapter, define a weakness.

5 In the context of this chapter, define a strength.

6 Briefly outline the procedure for performing a basic TOWS analysis.

7 Explain the term 'key success factor'.

8 Explain the term 'competitive advantage'.

9 Explain the term 'premium price'.

10 Briefly outline one of the two procedures for performing a more advanced TOWS analysis.

## Assignment questions

1 Choose one of the categories listed below that you have not studied before and identify the main companies operating in the category you have chosen. Identify the key success factors and possible areas of competitive advantage for the category you have chosen. Choose one of the main companies you have identified and perform a basic TOWS analysis of this company.

2 Choose one of the categories listed below that you have not studied before and identify the main companies operating in the category you have chosen. Identify the key success factors and possible areas of competitive advantage for the category you have chosen. Choose one of the main companies you have identified and perform an advanced TOWS analysis of this company using external environmental factors, strengths and weaknesses.

3 Choose one of the categories listed below that you have not studied before and identify the main companies operating in the category you have chosen. Identify the key success factors and possible areas of competitive advantage for the category you have chosen. Choose one of the main companies you have identified and perform an advanced TOWS analysis of this company using threats, opportunities, strengths, weaknesses and the four functions of business.

List of categories for TOWS analysis:

- Manufacturers of greeting cards.
- Manufacturers of pharmaceuticals.
- High-street financial institutions in the UK.
- Passenger road transport companies.
- Airlines operating on North Atlantic routes.
- Ferry companies operating out of English, Scottish and Welsh ports.
- Publishers of non-fiction books in the UK.
- Producers of frozen food in the UK.
- Footwear manufacturers whose goods sell in the UK.
- Furniture retailers in the UK.

## Reference

1 Johnson, G and Scholes, K (1999) *Exploring Corporate Strategy*, 5th edn, Harlow: Prentice Hall Europe.

## Further reading

Ambrosini, V with Johnson, G and Scholes, K (1998) *Exploring Techniques of Analysis and Evaluation in Strategic Management*, Chapter 8, Harlow: Prentice Hall Europe.

Bennett, R (1999) *Corporate Strategy*, 2nd edn, Chapter 5, London: Financial Times Pitman Publishing.

# 11 Stakeholder analysis

## Organisational context model

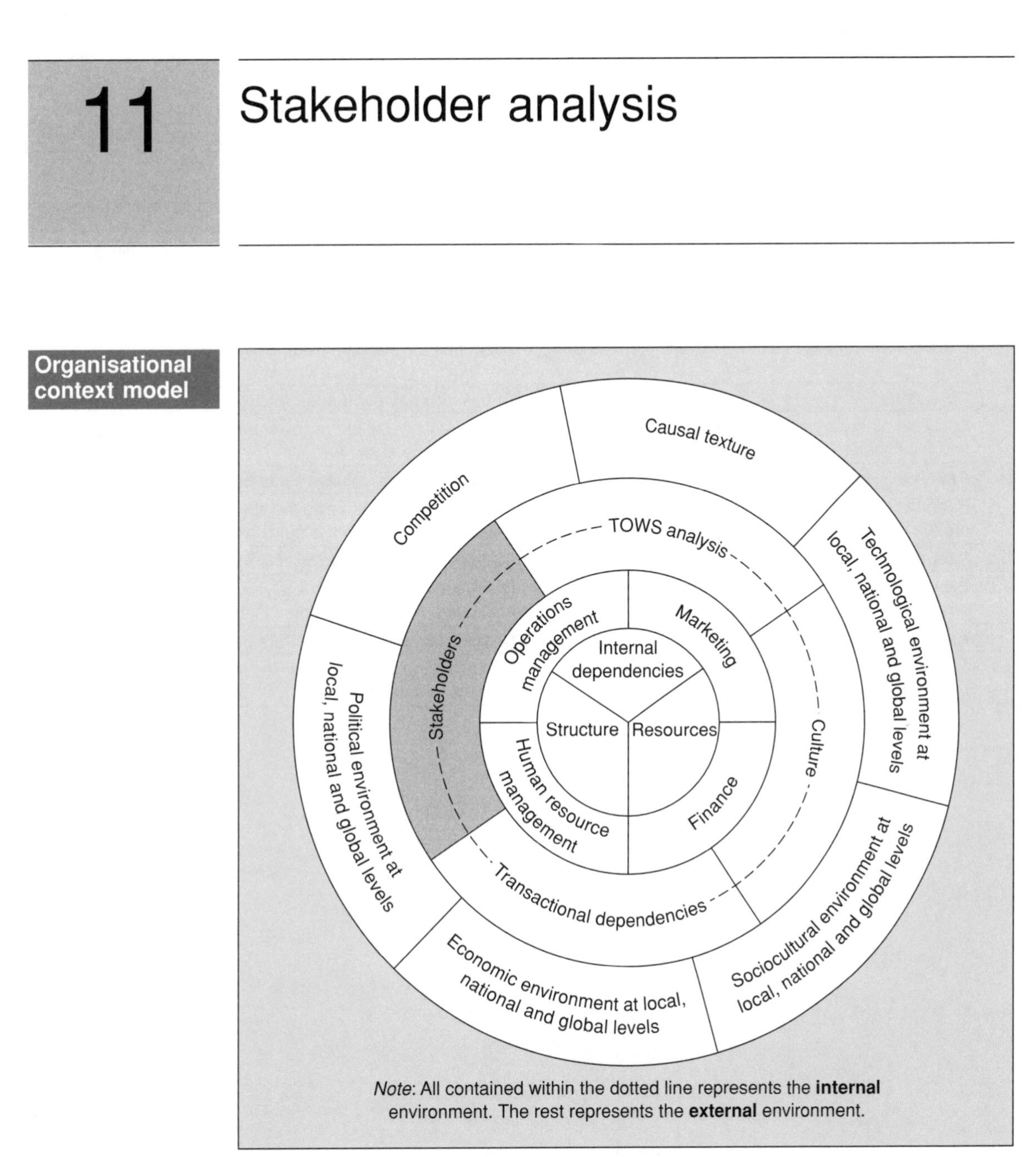

**Exhibit 11.1**

## Learning outcomes

While reading this chapter and engaging in the activities, bear in mind that there are certain specific outcomes you should be aiming to achieve as set out in Exhibit 11.2 (overleaf).

**Exhibit 11.2 Learning outcomes**

| | Knowledge | Check you have achieved this by |
|---|---|---|
| 1 | Define a stakeholder | stating in one sentence what a stakeholder is |
| | **Comprehension** | **Check you have achieved this by** |
| 1 | Understand why stakeholders can be important to organisations | explaining the role stakeholders play in organisations |
| 2 | Indicate who the key players are to an organisation | identifying the essential characteristics of key players |
| 3 | Explain how analysis of stakeholders is performed | illustrating the steps of stakeholder analysis |
| | **Application** | **Check you have achieved this by** |
| 1 | Apply the power and interest matrix to an organisation and stakeholders with which you are familiar | demonstrating use of the power and interest matrix on an organisation you know |
| | **Analysis** | **Check you have achieved this by** |
| 1 | Perform comparative stakeholder analysis for an organisation of your choice | comparing and contrasting the stakeholder analysis of an organisation in different circumstances |
| | **Integration** | **Check you have achieved this by** |
| 1 | Explain how stakeholder analysis may aid in the analysis of an organisation's environment | formulating links between analysis of an organisation's environment and stakeholders in the organisation |
| | **Evaluation** | **Check you have achieved this by** |
| 1 | Evaluate the relevance of stakeholder analysis to organisations today | evaluating the benefits and disadvantages of stakeholder analysis to organisations |

ENTRY CASE STUDY 11.1

FT

# When the pinstripe and bobble hat get hitched

**By Patrick Haverson**

The City has heard of shareholder action groups before, and the football world has plenty of experience of troublesome supporters' factions, but BSkyB's controversial £624m takeover bid for Manchester United has produced a unique blend of the two.

Shareholders United Against Murdoch is the product of this unusual marriage of the pinstripe and the bobble hat, and since the satellite broadcaster unveiled its agreed bid in early September the group has been making a thorough nuisance of itself. Only this week the football club was forced to issue a statement to the Stock Exchange, responding to an analysis of the takeover deal that SUAM sent out to United's institutional shareholders as part of its campaign to block the bid.

The people behind SUAM are a striking roll-call of the City, media world and academia. Its founders include Michael Crick, a respected author and reporter on *Newsnight*, the BBC's flagship current affairs programme, Richard Hytner, head of the Publicis advertising group and former chief executive of the Henley Centre for Forecasting, and Jonathan Michie, professor of management at the University of London.

United by their love for a club they have followed since their youth, and backed by funding from an anonymous wealthy fan, SUAM's members have taken the battle to BSkyB and the club's board.

A sophisticated media campaign has been conducted, all 30 000 of the club's shareholders have been contacted and lobbied, and City institutions approached for their support. With people like Andrew Salton, a City fund manager, on board, SUAM is in the position to communicate with institutional shareholders in their own language.

He says he joined SUAM for financial as well as personal reasons. 'I bought the shares as an investment and I think they'll do extremely well in the next five years. In the offer document they make a good case for how successful United has been and how rosy the prospects are – it is just not the sort of case usually made by a board that has agreed to sell out. I find it hard to understand why they should (accept the bid).'

SUAM has also consulted the Takeover Panel about the group's contention that the offer document sent to United shareholders was misleading because it failed to tell shareholders they had the right to reject the offer. As Mr Crick says: 'The bulk of United's individual shareholders have probably not got shares in anything else, and will not have come across offer documents before.'

SUAM failed to convince the panel of its case, but the group continues to snipe at the BSkyB deal from its lair in west London. The recent document sent to institutional shareholders arguing the BSkyB bid undervalued United included the supportive views of two stockbroking firms and Alex Fynn, a football marketing expert.

Mr Crick says the City has been taken aback by SUAM's campaign. For example, when the group visited the Takeover Panel 'they did seem to be surprised by how well we'd marshalled our arguments and the kind of people we were. Because we were normal football supporters they were expecting us to turn up in United scarves and hats.'

One high-ranking City banker has observed SUAM at close quarters and admits he has seen nothing like it before. 'It's unusual for shareholder action groups to develop in this way,' he says. 'Normally, it would develop through one or more institutions. These are individual investors who feel very strongly about the situation . . . and they are highly able people. I think they're doing a remarkable job. It puts the efforts of some highly-paid advisors in the shade.'

Anne Simpson, director of Pirc, the shareholder advisory group, is not surprised the BSkyB bid for United has produced such a committed shareholder pressure group. 'Football clubs are the original example of the stakeholder company. The supporters are stitched in as fans, shareholders and often members of local community.'

She believes a takeover of Manchester United does not just raise issues about how much the bidder should pay but also about whether it is in the best interests of supporters and the wider community. 'Fans are stakeholders and everyone from the DTI downwards is trying to think of the purpose of a corporation as broader than just creating shareholder value,' says Ms Simpson.

Mr Crick believes SUAM, which is working closely with official United supporters groups, has a chance of blocking the bid. At the very least, he is confident of convincing enough shareholders to reject BSkyB's offer and ensure at least 10 per cent of the shares remain in ordinary supporters' hands.

That would force BSkyB to retain a listing for United shares and continue listening to the views of the shareholder-fans. 'It would help preserve some independence for United,' says Mr Crick. 'It would mean we could carry on earning dividends and going to AGMs. It would not be great, but it would be something.'

*Source*: *Financial Times*, 24 October 1998. Reprinted with permission.

**Exhibit 11.3 Stakeholders' power and interest**

| Internal stakeholders | Stakeholder interests are: | Stakeholder power arises from: |
|---|---|---|
| Employees<br>Managers<br>Directors | • security of employment<br>• wage levels<br>• fringe benefits<br>• responsibility<br>• promotion prospects<br>• working conditions | • job grade or title<br>• position in organisational hierarchy<br>• personal reputation<br>• departmental reputation |
| Trade unions | • number of union members in the organisation<br>• same as its members (*see* list in box above) | • number of union members<br>• nature of bargaining (local or national) |
| Shareholders | • profit levels<br>• size of dividend payments<br>• capital growth in share price | • number of shares held |
| **External stakeholders** | | |
| Suppliers | • size and value of contracts<br>• speed of invoice payment | • location and availability of other suppliers |
| Customers | • quality of goods and services available<br>• prices and payment terms | • location of other suppliers<br>• quality of goods and services offered by other suppliers<br>• prices and payment terms offered by other suppliers |
| Competition | • quality of goods and services available<br>• prices and payment terms | • behaviour of other competitors |
| Financiers | • how promptly repayment of large and short-term loans occurs | • offering better deal (improved quality or better prices and payment terms) |
| Government | • payment of corporation tax<br>• implementation of legislation (e.g. competition and employment legislation) | • enforcing the legislation via the legal system if necessary |

## Introduction

Stakeholders are any individual or a collection of individuals with an interest in an organisation. Some stakeholders will be internal to an organisation and others will be external. Internal stakeholders include employees, managers, directors, trade unions and shareholders. External stakeholders include suppliers, customers, competitors, financiers, government and the general public. Various categories of stakeholder will affect or be affected by the organisation in diverse ways, hence stakeholders have different interests or stakes in the organisation. This is shown in Exhibit 11.3.

Stakeholders are also able to influence an organisation to act in their best interests. However, the interests of different stakeholder groups will vary and may even conflict with each other. For example, employees may seek high wages and above-inflation pay rises, while in contrast customers would prefer lower prices and lower costs, which are not possible if labour costs are high. The interests of stakeholders in an organisation and the way in which power is exercised by stakeholders are shown in Exhibit 11.3.

## Stakeholders and the organisation

An organisation's stakeholders will be important for an assortment of different reasons and to varying degrees, therefore different stakeholders will respond to the organisation and its behaviour in different ways. Stakeholders whose interests and expectations are met will tend to remain with the organisation. Unsatisfied stakeholders will leave or remain and use their sources of power in an attempt to persuade the organisation to meet their expectations or interests. This occurred in the entry case study to this chapter, when Manchester United fans who were against the takeover of 'their' club by BSkyB banded together to form Shareholders United Against Murdoch (SUAM) and clearly attempted to influence Manchester United, the regulatory bodies, institutional shareholders and the City to reject the BSkyB bid.

Stakeholders who experience a high level of satisfaction with an organisation will tend to demonstrate loyalty and choose to retain their position as stakeholders. For example, employees who feel that their well-paid jobs are secure and offer future prospects are likely to remain with that employer. In contrast, stakeholders who are disappointed with the organisation and its behaviour are more likely to relinquish their stake. The likelihood of an unhappy stakeholder withdrawing their stake in an organisation is increased if better opportunities and potentially greater satisfaction appear to be available by acquiring a similar stake in a different organisation. For example, shareholders in a company who feel that they are not gaining a good enough return on their investment may decide to sell their shares and invest the money in a company that will give a better level of return.

Alternatively, stakeholders who are unhappy with the organisation may decide to remain and attempt to change things. For example, unsatisfied shareholders may decide to try to influence changes to the organisation's leadership and strategies, with the aim of benefiting in the long run. To achieve this they

**Exhibit 11.4**
**Stakeholder diagram**

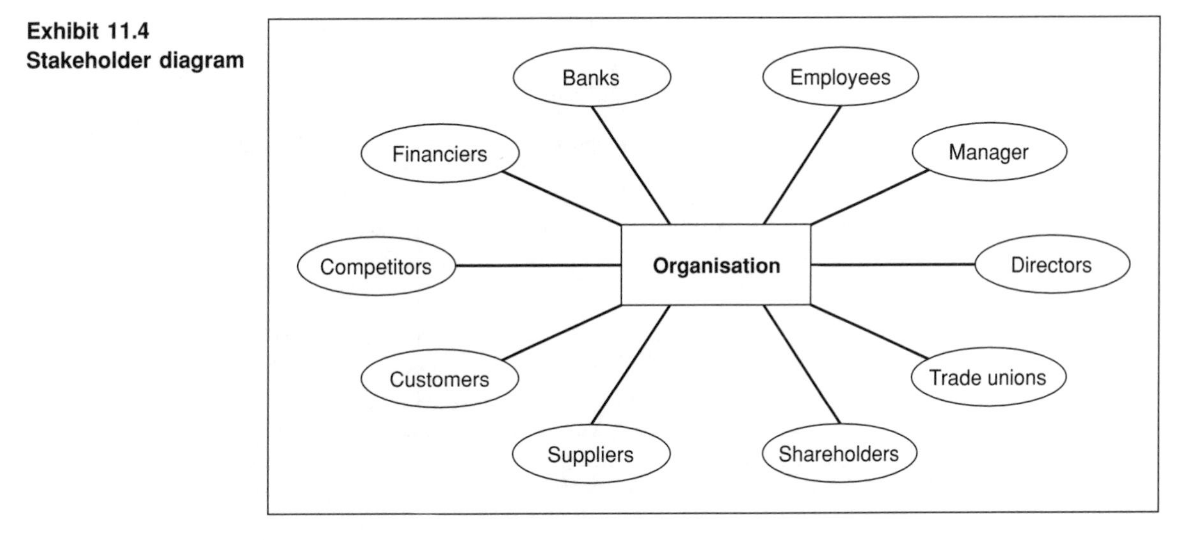

will have to be able to exert the necessary amount of influence on planning and decision making within the organisation. This requires a suitable combination of authority, determination and ability. It is usually large institutional investors who stand the best chance of being successful with this type of approach, as they have greater power than smaller investors.

## Analysing stakeholders

The analysis of stakeholders involves identifying who they are and considering their power and interest with regard to the organisation. Stakeholders can be identified by brainstorming and shown on a stakeholder diagram; *see* Exhibit 11.4 on p. 331. Once identified, the relative power and interest of the stakeholders can be mapped on to a power and interest matrix; *see* Exhibit 11.5.[1]

### Stakeholders with high power and high interest (category D)

Stakeholders with high power and high interest are key players in the organisation and are often involved in managing the organisation and its future. If key players are not directly involved in managing the organisation, then it is vital that they are given serious consideration in the development of long-term plans and the future direction of the organisation, as they have the power to block proposed plans and implement their own alternative agenda.

### Stakeholders with high power and low interest (category C)

Stakeholders with high power and low interest are those who must be kept satisfied. A good example of stakeholders who must be kept satisfied are institutional shareholders. Institutional shareholders will often remain compliant

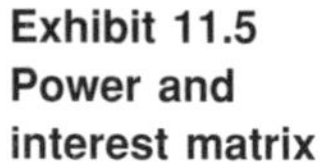
**Exhibit 11.5 Power and interest matrix**

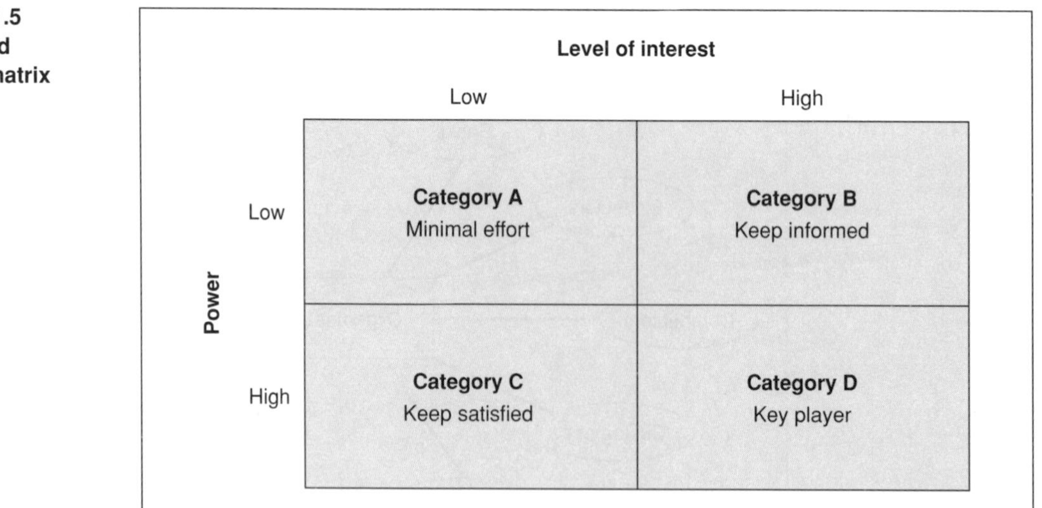


*Source*: Johnson, G and Scholes, K (1999) *Exploring Corporate Strategy*, 5th edition, Prentice Hall Europe. Reprinted with permission.

while they receive acceptable returns on their investment and are pleased with the organisation's management and activities. However, the ability of category C stakeholders to reposition themselves on the power and interest matrix into category D and become stakeholders with a continuing high degree of power and an increase in their level of interest should not be under-estimated. This occurs when category C stakeholders are not kept satisfied and feel that their interests are not being best served. Hence stakeholders with high power and low interest will increase their level of interest to make sure that their interests are met. The shift in position of unsatisfied category C stakeholders may impede an organisation's plans and prevent the expectations of key players or category D stakeholders being met as expected.

Therefore a canny organisation will ensure that the expectations of category C stakeholders are well met and the necessary adjustments made to meet changing expectations arising as the current issues facing the organisation change. This helps ensure that category C stakeholders do not feel that their interests are being marginalised at the expense of the interests of key players, category D stakeholders. Hence the repositioning of category C stakeholders should not be an unexpected occurrence if they are managed appropriately. This requires a good working relationship and open channels of communication to be developed between category C stakeholders, the organisation and key players or category D stakeholders.

### Stakeholders with low power and high interest (category B)

The stakeholders in category B are those with low power and high interest, who are able to exert relatively little power in influencing the organisation and its actions. However, these stakeholders have a high level of interest in the organisation and will voice their concerns if that interest is not being considered in a suitable manner. If category B stakeholders voice their concerns loudly enough and in the right way, e.g. via lobbying or petitions, they may be able to influence one of the powerful group of stakeholders in either category C or D and affect their behaviour. Therefore organisations need to keep category B stakeholders informed of the organisation's activities and decisions and in doing so convince them that their interests are being taken into account and considered seriously.

### Stakeholders with low power and low interest (category A)

Stakeholders with low power and low interest are those in whom the organisation need only invest minimal effort.

## The Automobile Association and its stakeholders

It should be recognised that the position of stakeholders on the power and interest matrix is dynamic and will vary over time according the current issues that the stakeholders have to consider. The situation in which the Automobile

Association (AA) found itself during April 1999 provides a good example of an organisation with groups of stakeholders who line up in a certain way due to a particular issue, in this case demutualisation.

The AA was founded in 1905 and by 1999 held around half the motor breakdown market, a market that was experiencing significant change. These changes included the acquisition of Green Flag by Cendant; the entry of the insurance company Direct Line into the market; and the RAC's expected trade sale or flotation. Therefore in April 1999 the AA considered its options with regard to retaining its mutual status or demutualising. It was rumoured that Ford had informally approached the AA with a takeover offer that would end the latter's mutual status. Other interested bidders were thought to include Centrica and a number of venture capitalists. The director-general of the AA, John Maxwell, initiated a strategic review to allow the AA to assess its options. The options available included demutualisation, a joint venture with a suitable partner or takeover by another company. The merchant bank Schroders was advising the AA.

The AA has annual sales of around £600m from its businesses, which include roadside service, publications and driving schools, and its value is estimated to be between £1bn and £1.5bn. Pursuit of the demutualisation option and stock-market flotation would give each full member of the AA a moderate windfall of £200–250. The AA has 9.5m members and 4.3m are full paying members who would receive the windfall payouts. However, excluded from the demutualisation windfall would be the 1.7m associate members, including the families of full paying members who benefit from the association's services; and the 3.5m members who are drivers of fleet cars with AA cover and drivers who received their AA membership as part of a package when purchasing a car.

**Exhibit 11.6 Power and interest matrix for the Automobile Association (AA)**

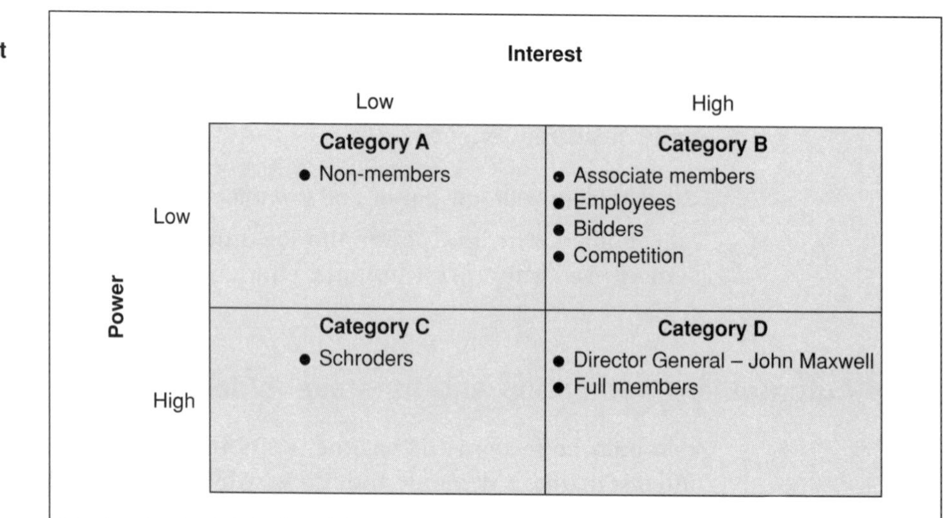

### The AA and stakeholders with high power and high interest (category D)

The key players were the director-general of the AA, John Maxwell, and his immediate management team carrying out the strategic review of the AA, as well as the full members of the AA; *see* Exhibit 11.6. John Maxwell and his management team were key players with high power and high interest, as their planning and decision making would determine their future with the AA; the future of the AA; the future of those who work for the AA; and the future of AA members. The full members would collectively decide if the AA was to demutualise. They might have chosen to support any demutualisation recommendations made by John Maxwell and his team, or reject them in favour of a bidder, such as Ford, buying the AA. The full members, for example, might have decided this if they were to lose confidence in John Maxwell and his management team and their ability to carry out the demutualisation successfully. Alternatively, full members might have taken the following view, expressed by one of their number in the press in April 1999:

> I got my membership when I bought a much-loved but temperamental MG Midget. If the AA does choose to demutualise, I would hope they would to pay a differential for members who have been with the AA longer. I might vote in favour of the move if they were going to pay me a £300 windfall but the down-side could be that if they become a corporate commercial entity, the cost of its services could soar.[2]

### The AA and stakeholders with high power and low interest (category C)

The merchant bank Schroders were category C stakeholders, as they had relatively little interest in whether the AA finally decided to demutualise. However, while in the position of corporate advisers to the AA, they were relatively powerful as they were able to advise and potentially influence John Maxwell and his management team.

### The AA and stakeholders with low power and high interest (category B)

The category B stakeholders, those with high interest and low power in the demutualisation issue, included associate members and employees. The associate members clearly had a high interest in whether or not the AA decided to demutualise. The primary concerns for associate members were the effect of demutualisation on the services they receive and the cost of associate membership. However as non-voting members, associates had no direct power to influence the outcome of any ballot on demutualisation. Equally, employees had a high interest in the future of the AA and would be concerned as to the effects of demutualisation. Potential effects of demutualisation on the AA could include its having to be more competitive and this being achieved via cost cutting and job losses. However, employees had no direct role in the ballot and would ultimately have to accept its outcome.

The stakeholder matrix suggests that category B stakeholders, high interest and low power, have to be kept informed, which is true of stakeholder groups like associate members and employees. In April 1999, the AA kept its members and employees informed by issuing the following statement to the media and via answerphones in its own offices:

> The AA has always kept an open mind about its structure as it pursues its prime purpose: to serve the best interests of its members. No decisions have been made in this respect.

However, also with high interest and low power were other stakeholders like potential bidders such as Ford and competitors like Direct Line and Green Flag. These were external stakeholders who had a great deal of interest in what the AA eventually decided to do, as their business and the marketplace they operate in would be directly influenced by that decision. Any organisation should be aware that any information it releases with the intention of keeping stakeholders such as employees and associate members informed will be in the public arena and therefore available to stakeholders such as competitors and potential bidders.

#### The AA and stakeholders with low power and low interest (category A)

The category A stakeholders are those with low power and low interest. For the AA, non-members fell into this category. They were unable to receive breakdown services from the organisation and had no influence over its demutualisation decision. However, it should be recognised that stakeholders' power and influence can alter over time. The opportunity of a £200–250 windfall might have encouraged some non-members to become members and move to category D, high interest and high power. This was perfectly possible, as the AA made it clear that it was not closing its doors to new members, nor was it expecting to distinguish between long-term and short-term full members:

> The AA has no intention of bringing the shutters down on membership. Everyone is as free to join the AA as they were before.[3]

> There is no distinction made among full members.[4]

If the number of new full members joining had been very large and there was no differentiation between new and longer-term members, then the value of the windfall paid to full members could have decreased. This could have pushed longer-term full members to seek to lobby or influence John Maxwell and his management team to distinguish between long- and short-term members.

### Stakeholder alliances and coalitions

When analysing stakeholders two points should be noted. First, people and organisations may belong to more than one category of stakeholder. Secondly, stakeholders and organisations may depend on one another, with the nature of the dependency varying according to the amount of power and/or interest the stakeholder has in the organisation. For example, if the director general of the AA, John Maxwell, favoured demutualisation, then he would have depended on the full-time members voting in large enough numbers for the demutualisation proposals. He would have had to recognise that full members might have been subject to influence by associate members, who may be related to full members. Similarly, some employees (category B) were also full members of the AA and how they were treated and informed as employees might have influenced their voting behaviour as full members. The employees might have felt that cost cutting and job losses were likely to result from demutualisation. Hence they might have sought to lobby and influence the voting full members to vote against a change in the AA's structure or to vote for a takeover rather than demutualisation if they thought their best interests would be served in this way. Equally, if associate members were concerned about the service they receive and its cost, they might have sought to influence full voting members, which would perhaps have been easy if the full voting members were family members. In addition, associate members and employees might have sought to influence John Maxwell and his management team directly, via letter-writing campaigns and petitions.

Therefore the arguments in favour of demutualisation had to focus on the benefits for full members (cash windfall and service levels at least maintained, preferably improved in some way); associate members (service levels at least maintained, preferably improved in some way); and employees, particularly those who were also full members (issues of job security and future operation of the AA for employees were crucial).

The members of the AA were balloted in August 1999 on the proposed sale of the AA to Centrica. The result of the ballot was announced in mid-September 1999 and showed 67 per cent of eligible members voted and 96 per cent of them voted in favour of the sale to Centrica.

## Comparative stakeholder analysis

It has been shown that stakeholder analysis can be used to identify the current position of stakeholders on the power and interest matrix. This analysis can be extended to considering the reaction, behaviour and position of stakeholders if a particular strategy or plan were to be implemented by the organisation.

A company that decided to implement a bold competitive move in May 1999 was the television company BSkyB. It announced that it was to give away free the set-top boxes that unscramble the digital signals and allow digital television to be watched. In 1998 BSkyB launched 140 digital channels and by May 1999 551 000 people had invested the £200 needed to view digital satellite television.[5] However, BSkyB's market research showed that consumers were reluctant to pay the set-up cost of £200 for the set-top box.[6] Hence it prepared to shoulder the cost of acquiring new customers, £155 per customer, by giving the boxes away free. This move took digital TV from an expensive luxury to being affordable to anyone who could pay the £40 installation fee and the monthly subscription, ranging from £7 for six channels to £32 for all channels, including three movie or sport channels.

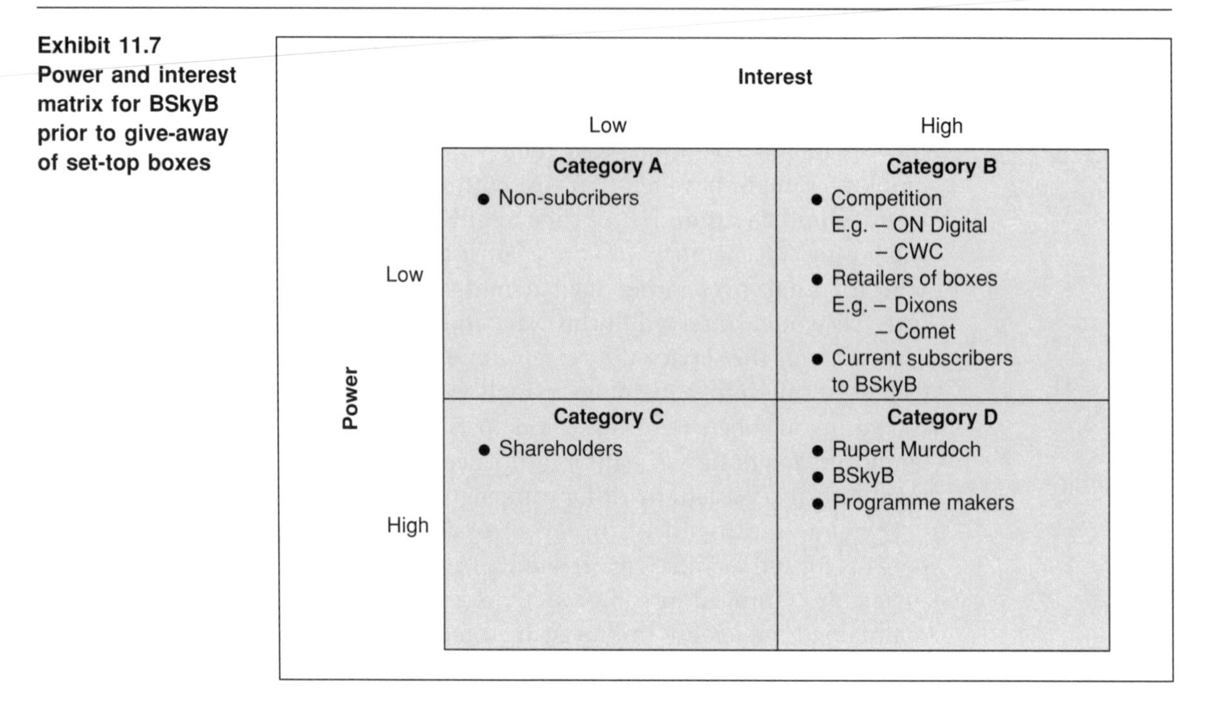


**Exhibit 11.7 Power and interest matrix for BSkyB prior to give-away of set-top boxes**

The free give-away of set-top boxes was an attempt by BSkyB to build up a customer base for digital TV. Prior to the announcement, the stakeholders in BSkyB would have lined up on the power and interest matrix as shown in Exhibit 11.7. In deciding to give away the set-top boxes free, BSkyB would have considered how stakeholders would react. The changes and positions that would be sought on the power and interest matrix are shown in Exhibit 11.8.

The important stakeholders in BSkyB in each category most affected by the free give-away were identified. BSkyB then took a number of actions to manage the stakeholders affected by this competitive move effectively, with the aim of achieving a line-up of stakeholders that would result in the give-away of set-top boxes being successful.

### Category A stakeholders, low power and low interest

The most obvious change sought by BSkyB in giving away set-top boxes was to convert non-subscribers, category A, into subscribers, category B stakeholders. This was achieved by offering the set-top boxes for free, with a £40 installation fee and subscription prices, as shown in Exhibit 11.8.

### Category B stakeholders, low power and high interest

In category B the important stakeholders were the existing BSkyB customers who had already paid out nearly £200 for a set-top box and would therefore have to be carefully managed. BSkyB wanted these stakeholders to remain in category B, hence the attempts to appease and retain these existing customers centred around holding their current subscription rates until September 2001; *see* Exhibit 11.9.

**Exhibit 11.8 Power and interest matrix for BSkyB after announcing the give-away of set-top boxes**

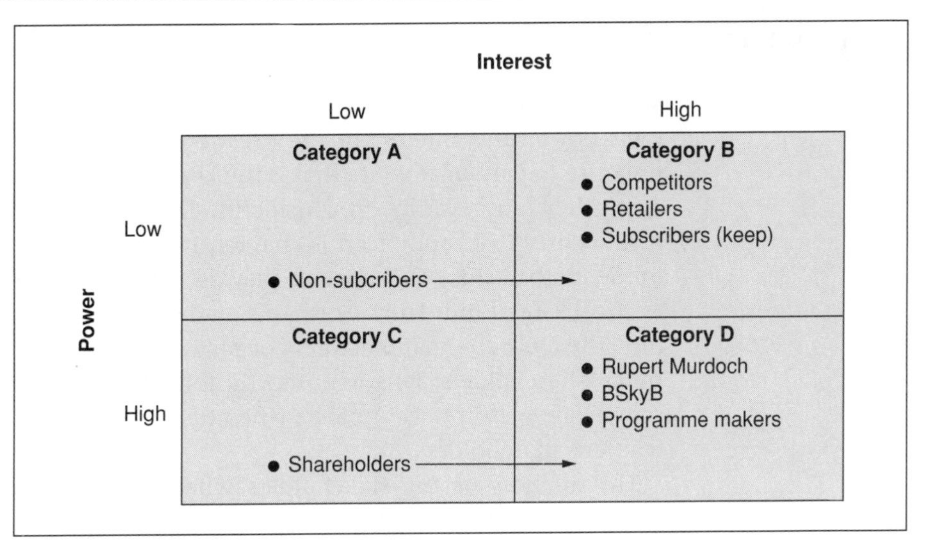


**Exhibit 11.9 BSkyB prices in May 1999**

| Package | New subscribers with free set-top box | Existing subscribers who paid for set-top box |
|---|---|---|
| 6 channels | £7 per month | £7 per month |
| 12 channels | £10 per month | £9 per month |
| 40 channels | £13 per month | £12 per month |
| All channels, including 3 movie or sport channels | £32 per month | £30 per month |

*Source*: *Daily Telegraph*, 6 May 1999.

### Category C stakeholders, high power and low interest

The shareholders, category C stakeholders, were very important in this bold competitive move to give away free set-top boxes. Objections from the shareholders would have meant that BSkyB would not have been able to afford this drive to get new customers, as the cost is borne by the shareholders in the short term, with the expectation of better returns in the longer term. The agreement of shareholders, particularly large institutional investors, meant that BSkyB could afford the free give-away. The shareholders of BSkyB have allowed the company to wipe out £315m in expected profits and to suspend dividend payments, with the immediate reward for shareholders of a leap in the share price from 542p to 607p, a 12 per cent increase, and more expected in the long term.

### Category D stakeholders, high power and high interest

The free give-away was aimed at keeping Rupert Murdoch and BSkyB as key players, with increases in their already high power and interest. For this to occur successful management of important stakeholders in the other three categories was crucial.

## Conclusion

Stakeholders in an organisation can be identified easily, but it is determining their power and interest in the organisation and how this may change in response to particular issues that is the key to analysing stakeholders and managing them successfully. In considering Manchester United, the organisation in the entry case study to this chapter, it is easy to identify the stakeholders in the football club. They include players, very talented individual players such as Ryan Giggs and Andy Cole; the manager, Alex Ferguson; the board that runs the club; the manufacturers of Manchester United merchandise; institutional shareholders; fans who mostly follow Manchester United via TV and the media; season ticket holders who attend a lot of matches; and finally fans who are shareholders.

The analysis of the stakeholders when a particular issue arises is more tricky. When BSkyB launched its takeover bid for Manchester United in the second half of 1998, Manchester United could clearly have predicted a tussle with the regulatory authorities, the OFT and MMC and a backlash from sections of the media not owned by Rupert Murdoch. A negative response from fans and from supporters who were also shareholders might have been predicted. However, this did not appear to have worried Manchester United and BSkyB when the bid was first announced. This saw the launch of a campaign group SUAM, Shareholders United Against Murdoch, comprised of small shareholders who were great fans of Manchester United. SUAM presented the case against the takeover of Manchester United by BSkyB and lobbied extensively for their cause among shareholders and city institutions in a very active and sophisticated manner. It was this alliance of small shareholders and their polished behaviour that was not predicted by the more powerful stakeholders. The identification and prediction of the actions of just such unpredictable stakeholder groups can make the management of stakeholders very good rather than just adequate. In the end, the fans and smaller shareholders got their way as the BSkyB bid for Manchester United was blocked by the government, which supported the report of the competition authorities, with some reservations. The trade and industry secretary, Stephen Byers, said:

> I accept their conclusions that the merger would damage the quality of British football by reinforcing the trend towards growing inequalities between the larger, richer clubs and the smaller poorer ones.[7]

In contrast, Kim Howells, the minister for consumer and current affairs, said:

> The day when politicians or Whitehall officials start to determine how markets should look is a death knell for competitiveness in Britain.[8]

## ETHICAL ISSUES CASE STUDY 11.2

FT

# Abbey to charge £5 for paying bills at branches

**By Christopher Brown-Humes**

Abbey National is to charge its own customers £5 per item for paying utility bills and other third party credits in its high street branches.

It is thought to be the first of the big banks to impose such a charge on its customers.

Abbey, which has 800 branches, said it wanted to encourage customers to pay their bills by telephone, post, direct debit or cash machine rather than at branch counters.

'We don't want customers to pay this charge – it is designed to encourage them to use the other methods, which are free of charge. These will allow customers to conduct such transactions more quickly and easily, and at the same time allow us to reduce queues in branches,' the bank said.

Critics said Abbey was unfairly restricting access to its facilities.

One said the charge was extortionate and 'far more than it costs Abbey to provide the service'. Another said a visit to the bank branch was a vital source of social contact for some older people.

Abbey denied the move would discriminate against poorer customers. It said all its current account holders automatically had access to the other payment methods.

The charge will be introduced on August 2.

One analyst said Abbey's move was part of an inexorable trend to move banking out of branches and into cheaper distribution channels.

Abbey dismissed suggestions that the move would lead to branch closures or job losses. It noted that direct debit was the most popular method for customers to pay their bills.

Abbey was embroiled in controversy several years ago when it introduced a £1 charge for customers who used branch counters to access their Instant Plus account.

The bank justified the move by saying that Instant Plus was a card-based account that was meant to be accessed using the bank's network of cash machines.

*Source*: *Financial Times*, 14 May 1999. Reprinted with permission.

**Questions for ethical issues case study 11.2**

1 List the stakeholders mentioned in the case study and their source of interest in the Abbey National.

2 Plot the Abbey National stakeholders you have identified on a power and interest matrix.

3 Do you think that the stakeholders affected should exercise their power and move their banking and personal finance arrangement to another bank or building society? Justify your answer.

## EXIT CASE STUDY 11.3

# Interflora

**By Claire Capon**

Interflora was founded in 1923 and is an association of florists owned by its members, which today is the largest UK flower delivery organisation, supplying four million bouquets annually, worth over £60m. The florists who belong to Interflora accept orders from around the world for flowers to be delivered in their local neighbourhood. Membership of Interflora allows co-ordination of flower deliveries and the organisation provides computer terminals for participating shops and clearing payments between its members. The updating proposals discussed in this case study proposed an increase in a florist's annual subscription charge for belonging to Interflora from £300 to £1750 and a reduction in the cost of each transaction from £2.99 to 60p.

Interflora and its affairs are managed by the members via an elected board of directors. Doug McGrath was identified by headhunters and in January 1996 was appointed chief executive of Interflora. He was recruited

**Exit case study 11.3 *continued***

to up date and overhaul the association, which was viewed by the board as a cosy trade organisation with traditional values. This was considered uncommercial and uncompetitive by Doug McGrath and the Interflora board. The updating was to take the form of the introduction of a corporate and commercial attitude to conducting business.

The updating of Interflora was announced at the organisation's 1996 Bournemouth conference. It was revealed by Doug McGrath and David Parry that Interflora was to have a mission statement and modern management thinking was to be introduced to the organisation to solve the problems faced by the organisation and the membership. The reasoning centred around the fierce competition faced by florists from garages and supermarkets, some of which offered home delivery of bouquets, e.g. Marks & Spencer and Waitrose. The view taken by Doug McGrath and the commercially minded chairman, Bristol florist David Parry, was that Interflora should ditch its traditional family values and act like a modern corporation.

Potentially, McGrath and Parry could have been a strong team that could have been of considerable benefit to Interflora, with McGrath's considerable business experience and Parry's extensive knowledge and experience of the florist trade. The view expressed by Caroline Marshall-Foster, editor of the trade publication the *Florist*, was that floristry was potentially a very profitable business and it is probably the opportunity for profit that McGrath and Parry saw.

If Interflora were to behave like a modern corporation, Doug McGrath saw opportunities for the centralisation of activities such as the purchase of flowers, which would produce economies of scale and make Interflora the largest purchaser of flowers in the UK. This would mean that all the flowers bought by Interflora would be purchased collectively on behalf of the members, giving it the power to negotiate better deals with suppliers than if members brought their flowers individually. The policy of centralised purchasing would be introduced along with standardisation of many other facets of the service provided by Interflora florists, such as opening times, service and corporate image, e.g. window displays.

David Parry and Doug McGrath left the Bournemouth conference convinced that their presentation of an updated and modern Interflora had been successful and that the conference delegates supported the proposed changes. However, they had not read the mood of the conference correctly. In fact, Interflora members saw the presentation of Interflora's image standardisation as dictatorial. This was illustrated by the views and actions of Bev Woods, an Interflora member and florist.

Bev Woods ran a florist shop in Leeds and did not view the mission statement and proposals on standardisation as suitable for her shop, as Interflora work comprised only a small proportion of the total amount of work undertaken by her business. A policy document was issued to all Interflora members outlining and reinforcing the conference proposals. The general view of the membership was that the proposals would be expensive for small florists, who make up a significant proportion of Interflora's members, and hence posed a financial threat to the long-term survival of small florist shops.

Realising that the proposals had not been as well received as he had initially thought, Doug McGrath decided to hold face-to-face meetings with Interflora members who were unhappy with proposals presented at Bournemouth and in the policy document. At one of these meetings, Bev Woods challenged the board over a small administrative change that she thought could be illegal. This challenge was met with a patronising reply, and in response the other Interflora members present displayed considerable anger at the manner in which Bev Woods was treated.

Hence Bev Woods and a fellow florist from London, Rose-Marie Watkins, joined forces and sought to gain the 250 signatures necessary to call an extraordinary general meeting (EGM) of Interflora. This involved the two women in stuffing and addressing 2600 envelopes by hand. The letters were posted so that they would arrive on the doormats of Interflora members the day after Valentine's Day and three weeks before Mother's Day in 1997, making use of a slight lull in the business calendar of busy florists.

Having realised that the proposals presented at Bournemouth had not been well received, Doug McGrath dropped plans for the centralised purchasing of flowers and other proposals that 50 per cent or more of Interflora's members disagreed with or felt neutral about. This was in line with a poll of members conducted by Interflora management. The poll was a telephone survey of a sample of 611 out of 2600 members. The results showed that 65 per cent of members agreed with the proposed changes. This was not accepted by Bev Woods and Rose-Marie Watkins, who thought that the views of all 2600 members should be taken into account.

The efforts and timing of Bev Woods and Rose-Marie Watkins paid off and by Mother's Day 1997 enough signatures had been gathered to force an EGM to be held in May 1997. This interrupted the programme of change that Doug McGrath and David Parry wanted in place by 1 June 1997. Bev Woods and Rose-Marie Watkins appointed a solicitor to help them prepare for the EGM, which was viewed as crucial if Interflora members were to retain their rights under the memorandum and articles of association, which were to have disappeared under the planned changes. Assistance in the battle against the board was forthcoming in the shape of Geoff Hughes, an ambitious Bristol-based florist, viewed by many as a political animal with a strategic outlook.

Geoff Hughes' primary aim was to remove Doug McGrath, David Parry and the rest of the Interflora board, as they were viewed as the drivers of the proposed change. This went beyond Bev Woods' original intention, which

**Exit case study 11.3** ***continued***

was to see the structure of Interflora change, as she thought that it allowed board members too much power. Despite being over-ruled, Bev Woods remained very much an active figurehead in the campaign to oust the Interflora board. The aims pursued by Bev Woods, Rose-Marie Watkins and Geoff Hughes were to diminish board support and appoint a caretaker board, a proposal which sent shockwaves through the organisation.

Bev Woods and Rose-Marie Watkins were determined to win and prepared very carefully for the EGM, rehearsing their speeches and actions. At the EGM they, Geoff Hughes and other Interflora members all had board members to mark. The turnout at the EGM was substantial, with 1200 florists coming to Warwick University, three times the expected number. In the electric and gladiatorial atmosphere of the EGM, Bev Woods spoke first and made an inspirational speech designed to get the support of non-committed Interflora members. The speech was even much admired by the opposition, Doug McGrath.

Rose-Marie Watkins employed cunning and laid a trap for David Parry to fall into. His role at the EGM was to chair the meeting and hence remain impartial. In her speech, Rose-Marie Watkins quoted from a recent *Daily Telegraph* article citing David Parry as stating that approximately 1000 Interflora members would cease to be members under the updating proposals. When this point was made, David Parry responded by rising to his feet to correct the remark by making a statement and in doing so breached his position as impartial chairman. This was raised as a point of order by Bev Woods, who indicated to him that by making a statement he was not being fair to both sides. The Interflora members present applauded. David Parry moved on to arguing with Geoff Hughes and the Interflora board continued to manage the meeting badly. A succession of Interflora members made speeches supporting the proposals of Bev Woods, Rose-Marie Watkins and Geoff Hughes.

There followed a much-needed lunch break. During lunch one Interflora member complained about the cold quiche buffet and in talking to the restaurant manager discovered that the amount paid for the lunch was £6 a head less than Interflora members had been charged. This was raised as a question in the afternoon session and it was suggested that the Interflora board was making money from its own members. This heightened the anger of the already hostile Interflora members and the meeting degenerated into an irate protest against the board. It was later clarified that the extra £6 a head was to cover the expenses of drinks, VAT and other extras.

The meeting was adjourned and voting took place, with board members waiting an anxious two hours for the results. These results removed all 13 board members, leaving Bev Woods, Rose-Marie Watkins and Geoff Hughes feeling ecstatic at their victory. However, a surprise was in store for the two women. The election of the new caretaker board saw their friend and ally Geoff Hughes moved into the role of chairman. It was from this position that he declared that he had never disagreed with business ideas of the previous board and he thought them appropriate.

In consequence, Geoff Hughes was not re-elected as chairman of Interflora in 1998. Interflora decided not to appoint a florist and Martin Redman, an accountant, was appointed chairman, not chief executive, of Interflora. There had been a radical change of personalities on the board of Interflora, but not of the aims that it was seeking to achieve.

*Sources*: Wolffe, R (1997) 'Interflora board ousted in protest over restructuring, *Financial Times*, 12 May 1998; 'Guns "n" Posies', *Blood on the Carpet*, BBC2, 10 February 1999.

**Questions for exit case study 11.3**

1 Identify the stakeholders in Interflora before the 1996 Bournemouth conference and plot them on a power and interest matrix.

2 Consider the situation that Interflora was in by the time of the EGM in May 1997. Identify any additions or deletions to the Interflora stakeholders you identified in answering Question 1. Plot all the stakeholders you have identified in Interflora as they would appear in May 1997, immediately after the EGM at Warwick University.

3 What lessons are to be learnt from the Interflora experience with regard to identifying and managing stakeholders?

## Short-answer questions

1 Define a stakeholder.

2 Define a key player.

3 Identify a stakeholder in an organisation and their source of interest.

4 Identify a stakeholder in an organisation and their source of power.

5 Briefly summarise the expected behaviour of satisfied stakeholders in an organisation.

6 Briefly summarise the expected behaviour of unsatisfied stakeholders in an organisation.

7 What should organisations seek to do with stakeholders who have high power and low interest?

8 What should organisations seek to do with stakeholders who have low power and high interest?

9 What should organisations seek to do with stakeholders who have low power and low interest?

## Assignment questions

1 It is late April and you work as a manager for a holiday company in charge of all its resorts and business in southern Spain. The company sells package holidays to the Mediterranean, France, Spain, Italy, Greece and Cyprus. The first plane of holiday-makers for the summer season arrives in Nipas in southern Spain. It is the first year the company has sent holidaymakers to the resort. The holidaymakers arrive late one night after a bumpy flight with bad turbulence. Within 24 hours there has been a significant list of complaints, about building work around the hotel pool and the hotel food portions being small, and two guests from separate parties have complained of upset stomachs.

Identify your stakeholders, plot them on a power and interest matrix and decide what action you are going to take to appropriately manage each of the stakeholder groups you have identified. Present your findings in a 2000-word report.

2 Choose and research an organisation. Identify all its stakeholders and plot them on a power and interest matrix. Comment on how and why you think the power and interest of the stakeholders will change in the next 12 months and the next five years.

## References

1 Johnson, G and Scholes, K (1999) *Exploring Corporate Strategy*, 5th edn, Harlow: Prentice Hall Europe.
2 Jagger, S (1999) 'AA ponders its road to the future', *Daily Telegraph*, 24 April.
3 Ibid.
4 Ibid.

5 Newman, C (1999) 'BSkyB signals new phase for digital customers', *Financial Times*, 6 May.
6 Potter, B and Levi, J (1999) 'BSkyB raises digital stakes', *Daily Telegraph*, 6 May.
7 'Minister rekindles controversy over football takeovers', *Financial Times*, 19 April 1999.
8 Ibid.

## Further reading

Ambrosini, V with Johnson, G and Scholes, K (1998) *Exploring Techniques of Analysis and Evaluation in Strategic Management*, Chapter 10, Harlow: Prentice Hall Europe.

Bell, E (1998) 'Winner takes all', *Management Today*, September (provides further information on BSkyB & ON Digital).

Johnson, G and Scholes, K (1999) *Exploring Corporate Strategy*, 5th edn, Chapter 5, Harlow: Prentice Hall Europe.

White, J (1999) 'United's hard sell', *Management Today*, February (provides more information on Manchester United and the BSkyB bid).

# 12 Managing a changing environment

**Organisational context model**

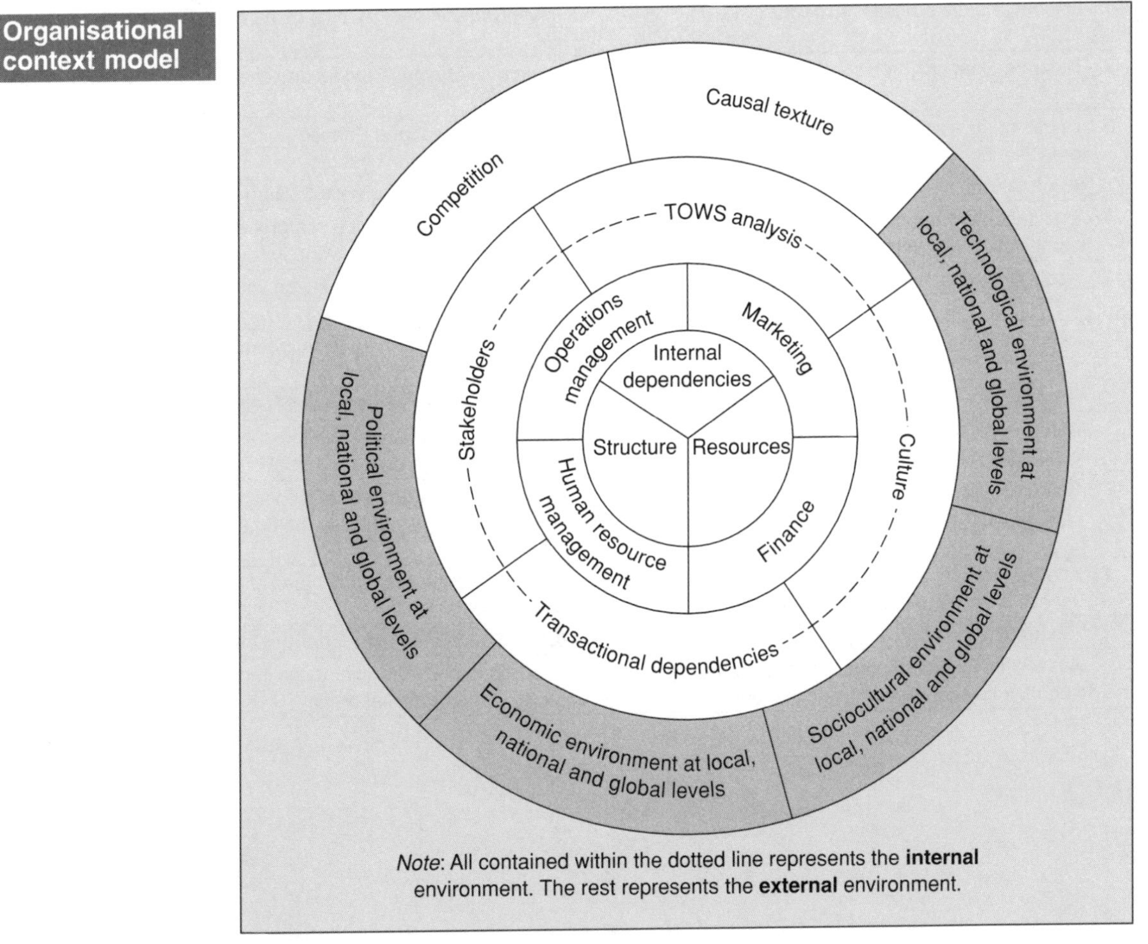


*Note*: All contained within the dotted line represents the **internal** environment. The rest represents the **external** environment.

**Exhibit 12.1**

**Learning outcomes**

While reading this chapter and engaging in the activities, bear in mind that there are certain specific outcomes you should be aiming to achieve as set out in Exhibit 12.2 (overleaf).

**Exhibit 12.2 Learning outcomes**

| | Knowledge | Check you have achieved this by |
|---|---|---|
| 1 | Recognise the demand for change | reviewing the reasons for change |
| 2 | Recognise the different types of change that organisations can undergo | reviewing the different sorts of change that organisations have to manage |
| 3 | Recognise the causes of organisational change | specifying the basis for change in an organisation |
| | **Comprehension** | **Check you have achieved this by** |
| 1 | Understand the process of change | identifying different models of the change process |
| 2 | Illustrate Greiner's model | sketching Greiner's model and explaining what it represents |
| 3 | Understand what is required for successful organisational change | explaining Deal and Kennedy's seven elements for successful cultural change |
| | **Application** | **Check you have achieved this by** |
| 1 | Demonstrate the process of change models on an organisation with which you are familiar | illustrating the relevance of a process of change model to an organisation you know |
| 2 | Apply Greiner's model to an organisation | illustrating how an organisation you know fits or does not fit Greiner's model |
| 3 | Apply Deal and Kennedy's seven elements for successful cultural change to an organisation you know | illustrating what an organisation must do to achieve successful cultural change |
| | **Analysis** | **Check you have achieved this by** |
| 1 | Analyse the change process that an organisation has experienced using the change process models | comparing and contrasting the analysis produced by the different change process models |
| 2 | Anticipate the future structure of an organisation using Greiner's model | identifying an organisation's position on Greiner's model and assessing its ability to manage itself through periods of revolution |
| 3 | Analyse, for an organisation you know, its ability to survive a period of cultural change | anticipating the likely outcomes of cultural change for the chosen organisation |
| | **Evaluation** | **Check you have achieved this by** |
| 1 | Assess the impact of major change on organisations and their key departments and stakeholders | identifying and determining the impact of change on all parts of the organisation |

ENTRY CASE STUDY 12.1 

# Shoemaker finds one size does not fit all

**By William Lewis**

For generations of children forced by their parents to dress sensibly for school the unthinkable is happening: Clarks shoes have actually become cool. The company's 'Wallabees' are to be seen on the feet of the Verve rock band on the cover of their latest album; Clarks desert boots, meanwhile, are worn by Oasis.

There have been changes, too, for the shareholders of the privately owned, UK-based C&J Clark, manufacturer and retailer of the shoes.

Under the leadership of Timothy Parker, who was appointed chief executive in 1995, and Roger Pedder, who became chairman in 1993, Clark has bounced back to financial health, taken a lead position in the UK shoe market and is now focused on achieving rapid international expansion.

Following factory closures and staff cuts announced in 1996, Clark reported pre-tax profits of £35m for the year to January 31 1998, against losses of £409 000 the previous year – including the £30.4m cost of closing a number of factories. For several years shareholders have been squabbling among themselves about whether Clark should remain a private company or float on the London Stock Exchange. Mr Parker and Mr Pedder are conducting a review which could lead, within two years, to the board recommending to shareholders that a flotation takes place.

Nowhere is the change at the company more evident than at Clark's North American operations, previously based in Philadepia and now headquartered in Boston, Massachusetts. Things had got so bad that Mr Parker wrote in Clark's annual report for the year end January 31 1996 that the wholesale operation of Bostonian, the US shoe company it acquired in the early 1980s, had had 'a disastrous year'.

Mr Pedder added that 'profits effectively collapsed' in the US due to 'a combination of over-optimism, poor marketing and a failure to control the sales function'. In the year to January 31 1996, the US operations achieved a trading profit of just £700 000 compared with £3.6m in the previous year.

Looking back, Mr Pedder says the problems were due in part to the management structure in the US. Clark had three US shoe businesses: Bostonian, Hanover and Clarks England. In the early 1980s, Bostonian was merged with Hanover, another shoe wholesale, retail and manufacturing operation. However, Clarks England, the direct off-shoot of the UK business, was kept separate.

Bob Infantino, who had been managing the Clarks England retail, wholesale and manufacturing operations, was appointed at the end of 1995 to head a new grouping known as the Clarks Companies North America. Clarks England, Bostonian and Hanover now report directly to him. 'It was a messy structure so Tim made the obvious decision to put both brands under the command of Bob,' Mr Pedder said.

A loss-making shoe factory in Pennsylvania has been shut, the company's headquarters moved to Boston and while the Bostonian and Clarks England brands continue to be marketed separately, some of the traditional features of Clarks shoes are being incorporated into the Bostonian shoe range, and vice versa.

Several of the company's loss-making stores have been replaced with shops with livelier interiors to appeal to a younger market. The shoes themselves have been redesigned, with the US staff given the freedom to develop their own version of traditional Clarks shoes.

Mr Infantino says the company no longer focuses on exporting to the US shoes that have been designed for the UK market. Clark North America's designs reflect the facts that shoes are worn more tightly in the US compared with Europe and that the average size is also larger. While Clark-owned stores and others are still selling traditional Clarks shoes, Mr Infantino says that US sales are increasingly coming from ranges adapted to US tastes.

'We were an organisation buying shoes from English factories and trying to sell them here,' Mr Infantino says. 'The customer did not want them but we were force-feeding them into the stores,' he says.

These changes helped Clark North America achieve profits before exceptional items of £10.96m in the year to January 31 1998, more than doubling the figure achieved the year before. Sales, at £157.6m, were up by 14 per cent.

Unlike in the UK, where Clark is primarily a retail operation, the success achieved in the US has derived mainly from the rapid growth of Clarks and Bostonian wholesale businesses. Questions remain about the profitability of the US retail portfolio and there are plans to prune the number of Hanover stores, while increasing the number of Bostonian and Clark stores.

Mr Pedder now talks about a 'fundamental shift' in Clark's focus on non-UK revenue growth. 'I would like to see 60 per cent of sales overseas within five years,' he says. With non-UK sales currently at 40 per cent of the overall sales last year, that implies rapid growth both in the US and Clark's operations in east Asia and Australia. However, Mr Pedder describes the ambition as his 'legacy' and wants to see a 50–50 split between UK and non-UK by 2000.

To achieve this, Mr Pedder says, Clark is moving to a more flexible structure, with a 'loose holding company with a big UK base that is primarily a retailer, and a worldwide organisation based on wholesale marketing companies'. He describes the linking features between the various regional companies as common ownership, a common brand and a common set of brand ethics.

'We want to be the number one branded casual footwear company in the world,' Mr Pedder says. 'So we have to produce flexible, comfortable, stylish – although not necessarily fashionable – excellent value-for-money and robust shoes for the different markets in which we operate. It does not matter if the shoes look or feel different if the linkages are intact.'

*Source: Financial Times*, 16 June 1998. Reprinted with permission.

## Introduction

The first 11 chapters of this book have been concerned with analysing organisations and their environments (internal and external). Therefore in conclusion this chapter examines how organisations cope with change arising from those environments and examines some tools and techniques that allow organisations to plan for and manage change.

## The demand for change

The demand for change in an organisation is caused by a variety of internal and external factors. External environmental factors driving change include competition, markets and customers. In comparison, internal drivers of change include employees and organisational departments. An organisation that effectively scans its external environment (*see* Chapters 1, 2, 3 and 6) will be in a strong position to deal with change and related issues. Organisations have to assess the outcome of carrying on with current tasks and activities or decide to plan and implement change to allow the organisation to develop from its current position. The decision to implement change will depend on whether the organisation wants to maintain or increase market size. If an organisation is seeking to maintain market size, it should realise that this is a difficult task in declining markets and an easier task in stable or growing markets. Alternatively, if an organisation is aiming for market growth, it should recognise that this is most realistic if markets are growing, but still possible if markets are stable.

The decision to implement change requires an assessment of the resources available and those required. The difference between the two should be evaluated (*see* Chapter 4). Stakeholder behaviour, alongside current and changing expectations, should also be considered when planning for change (*see* Chapter 11).

## Types of change

The changes that organisations choose to implement will cover many different aspects of organisational life. These changes can include alterations to organisational size, structure and culture, as well as changes to operational activities and the roles that people undertake in an organisation. Change is either reactive or proactive. Reactive change is where the organisation reacts to an event that has already occurred. In contrast, proactive change is where an organisation plans and prepares for expected and anticipated events and maybe even how to deal with unexpected events.

Unexpected events include shock occurrences, such as a chemical company experiencing a leak or explosion or a food manufacturer suffering contamination of food or drink products with glass or poison. These types of events are very rare and unpredictable, but organisations can plan for the steps they will take in such circumstances. The planning should cover what action should be taken and who in the organisation is responsible for taking it. The issues that should be covered include dealing with the event immediately it happens, for example recalling all food or drink products that might be affected and offering a full refund, which is important if the dent to consumer confidence is to be minimised; liaising with the emergency services and investigating authorities, important in the example of the chemical company leak or explosion; dealing with those members of the public affected; and liaising with the media, the effectiveness of which is likely to affect public perception of the organisation and consumer confidence.

Major organisational changes in size and structure will affect hierarchical and reporting relationships, along with communication and decision-making systems. An organisation that decreases the number of middle managers it employs will also be reducing its size and that of its hierarchy. Hence the structure of the organisation will be flatter and more communication will need to occur between those at the top and those further down. At the same time, decision making will have to be either decentralised to those further down the organisation or centralised with those at the top.

A change in organisational culture, which is difficult to achieve, is likely to involve changes in personalities and the position these people occupy in the organisation. This means that leadership styles and the way in which people are motivated will change. Different organisational cultural styles are examined in Chapters 4 and 5. For employees of an organisation undergoing change, the changes will be felt in the tasks and activities they perform on a daily basis as part of their jobs. This may include changes in the way the job is carried out. The content of a particular job may change completely or only slightly. Technology may alter the way someone's job is done and the way work is co-ordinated in the organisation.

## Causes of change

The demand for change is caused by shifts in the external and internal environments, as outlined above. External environmental factors were discussed in earlier chapters. The PEST issues – political, economic, sociocultural and technological – are all drivers or causes of change that all organisations face. The influence an organisation has over these drivers of change is often limited, as they usually arise from another organisation or development over which the company has no or limited influence. For example, the Bank of England may be lobbied extensively by businesses to decrease interest rates, but may choose to take no notice and either not decrease rates or reduce them by a smaller amount than was lobbied for. In such an example, the outcome may be a 0.25 per cent reduction in interest rates instead of a much lobbied for

0.5 per cent decrease. This demonstrates the limited influence that businesses may have on changes directly affecting them.

The actions of competitor companies or organisations also result in businesses having to manage change. The introduction of a price-cutting strategy by a competitor is clearly an event that requires an immediate reaction if a company is to retain market share. In May 1999 when BSkyB announced that it was to give away digital set-top boxes for free, one of its main competitors, ON Digital, immediately announced that one of its set-top boxes would be given away free to anyone buying a new television set for more than £200. It subsequently decided to give away a set-top box to every customer. However, there is also a longer-term issue of change that needs to be addressed: how is the organisation going to ensure that it is not taken by surprise again? And how is it going to ensure that a price war does not develop? Price wars are dangerous to the participating businesses, as if prices reach a low level, such that no profit is being made and possibly costs are not covered, then the companies involved may become insolvent and cease to trade. In the case of BSkyB, the cost would be recouped in the very long term from receipt of increased subscriptions for digital TV channels.

The introduction of a new technology or technologically advanced product by a competitor will require a considered response that may take time to develop. Dyson's introduction of the bagless vacuum cleaner took the market by surprise and the company rapidly seized market share from other vacuum producers. Manufacturers of traditional vacuum cleaners took several years to develop a noticeable and competitive response to Dyson.

Change in the attitudes, expectations and behaviour of stakeholders creates further change that organisations have to deal with. Internal stakeholders such as employees and trade unions, along with external stakeholders including competitors, customers, financiers and shareholders, are stakeholder groups who are players in the change process. Demands from employees and trade unions for more pay and different wage and salary structures may mean that the rewards system operated by the organisation changes, although this is unlikely to happen in isolation. The likely outcome would be that a deal is struck over working hours or productivity in return for improved wages and conditions of employment.

A pessimistic scenario would be a company performing badly in a declining market, finding change thrust on it by the actions of its financiers calling in loans. The outcome would be a falling share price resulting from lack of market faith in the company's shares and a glut of shares on the stock market as many shareholders are selling. An organisation in such an unfortunate financial situation may seek to consolidate by closing down activities and outlets that are not profitable and not core to its business. This will mean changes in the structure of the overall organisation and in the size of the workforce. However, a company that anticipates the possibility of that scenario can attempt to change and avoid financial ruin. This is done by seeking to improve the financial performance of the business and also continuing to satisfy its stakeholders.

The chemical company ICI, which was traditionally known for supplying bulk chemicals to industry, realised that it was becoming less and less competitive in the industrial bulk chemical market in which competition was price based. Therefore in 1997/8 ICI sold its bulk chemicals businesses and concentrated on acquiring brands that would provide a much better return, such as Polyfilla and Cuprinol wood preserver. This move was well received by shareholders and other investors, as the share price rose and profits improved.[1]

## The process of change

In this section an overall view is taken of the process of change and the issues affecting the managers and stakeholders involved.

Lewin[2] suggests that change is the outcome of the impact of driving forces on restraining forces, more commonly known as force field analysis; *see* Exhibit 12.3. This can be thought of as the status quo that is under pressure to change. The resulting change is a direct outcome of either the driving forces or the restraining forces being more powerful. It is normal for the driving forces of change to have economic attributes. These economic attributes may arise from the external environment as a result of macroeconomic changes or because of internal issues, e.g. the need to cut costs and improve profit margins. In the entry case study to this chapter the economic drivers for change in Clark's US operations arose directly from internal issues, namely the trading profit of £700 000 in the year ending 31 January 1996, compared with £3.6m profit in the previous year.

In most organisations undergoing change, at least part of the workforce will be dedicated and faithful to the existing work practices. The news that these practices will have to change will mean that staff feel concerned about their jobs and future with the organisation. Their reluctance to accept and adopt suggested changes forms the restraining forces that contribute to the unbalancing of the status quo. The restraining forces will seek to persuade the organisation to discontinue or alter the recommended changes and counteract the driving forces of change.

Managers in organisations experiencing change obviously need to be aware of both driving forces and restraining forces. The managers involved should seek to communicate strong justification of the changes by offering a clear

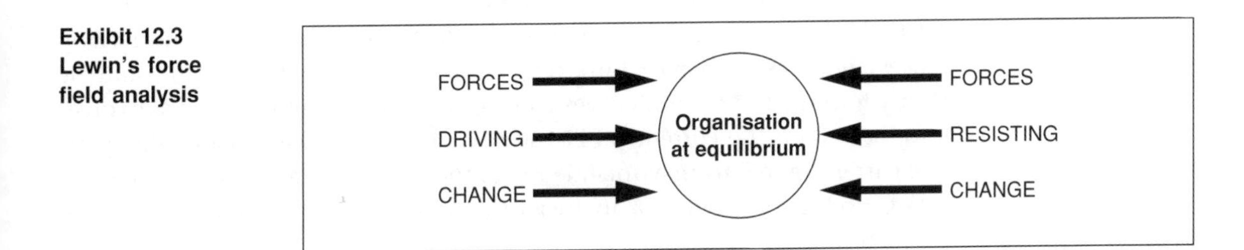


**Exhibit 12.3 Lewin's force field analysis**

and unclouded explanation of the reasons for the changes and any advances in employee empowerment that will result. This was clearly the case with Clarks, as there was a shift in the way shoes were produced for and sold to the American market. American staff were given the freedom to redesign and adapt UK designs for the US market. This resulted in American staff selling US-designed shoes to American customers, instead of thrusting UK-designed shoes at them.

However, in planning such significant change a careful balance needs to be struck between providing too much information, which may raise fears and queries that cannot yet be dealt with, and providing inadequate information, which leaves people feeling that they have been told nothing and are being kept in the dark. Hence change requires very strong and effective leadership from the managers involved, particularly if structural and cultural changes are to occur. The leadership in any new working teams or groups will have to be particularly strong and focused if the implemented change is to be successful. In the entry case study to this chapter the clear and strong leadership at Clarks was provided by chief executive Timothy Parker and chairman Roger Pedder in the UK and Bob Infantino in the US.

A strong and effective change manager will disperse resistance to change by highlighting early achievements resulting from the change, which will also help maintain the momentum of the change programme and prevent it suffering from setbacks and periods of slow progress. Recognition that not everyone will support the change and that the feelings of those who are likely to be hurt by it require sensitive handling will both help the change process to be managed successfully.

## The change process model

The Lewin model[3] was developed to identify and examine the three stages of the change process. The first stage is unfreezing current attitudes and behaviour; *see* Exhibit 12.4. This unfreezing stage takes the view that if attitudes and behaviour are to change, then old behaviour must come to be regarded as unsuitable and must stop. Hence the requirement for change must be appreciated directly by the people to be affected by the change, as it cannot be imposed on them if the change process is to succeed. The unfreezing process may be achieved by realisation among the people involved that change is required for some reason. Common driving forces of change often result from commercial pressures, which include the need for increased sales or market share, better profitability or more efficient production. The unfreezing process requires employees affected by the change to understand and be clearly informed of the difficulties confronting the organisation. This involves information on the current problems being communicated to all employees. The current problems may include issues such as reducing market, poor quality of product or service, or inefficient production levels when compared to competition.

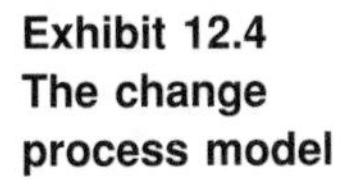
**Exhibit 12.4**
**The change process model**

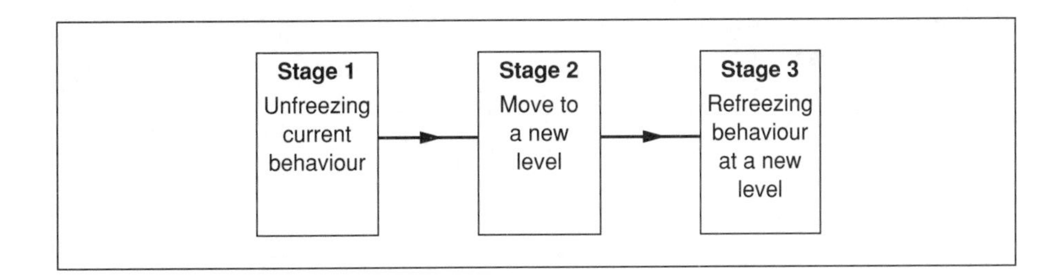


The second stage of Lewin's change process model is moving to a new level. A search for answers to the difficulties faced by the organisation has to occur. This involves reviewing alternative solutions, the examination of changing values and culture within the organisation, and an assessment of the organisational structure that would most suit the changing organisation. The people affected by the change should continue to receive regular updating communications about possible answers and solutions to the difficulties faced by the organisation. These informing and updating communications should be both verbal, via meetings, and written, for example via a newsletter. The use of meetings allows people's ideas and views on the proposed changes to be gathered and their questions, queries and fears to be addressed. Issuing a newsletter allows what has been discussed at meetings to be confirmed and is also a useful summary, particularly for anyone who was not able to be there. Meetings should be organised and structured so that all those affected by the change have the opportunity to be involved in the debate and discussions. If the change is significant and large numbers of people are involved, then a series of meetings covering different departments, sections or divisions will be required throughout the period in which the change occurs.

Solutions and answers to the difficulties faced by the organisation need to be developed and the essential changes planned. Hence there is a need to continue communicating with those affected by the change. Once this has been achieved the implementation of the planned changes can be arranged. Then the chosen solutions can start to be implemented, which may necessitate running new and current systems and methods of working in parallel. This allows the newly implemented solutions to be assessed as working satisfactorily, before the old method or system of working is withdrawn. There should then be a final review and tweaking of systems and methods of working to ensure that the required level of work and satisfaction is being achieved by the people involved.

The third stage of Lewin's change process model is refreezing attitudes and behaviour at the new level. This takes place once acceptable solutions have been found. The refreezing stage involves positive reinforcement and support for the implemented changes. This can be done by highlighting improvements in difficult areas that have occurred as a result of the changes, for example an upturn in sales or improvement in quality, which could include testimony from a satisfied customer reproduced in a company newsletter. People from

**Exhibit 12.5 The change process by Burnes**

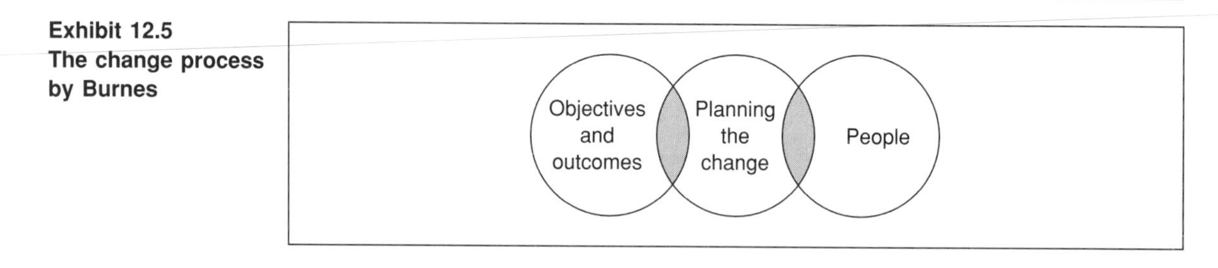


other parts of the organisation not directly affected by the changes also need to be informed of what has been altered.

An alternative view of the change process is presented by Burnes.[4] He suggests that the change process consists of three interlinked elements: objectives and outcomes; planning the change; and people; *see* Exhibit 12.5.

## Objectives and outcomes

In assessing the objectives and outcomes, Burnes presents four phases that organisations should work through in deciding if change is appropriate and the desired objectives and outcomes. First, a trigger for organisational change should exist and fall into one of the following broad categories: the requirement for change and/or better performance is a core theme of organisational strategy; serious problems exist with current performance; or opportunities exist that will provide the organisation with a greater return than that currently achieved.

Once organisational change has been triggered, the second phase, the remit, is entered. Drawing up the remit is about establishing clear agreement concerning where the accountability and authority to instigate change exist in the organisation. This needs to be undertaken prior to proceeding further with the change process. The accountability and authority to instigate the change process rest with senior management in a bureaucratic and hierarchical organisation, but will be devolved in an organisation with a flatter structure and more modern approach to management. The remit also needs to outline the objectives of the change programme along with the time in which the change can be expected to be implemented and achieve its objectives and outcomes. Finally, it is very necessary for any remit concerning change to make crystal clear that a full range of options needs to be examined and evaluated in deciding what change should take place, not just one or two options.

Once the remit is established, an assessment team should be confirmed to explain the difficulty or opportunity facing the organisation; explore possible options; discuss the problems/opportunities and options with interested parties; and make recommendations about the way forward. The assessment team will usually contain representation from a wide range of people, for example those affected by the change, specialist staff (such as finance, personnel), senior management and in some cases a change agent. This is an outside consultant or facilitator who promotes and stimulates the change process.

The assessment team will seek to explain the difficulty or opportunity facing the organisation by gathering information and speaking to those directly and indirectly involved. This is followed by establishing possible options for dealing with the difficulty or opportunity. The options or solutions that offer the greatest benefit and are practical should be identified. The difficulties and possible solutions need to be discussed with the stakeholders affected by the change process. Finally, the assessment team should present its recommendations to those responsible for making the final decision on what change will occur. Those responsible may be senior management and/or those affected by the change, and they will decide to accept, reject or modify the proposals.

### Planning the change

Once the remit for change and an assessment of the change required have been settled, the planning of the actual change to take place and its implementation are undertaken. The initial stage is to confirm the change management team, which will usually include many of the people who were members of the assessment team. Where and to whom in the organisation the change management team reports, and how it gains access to resources to foster change, need to be extremely clear, as during a period of change these can become ambiguous.

The function of the change management team is to devise a schedule for change, specifying the tasks and activities that need to occur if successful change is to result. The tasks and activities should link directly to the objectives and outcomes of the change and be expressed in very focused terms if they are to be carried out accurately and within the desired timescale. This requires the support of those in favour of the change to be harnessed and used to progress the change programme. In a programme of major change it is important to review events periodically and assess if the tasks and activities have been correctly carried out and are in fact achieving the objectives and outcomes of the change. In addition, such reviews are likely to yield lessons for future change programmes. Finally, most programmes of major change will require some form of training to allow those affected by the change to cope and perform well under the new working conditions. The training required can take many forms and may include skills development to cope with a quality programme that has been implemented or technical training to allow staff to operate new equipment, machinery and computers.

### People

The third and final element of Burnes's change process model is people. Whatever the change undertaken, people are normally part of the change process and therefore have to be given consideration. Burnes proposes that in implementing change there are three people-related issues that must be handled if the change is to succeed and the organisation flourish. People must be responsive to the need for change; they should be involved in the change process; and the impetus resulting from the change should be sustained.

Ensuring that people are responsive to organisational change is difficult, as it involves moving from the familiar to the unfamiliar. Therefore organisations need to ensure that the sources of resistance to change and how to deal with them are understood. This can be done by making sure that employees are familiar with company plans and the pressures faced by the organisation from, for example, customers, competitors and suppliers. In this context, the focus of the information provided to employees should be to show that change is undertaken to secure, not jeopardise, the future of the organisation. Accordingly, the difference between current and desired future performance should be outlined in a manner that allows employees to relate the proposed change to themselves and their individual/group/section performance. This allows employees to think about how such improvement may be made and to contribute their own ideas to the change process. The successful achievement of change to meet the intended objectives and outcomes can be publicised to promote further change or change in a different part of the organisation, as required. In doing this mistakes do not have to be swept under the carpet, but should be learnt from for future change projects.

The management of a changing organisation also has to be responsive to change and understand the anxieties and apprehension felt by those affected directly by the changes. Significant concern about change in an organisation may indicate that all is not right with the proposed alterations and further consideration of the suggested changes may be needed. The resistance to change that arises from anxiety and apprehension may be at least partly overcome by involving the people affected by the change in the change process. Most change programmes are long and complex and if people are to be properly involved, they have to be so from beginning to end, in the development, planning and implementation of the changes.

Involvement should start with clear and regular communication with those affected. The communication process will involve the assessment team and the change management team and should seek to explain the context of the proposed change, its details and its consequences. The communication process should be two way. The people affected by the changes need to listen and take on board what is planned, and it is equally necessary for the change management team to listen to those affected and their views and ideas about what it is appropriate to change. This may result in the change management team reviewing their own ideas and assumptions about what change is necessary. They should gain assistance in the change process rather than resistance to it. This can be encouraged by making every effort to involve those most closely affected and giving them responsibility for the change project.

Once the change has been developed, planned and implemented, the challenge for the organisation is to sustain its impetus. This is important, as if the impetus created by change is not sustained then there exists a danger that people will revert to old behaviour and past ways of doing things. In order to sustain the impetus, the organisation should consider several issues, starting with the provision of additional resources, both financial and human. These would provide clear support for staff striving to uphold previous levels of output for the duration of the turbulent period of change. The change management team should also be given support and not be ignored, as they have been charged with motivating others and dealing with problems associated with the change process. If support for the change management team is not forthcoming, its members are likely to become disheartened and demotivated and will no longer be in a position to support and motivate others.

The development of new skills and knowledge among staff and leadership styles among managers are frequently required as part of the process of change. Backing for the process can also be provided by the organisation's meeting the challenge of allowing staff and management to gain the required knowledge, skills and leadership style in a non-threatening and encouraging manner. Sustaining the impetus of change can be reinforced by rewarding behaviour that supports the change and the new way of doing things. The rewards can be monetary or simply praise for achieving success in the changed organisation.

The process of change is an intricate blend of setting objectives, planning the change and managing the people involved, with all three areas involving gathering information and making decisions about what is most important. However, in many instances change will continue to occur long after the actual programme for change has been successfully implemented. This can become a pattern of continuous improvement and change.

## Changing organisational structure

Structure and culture are the two most significant things that an organisation can change about itself. This section will look at changing structure and the next section at changing culture. The five basic organisational structures are identified and discussed in Chapter 4. Thompson[5] identifies four determinants of organisational structure: size, tasks, environment and ideology; *see* Exhibit 12.6.

### Size

The size of an organisation will influence which structure is most suitable. The most suitable structure is that which allows the best and most effective communication and co-ordination within the organisation. A large organisation or one that has grown in size in terms of more and bigger markets, a greater range of products and services, an increasing number of employees or a greater number of factories or outlets will require a different structure from a small business that has just opened its first factory or outlet. The centralised simple and functional structures are suitable for smaller organisations and offer a number of advantages and disadvantages, but these cease to be appropriate once an organisation has expanded significantly; *see* Chapter 4.

**Exhibit 12.6 Determinants of organisational structure**

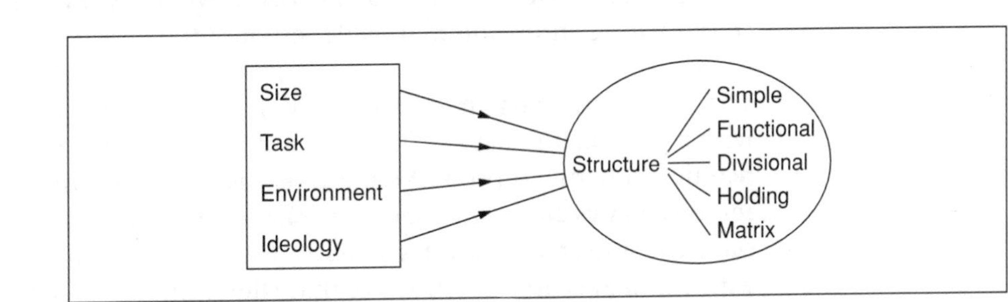

The question of structure can be illustrated by ICI and the dilemma the company faced in March 1998. It had just sold off its bulk chemicals businesses, retaining its paints business with the ICI-Dulux brand, and had acquired other branded products such as Polyfilla. The debate centred around whether to keep the paints business together with the other branded goods or to separate it. ICI's view was that there were considerable benefits in keeping its paints business and the Dulux ICI brand together with its other branded products. However, one of its main rivals, Kalon, which specialised in paints, gained a much better profit margin than ICI from paint. The alternative view was that paint is a bulk chemicals business and there is no benefit in keeping it with branded goods, so it should be spun off into its own separate division. This was the approach of another of ICI's rivals in the paint business, Courtaulds, which had done just that. If ICI kept paints and branded chemicals together and the business did well, shareholders would be delighted; but if profits were not good and shares performed poorly, then shareholders were likely to question the structure of the company.[6]

**Tasks**

A large and complex organisation with both interlinked and dissimilar tasks and activities will have a crucial need for effective co-ordination and communication. Where an organisation's tasks and activities are complex and interlinked, the most suitable structure will be decentralised such as the divisional or matrix structure. In contrast, if all the tasks and activities carried out by the organisation are dissimilar and unconnected, then the holding company structure may be appropriate; *see* Chapter 4.

**Environment**

The external environment of an organisation exerts pressure on it to change. These external environmental pressures result from all areas of the external environment, typically including political, economic, sociocultural and technological pressures (*see* Chapters 1 and 2). Other external environmental pressures arise from changing customer demands and market size, as well as the behaviour of competing organisations (*see* Chapters 3 and 6). The key issue for organisations to consider is their ability to prioritise and select which pressures to respond to and the degree and speed of response required.

An organisation with a centralised structure will be much less flexible and therefore less able to respond rapidly to major change. In contrast, an organisation with a decentralised structure will be able to react more quickly to major change if required to do so, but the disadvantage is the greater difficulty in co-ordinating an organisation-wide response to change.

**Ideology**

The logical argument here is that the longer an organisation has operated with a particular structure, the more difficult it will be to change it. The difficulties will arise as people who are very familiar with their specific jobs and responsibilities are likely to be set in their ways and wary of change. This will have to be overcome, as described in the section on the process of change earlier in this chapter. A structure that allows the complexity, similarities and dissimilarities in the organisation's tasks and activities to be managed so that it operates both efficiently and effectively should be aimed for. This may mean that a new structure has to be designed that uses the best ideas from the five basic organisational structures.

## Greiner's model

The changes that an organisation's structure may undergo are summarised in Greiner's model; *see* Exhibit 12.7. This relates growth rate, age and size of the organisation to five phases of growth and development.[7] Each phase of growth consists of an evolutionary stage and revolutionary stage. Hence it should be clearly understood that 'each phase is both an effect of the previous phase and a cause for the next phase'.[8] An evolutionary stage is one in which there

**Exhibit 12.7 Greiner's model**

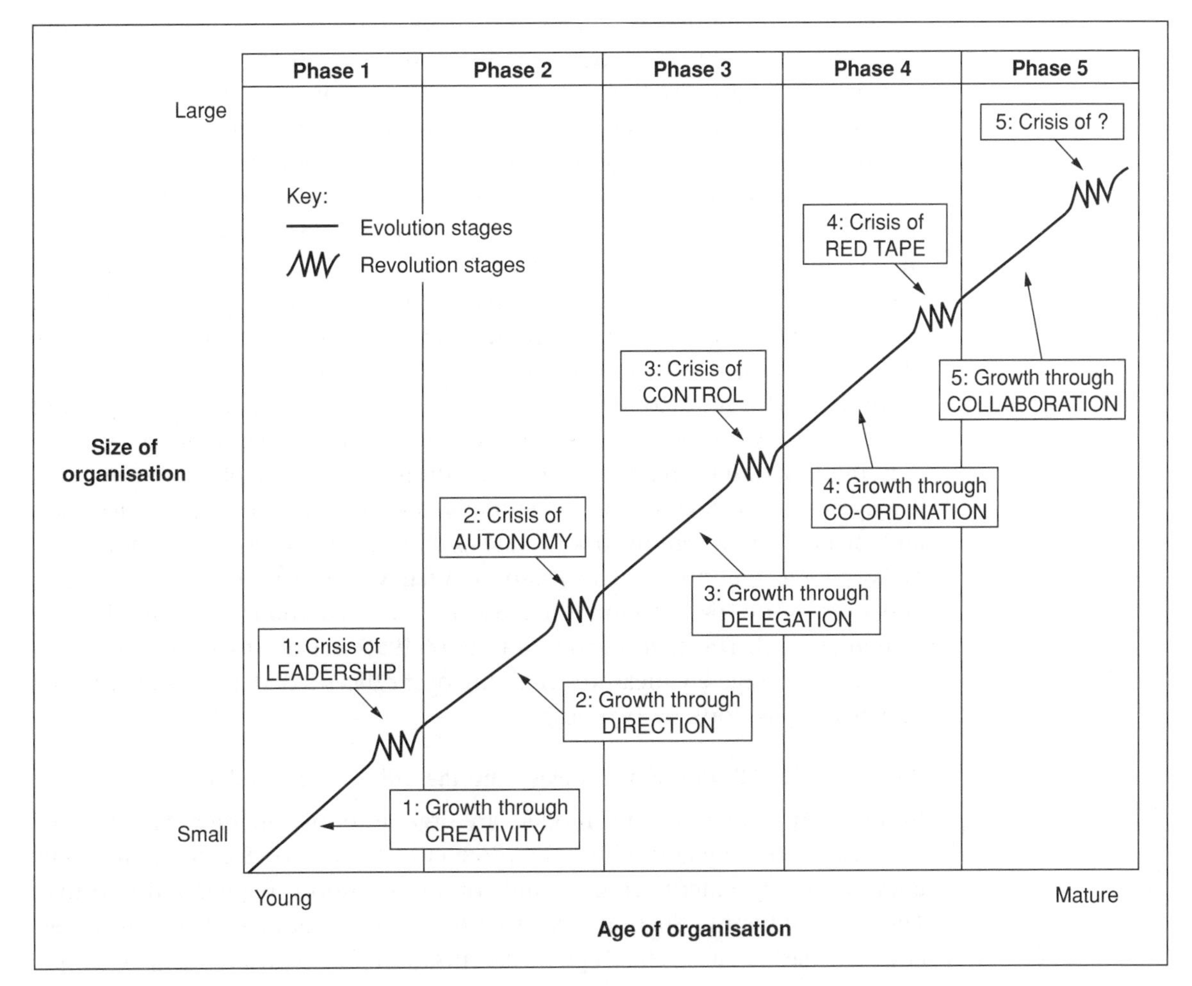

is a dominant management style that is successful. In a revolutionary stage there is a dominant management problem that has to be resolved for the organisation to continue to grow and move on to the next evolutionary stage. The rate of growth is indicated by the steepness of the line on Greiner's model, a steep line indicating high growth and a gradual line indicating slow growth.

The key impact of an organisation's age is that the older an organisation, the more likely attitudes and behaviour are to be engrained, resulting in a potentially greater resistance to change. The larger the organisation, the more extensive the task of ensuring that communication, co-ordination and interdependence of the organisation's activities are achieved. An increase in both age and size of an organisation usually suggests that it has grown steadily over a number of years, which is termed a stage of evolution. During the stages of evolution (*see* Exhibit 12.7) the style of management behaviour continues to be more or less stable, with only minor adjustments required to ensure that the organisation is able to perform its tasks and activities effectively. In contrast, periods of great flux and upheaval in an organisation's life are referred to as stages of revolution.

A stage of revolution is when old management and work practices are put under pressure from new requirements brought about by the increasing age and size of the organisation. The stages of revolution are shown on Greiner's model as crises with which the organisation has to cope if it is to pass successfully into the next stage of relative stability or evolution.

In the summer of 1999, the UK Passport Agency was in the midst of a period of revolution, with the installation of a new computer system and change in the law requiring all children under 16 who were not currently on a parent's passport to have their own passport. This created a crisis for the Passport Agency, as the number of applications for passports rose and waiting times for the issue of new passports also increased, from 6 to 8 weeks to as much as 15 weeks in some cases. This resulted in people panicking and believing that they were not going to receive their new passport in time to go on their summer holiday. If an organisation is unable to deal successfully with a stage of revolution, then it may be that it stagnates and eventually goes into decline, losing customers and market share, selling out-of-date products and services and having to make people redundant. The Passport Agency is the only issuer of British passports in the UK, therefore the government was anxious to ensure that the system worked. The crisis of the summer of 1999 was dealt with by the introduction of emergency measures enabling post offices to extend the expiry date of existing passports by two years.

#### Phase 1 – Growth through creativity and the crisis of leadership

In phase one of Greiner's model, an organisation's energies are directed towards developing and selling the product or service. Hence minimal energy goes into management activities and communication is regular and informal. The management of such an organisation reacts to feedback from customers and the marketplace. This type of organisation will typically adopt a simple structure (*see* Chapter 4). As the size and age of the organisation increase, a crisis of leadership will occur. The organisation's leader will be under considerable stress and unable to manage a growing organisation on their own via informal communication and co-ordination, which is the crisis of leadership. There will be a clear need for greater management expertise in the organisation if the workforce and resources are to be handled effectively.

#### Phase 2 – Growth through direction and crisis of autonomy

The crisis of leadership is resolved through 'professionalising' the organisation. This involves any number of the following occurring: specialised managers being introduced to the organisation; restructuring, typically the functional structure being adopted; implementation of formal control and communication systems; and the development of a decision-making hierarchy. Professionalising of the organisation helps iron out the difficulties arising from running a growing organisation very informally and allows the organisation to continue to grow in age and size. After this further stage of evolution, 'growth through direction', another crisis occurs, this time the crisis of autonomy. The lack of autonomy results in employees feeling restricted by centralisation and the formal control systems that have been implemented. There is also a lack of opportunity for employees to act on their own initiative.

#### Phase 3 – Growth through decentralisation (divisional) and crisis of control

The crisis of autonomy is handled by a move towards decentralisation, which gives more responsibility to plant and sales managers and provides an opportunity for employees to act on their own initiative. This is supported by implementing a structure such as the divisional structure, which uses profit centres and profit sharing schemes or bonuses to motivate managers to perform well. Short- and medium-term planning is left to middle management and their workforce, while senior managers deal with long-term or strategic planning. The crisis in this phase is one of control. The move to decentralisation results in management's feeling a loss of control over an organisation with a complex and diverse range of products and services in different and unrelated industries.

#### Phase 4 – Growth through co-ordination and crisis of red tape

The crisis of control is tackled by the introduction of more formal planning procedures and an increase of staff in roles concerned with company-wide control and management of the workforce and resources. Critical functions in the organisation will be centralised, while decentralised units that show some relation or similarity will be merged into product groups. This should allow managers in the organisation to take a corporate-wide perspective, rather than merely a local view of their own department or division.

The crisis arising from greater co-ordination is one of red tape. This is indicated by a lack of confidence in the relationship between staff in the organisation and headquarters, from where many of the co-ordination systems will have originated. The co-ordination systems may no longer support the local market conditions in which employees have to operate. In summary, there is an expansion of systems that is too extensive and rigid to allow an organisation still increasing in size and age to operate effectively. Innovation, which can aid organisational growth, will also be inhibited.

### Phase 5 – Growth through collaboration and the ? crisis

The red tape crisis is overcome by an increase in team and cross-functional activity, which usually involves the organisation restructuring to a matrix structure. The greater flexibility of the matrix structure allows team working and cross-functional activity to occur naturally. The strong co-ordinating management of the previous phase is replaced by social control and self-discipline of both individual workers and the teams in which they operate.

Greiner anticipates that Phase 5 may be concluded by a crisis of 'psychological saturation', where employees are exhausted by the demands of teamwork and the need for innovation. In conclusion, he suggests that organisations will adopt a dual structure. First there will be a 'habit' structure within which employees carry out their routine, day-to-day work. Secondly, there will be a 'reflective' structure for 'stimulating perspectives and personal enrichment'.[9] This will be a structure that allows employees to refuel and may include things like flexible working hours, revolving jobs and sabbaticals.

In conclusion, managers should be able to recognise the phase of Greiner's model in which their organisation is currently and its associated organisational structure and features, and hence be able to identify the next phase of growth. This should encourage the development of the key skills and strengths required to get the organisation through the next crisis or stage of revolution and to continue success in the next stage of evolution. Moreover, progression from one phase to the next is not automatic and managers must consciously act to move the organisation through the rough revolutionary stage to the next higher evolutionary stage.

## Making cultural change successful

Deal and Kennedy[10] outline seven elements required for successful cultural change to be achieved.

### Position a hero in charge of the process

Deal and Kennedy define a hero as a high achiever in the organisation and someone who personifies the organisation's cultural values and hence provides an explicit role model for employees. Heros 'show every employee "here's what you have to do to succeed around here." '[11] A hero put in charge of the change process will have to believe strongly in and be committed to the proposed changes. The person or hero in charge of the change process needs to inspire belief in and commitment to the change among the affected workforce.

### Recognise a real threat from outside

Major cultural change in organisations requires sound reasoning before the change process can be initiated, as well as the appointment of a hero. An organisation's external environment may alter to such an extent that the culture of the organisation and the external environment no longer match one another. The more significant the threat posed to the organisation by the mismatch between its external environment and culture, the more likely it is that the culture can be successfully changed.

### Make transition rituals the pivotal elements of change

The involvement of the people to be affected by change in the change process is a common recommendation of both academics and practitioners. Deal and Kennedy suggest a 'transitional ritual' or stage. This is where old ways of doing and organising things cease and new working relationships are established. This is a period of change in which people are encouraged to adopt new work patterns without rushing, while at the same time resisting the temptation to return to the old ways of working. Eventually the new working patterns and relationships become established as the norm.

### Provide transition training in new values and behaviour patterns

New working practices and relationships need help to become established. Hence a programme of change will have to be available to all the employees affected. A culture change programme should focus on new values, new behaviour and new language if new working practices and new relationships are to become permanent and last in the long term.

### Bringing in outside shamans

An organisation experiencing cultural change needs to drive the change from inside, hence the need for good management and clear direction. However, an outside 'shaman' or consultant can be useful in helping the people affected by the change span the gap between the two different organisational cultures. This can be done by defusing the friction and strife and helping those affected by the change see that the way forward suggested by the change in culture can work successfully.

### Build tangible symbols of the new direction

People in an organisation affected by cultural change need to see and feel the effects of the change if they are to consider moving forward with the organisation and its cultural change. A good example of this would be a well-managed alteration to the structure of the organisation, as this would send a clear and tangible message concerning the direction in which the organisation was heading.

### Insist on the importance of security in transition

Proposed change in an organisation will always create uncertainty and as such needs to be minimised. The greatest uncertainty that people will feel is that surrounding the security of their own jobs, and this needs to be made clear and dealt with swiftly. Those people who are staying with the organisation need to be clearly informed that this is the case. Equally, those who are to be bought off or made redundant also need to be told. Dealing with the issue of job security in an unambiguous way is an crucial part of effective change.

Change is sometimes required in an organisation. Its result can be good or bad. The certainty of change is that it is usually risky, expensive and time consuming. However, if the managers involved are sensitive to the organisational culture, then the change process can be managed successfully.

## Conclusion

This chapter gave an overview of organisational change in its earlier sections. It then went on to examine two entities that are explored in earlier chapters and are also often changed by organisations, namely structure (Chapter 4) and culture (Chapter 5).

## ETHICAL ISSUES CASE STUDY 12.2

FT

# NatWest 'must change culture to survive'

**By George Graham**

National Westminster Bank will not survive unless it changes the way it works, Derek Wanless, group chief executive, has told the bank's top 200 managers.

Unveiling plans to overhaul the bank's culture earlier this month, Mr Wanless told managers: 'There will be no second chance. If we cling to our hierarchical process-driven structure, we will not survive.'

He said that henceforth NatWest must abandon internal hierarchies and put its customers first. Years of undemanding customers had left the bank with the attitude that it could take them for granted. 'The way we work here has got to change, starting with our attitude to customers.'

His stark warning came after a year in which NatWest took a battering amid losses in its investment banking division and doubts over its strategic direction.

The need for reform was evident at NatWest's annual meeting yesterday, as shareholders and customers took it in turns to complain.

'As a private customer, I get the feeling that the bank no longer wants us,' complained one customer, angered by the closure of her local NatWest ATM. 'I'm speaking for your 6.5m customers: we are not your dairy cows for you to milk us,' grumbled another annoyed shareholder.

However, the bank rebuffed an attempt by Rory Murphy, general secretary of the NatWest Staff Association trade union, to win election to its board.

Mr Wanless's efforts to galvanise NatWest have brought charges within the group of a short-term approach to investment decisions.

At the senior executives meeting he retorted: 'We have two choices. We can continue to produce poor results and continue the inexorable slide downwards. Or we can score some goals.

'The changes I am looking for are not optional – not negotiable, not nice to have; they are absolutely vital to the future of the group.'

Mr Wanless said NatWest had preserved structural and hierarchical barriers which had made it more difficult to work as a team across the group. 'Those barriers must be brought down.'

Most of NatWest's difficulties last year lay in NatWest Markets, the investment banking division, much of which has now been sold.

Problems also arose in Coutts, the private banking arm; but the main retail banking division performed well, with profits up 15 per cent to £962m and a return on equity of 27.6 per cent.

Mr Wanless said NatWest's 'biggest sin over recent years was allowing ourselves to be distracted from really focusing on developing one of the world's greatest banking franchises'.

*Source*: *Financial Times*, 22 April 1998. Reprinted with permission.

**Questions for ethical issues case study 12.2**

1 'As a private customer, I get the feeling that the bank no longer wants us' complained one customer angered by the closure of her local NatWest ATM.

Do you agree with the customer? Use Lewin's force field analysis to summarise the case for and against change.

2 If you were Derek Wanless, group chief executive of NatWest, how would you apply Burnes's change process model to NatWest?

## EXIT CASE STUDY 12.3

FT

# Broadcasting the value of creativity

**By John Gapper**

Sir Christopher Bland sits on a white sofa in the chairman's office of the British Broadcasting Corporation, and considers the essence of the world's best-known public broadcaster. 'Its history creates its character. It is in the woodwork and the air conditioning,' he reflects.

There is quite a lot of woodwork in Broadcasting House, the headquarters of the BBC since 1932. Amid the Eric Gill statues and the wood-panelled offices in which the corporation's top brass work, it is hard to forget the BBC's unique history and public service values.

Sir Christopher has not always been part of this woodwork. After an early career in the army, he chaired London Weekend Television, the commercial broadcaster, for 10 years. He was made BBC chairman in April 1996, and retains an outsider's clarity about its inessential aspects.

'The most important thing is not organisation structures and charts, fascinating though they may be, but to see if creativity is flourishing. I was a management consultant for a long time, and I don't think management structures are anything other than ephemeral,' he says.

That might seem no more than commonsense, but in today's BBC it has a revolutionary ring. Sir John Birt, the BBC's director general since 1993, is known for ignoring internal scepticism and external controversy to implement sweeping organisational reforms.

But there is a different tone from the top of the BBC these days as it faces yet another review of licence fee funding. It is also trying to damp discontent about its five-year investment of up to £1bn in new digital channels such as a 24-hour news channel while cutting costs elsewhere.

This week, the BBC will publish a new statement of its guiding principles, deliberately placing far greater weight on its traditional public service obligations than on commercial ventures such as its international joint venture with the US cable network Discovery.

It has faced criticism that the public service dog is being wagged by the tail of its ambitions to become a global broadcaster. 'I don't believe that is happening, and this gives us an opportunity to restate very clearly which is the tail and which is the dog,' says Sir Christopher.

Yet even Sir Christopher admits the BBC is treading an increasingly delicate line by setting up an assortment of new public service and commercial channels, while at the same time trying to preserve its traditional strengths on public service television channels and radio stations.

Fears that it was compromising its core television and radio service erupted recently when it lost rights to live test match cricket to Channel Four. Opponents claimed it was stretching itself too thin with ventures such as its loss-making international news channel BBC World.

Sir Christopher's first response to the claim that it is no longer spending enough on UK sports rights is to rebut it. 'I can understand that feeling, but it is wrong. There is no single segment of the audience that we should try to serve regardless of the costs involved,' he insists.

But he goes on to acknowledge that the BBC is stretched to its limits by the new-found range of its commitments. 'The BBC is a powerful and well-organised force, but you can see a time when we could not take on much more. I think we are close to that,' he says.

'I would say the BBC is at full stretch. I think that is how organisations function best, but if somebody said to us: "Why don't you start three more channels?," I do not think we would want to take that on, even if the government offered to fund them,' he says.

There is some sense to adopting such a tone at the moment. The government will shortly set up a review panel to consider the level of licence fee funding for the second half of the BBC's 10-year Royal Charter, which was renewed by the last government from 1996.

The BBC wants the government to set the licence fee as high as possible. But Chris Smith, the culture secretary, is keen for it to commit itself firmly to the primacy of producing UK public service television rather than becoming a commercially valuable global brand.

The rhetoric is easy enough, but the BBC faces a struggle to prove it is still needed in practice as commercial channels proliferate. General entertainment and news channels like BBC1 face attacks from channels providing everything from 24-hour news to children's programmes.

**Exit case study 12.3 *continued***

As he admits, one conclusion might be that the BBC should simply provide services that fall in the gaps left by commercial stations – such as education and in-depth news. But he rejects such a role, saying the BBC must provide something for all those who pay £97.50 a year.

'The whole point is that it is universal, and that will become more important, not less. If you look at why the BBC is successful and the envy of the world, that is because it is big and independent, and funded through a mechanism that guarantees independence,' he says.

Instead, he says it must provide higher quality across a range of services. 'The BBC has to do something not just different, but better, than commercial competitors. Everything we do ought to be justified by a comparison with what the market can provide,' he insists. On such criteria, there are clear weak spots in the BBC's current range of services. Its News 24 channel – now broadcasting on digital television – is widely seen as falling below the standards of the main BBC bulletins, and those of its main British competitor, Sky News.

Sir Christopher tacitly admits that News 24 is a weak spot. 'It is a good service, and getting better. It takes time to get established and develop the right tone. It would be appropriate to examine whether it is genuinely doing things that are different in three or four years' time.'

Apart from defending the BBC's new services, Sir Christopher's essential role during the next year will be to help choose a new director-general to succeed Sir John. Sir John steps down in April 2000, and Sir Christopher will lead a selection process involving other governors.

The job of director-general is unusual in combining the functions of editor-in-chief and chief executive of the corporation. Indeed, some argue that the complexity of the BBC means that it would be more sensible to split the roles between a new director general and a deputy.

Sir Christopher says he is unimpressed by power-sharing. 'Organisations need leaders. If it is hard to find somebody who is all those things, very well, but that is our task because no organisation can easily be run by two or three people with complementary skills,' he says.

Beyond this – and emphasising that there are strong candidates from inside the BBC as well as outside – he will not go. But he emphasises that Sir John's successor will need many skills to deal not only with outside pressures but with the 'byzantine internal politics' of the BBC.

Despite years of organisational change and 'Birtist' managerial revamping, the chairman is convinced that the BBC's old recalcitrant heart still beats strongly. This does not seem to displease him particularly: 'Organisations are not primarily matters of charts and boxes.'

*Source*: *Financial Times*, 1 December 1998. Reprinted with permission.

**Questions for exit case study 12.3**

1 Identify:
- the demand for change at the BBC;
- the types of change occurring at the BBC;
- the causes of change at the BBC.

2 Apply both Lewin's and Burnes' change process models to the BBC. Explain and justify which model is most useful in helping you understand change at the BBC.

3 Identify where the BBC is on Greiner's model, today and in five years' time. Name the likely type of structure of the BBC in five years' time. How would the change process model you have chosen in answering Question 2 help the BBC achieve this?

4 Is there an alternative model that would help you in thinking about possible future structures of the BBC. Explain and justify your answer.

## Short-answer questions

1 Where does the demand for change in organisations arise?

2 Briefly summarise the causes of change.

3 Briefly explain what is meant by force field analysis.

4 List the three stages of Lewin's process change model.

5 Write down Thompson's four determinants of organisational structure.

6 In one sentence, define Greiner's model.

7 Name the five phases of Greiner's model.

8 Explain what a period of revolution is in Greiner's model.

9 Explain what a period of evolution is in Greiner's model.

10 List Deal & Kennedy's seven elements for the achievement of successful change.

## Assignment questions

1 There are two models for change presented in Chapter 12, Lewin's the change process model (Exhibit 12.4) and Burnes' change process (Exhibit 12.5). Choose *one* of the models of change. Compare and contrast your chosen model with another model of change from the literature (not Burnes or Lewin). Present your findings in a 2000-word essay.

2 The work of Thompson identifies four determinants of organisational structure. Review the literature and present at least two alternative views on the determinants of organisational structure. Present your findings in a 2000-word essay.

3 The work of Deal and Kennedy identifies seven elements required for successful cultural change. Review the literature and present at least two alternative views on successful cultural change. Present your findings in a 2000-word essay.

## References


1 Taylor, R (1998) 'ICI tries to get the right chemistry between old and new', *Financial Times*, 27 March.

2 Lewin, K (1951) *Field Theory in Social Science*, Harper and Row, quoted in Thompson, J L (1997) *Strategic Management*, London: International Thomson Business Press.

3 Lewin, K (1947) 'Frontiers in group dynamics: concept, method and reality in social science', *Human Relations*, 1, quoted in Thompson, op. cit.

4 Burnes, B (2000) *Managing Change*, 3rd edn, Harlow: Financial Times Prentice Hall.

5 Thompson, op. cit.

6 Taylor, op. cit.

7 Greiner, L (1972) 'Evolution and revolution as organizations grow', *Harvard Business Review*, July/August, reprinted May/June 1998.

8 Ibid.

9 Ibid.

10 Deal, T and Kennedy, A (1988) *Corporate Cultures*, London: Penguin Business.

11 Ibid.


## Further reading


Balogun, J and Hope Hailey, V with Johnson, G and Scholes, K (1999) *Exploring Strategic Change*, Harlow: Prentice Hall Europe.

Davidson, A (1999) 'The Andrew Davidson Interview', Christopher Bland/BBC, *Management Today*, June.

Littlefield, D (1999) 'Kerry's heros', *People Management*, 6 May.
Parker, G (1998) 'Fear and loathing in Smith Square', *Management Today*, October.
Pascale, R and Millemann, M (1997) 'Changing the way we change', *Harvard Business Review*, November/December.
Senior, B (1997) *Organisational Change*, London: Financial Times Pitman Publishing.
Webster, G (1998) 'Changing places', *Management Today*, August.
Whitehead, M (1999) 'Hat trick', *People Management*, 29 July.

# Future trends

## Introduction

This final chapter of *Understanding Organisational Context* briefly highlights and examines some of the current trends (*see* Exhibit A1) that are relevant to businesses at the start of the new millennium.

## Twenty-first-century managers

The successful manager in the twenty-first century will require a different set of skills from those required by a manager in the 1960s and 1970s. Managers of the 1990s had to embrace new computer-based technology, such as bigger and faster computers, running ever more sophisticated software, along with the Internet, e-mail and video conferencing, all of which altered the way people work (*see* Chapter 2).

The introduction and development of technology made working from a remote location all the more possible. The executive with a laptop computer can be linked to fellow workers, clients and the office from anywhere. Therefore the

**Exhibit A1**
**Future trends**

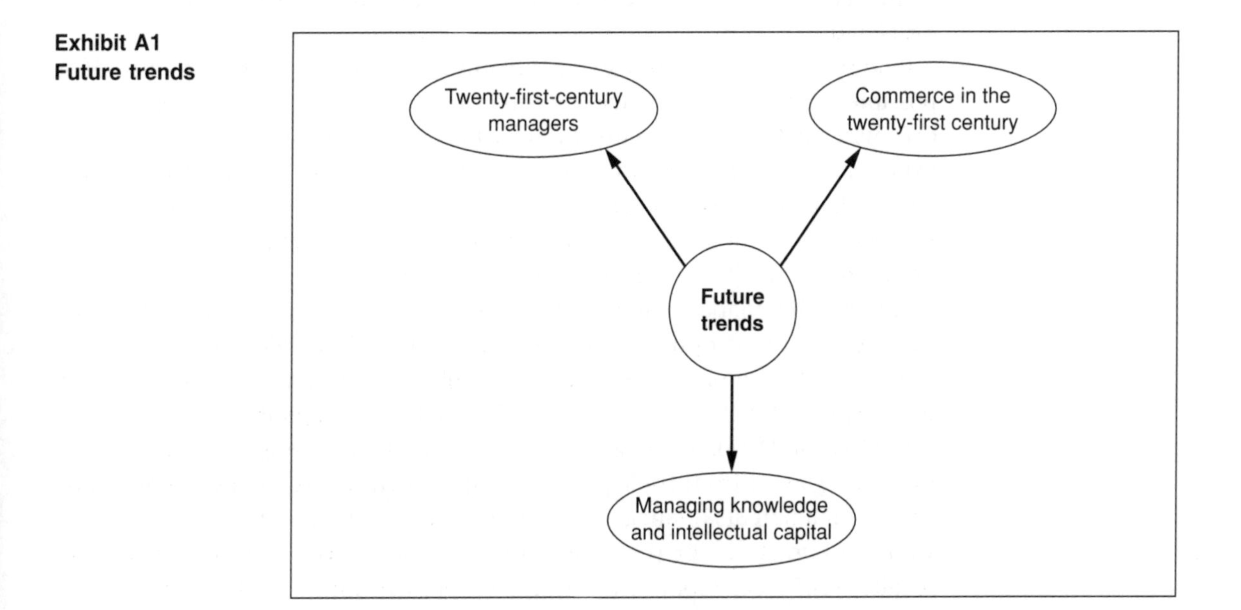

manager of the twenty-first century is more likely to work from home at least some of the time, possibly with an office which they can use on an occasional basis. A discussion on home working appears in Chapter 9.

However, the advent of technology does not mean that office-based working and meetings are going to disappear altogether, as the trust and respect of the others with whom one works and does business are still going to be needed. The attributes of trust and respect are always going to be best established by face-to-face meetings where eye contact occurs and body language can be read. The technology and electronic media will be used for a significant amount of routine communication in the future.

## Managing knowledge and intellectual capital

Knowledge and intellectual capital rest in the brainpower, experience and flair held by the people who work for an organisation. The phrase 'knowledge management' is used to suggest that knowledge is shared between people in an organisation, such that greatest benefit is gained from the knowledge for the organisation and its employees.

One important knowledge activity is the identification and development of informal networks through which knowledge can be shared. These informal networks are cross-functional in nature and usually comprise around 50 to 300 people with shared interests. The people in the informal network will freely share their knowledge and expertise with one another.[1] For example, at Ford there exists a virtual community numbering several hundred employees that focuses on newly breaking technologies.[2] The existence of such communities can be confirmed by observing direct or electronic conversations between members, which allows the level of participation in knowledge sharing to be measured.

If knowledge that exists in an organisation can be identified, then it is possible to map it and make it visible to others. This allows others in the organisation to identify who knows what. Corporate skills inventories and expert databases are example of such maps. These are difficult to maintain and update and this problem grows accordingly as organisations increase in size. However, the importance of having an accurate picture of the skills and knowledge that do exist in the organisation also increases.

Finally, the management of knowledge or intellectual capital does not have to rely on a sophisticated computer system. The DIY chain B&Q has a reputation for employing significant numbers of people aged over 50, staff who represent part of the company's intellectual capital. This intellectual capital is the DIY knowledge and experience of this group of people, which has been gained from their previous employment or experience as keen and successful amateurs. Managing the intellectual capital involves exploiting this valuable knowhow for the benefit of the organisation, staff and customers. Ideally, staff receive pay and a fulfilling job, customers receive sound advice and the organisation receives improved customer loyalty and profits, which helps increase shareholder value.

## Commerce in the twenty-first century

E-commerce consists of transactions between organisations or individuals and organisations using electronic networks. It concerns physical goods such as books and CDs and intangible goods such as financial service products, which are often sold via the Internet. Witness the success of Amazon, the biggest bookseller on the Internet.

Retailing via the Internet is mainly an activity of the medium to well-off sectors of the population, estimates suggesting that Internet shopping will be worth £300m by 2003.[3] These people tend to have access to computers at work, their children use computers at school and they have a computer in the family home. It is also suggested that, in these early days of e-commerce, Internet shopping tends to be done by those who have more disposable income than time. However, in the future virtual shopping will become available to everyone, as access to the Internet widens. This does not mean that everyone will need a computer in their home, as access to the Internet will become available via digital television. For example, an advert for a product will appear on television and all the user will have to do is click on the web address scrolling across the bottom of the screen to find out more about the product, purchase it and pay for it. The skill levels required to be able to do this will be virtually no more complex than using teletext, something which most people have mastered.

Hence access to the Internet for most consumers will become a reality in the early part of the twenty-first century in much the same way that connection to the telephone system for nearly all people became a reality in the middle years of the twentieth century. Barriers that need to be overcome to assist the expansion of access to the Internet include much wider availability of free local calls and the publication of website directories to enable people to find the site they want easily without searching for a long time.

In contrast, those who enjoy the physical experience of going shopping, seeing and being able to handle the goods before a purchase is made may be disappointed by the concept of Internet shopping. This is expected to change as Internet shopping becomes more interactive, whether accessed via a computer or digital television. The two other areas where Internet shopping may have an impact as its availability widens are in taking the place of traditional catalogue shopping, and repeat purchases where the purchaser knows what they are buying and the need to examine the product is minimal. For example, a proportion of the weekly shopping basket from the supermarket consists of repeat purchases of staple foods and goods bought on a regular basis. It is already possible in some areas of the country to order your groceries via the Internet from one of the major supermarkets, with home or office delivery or collection of your goods, ready, packed and waiting, from a pick-up point at the supermarket.

Successful Internet retailers will use the Internet to develop one-to-one relationships with individual customers. For example, the supermarket shopper on the Internet will be able to have their most commonly purchased groceries stored on an electronic shopping list and be asked if they wish to purchase individual items on the list today or an alternative that is on special offer.

The Internet is efficient in terms of time and distribution for goods such as books and CDs. Purchasing a book previously required the individual to go to the bookshop, locate the book, if it was in stock, or order it if it was out of stock. If out of stock, then the bookshop had to obtain the book from its warehouse or distributor. This could take several days. The customer then had to come into the shop to collect the book. In contrast, using the website of a company like Amazon, the customer can log on from home and check availability and price, order and pay electronically and wait for the book to be sent to their home. The development of the website for the Cambridge bookshop Heffers grew out of e-mail orders for books. The website was expensive to develop, but had recouped its costs within a year.[4]

In Europe the development of the Internet will allow trade between businesses in different countries to be done easily and all in one currency, the euro (*see* Chapter 8). Companies located in the UK, which has not entered the euro zone, will still buy and sell in euros and open the appropriate bank accounts to deal with the revenue they generate. The same will also occur eventually for retail services offered on the Internet, only it will be more difficult for individuals in the UK to trade in euros as what they pay will be in pounds and will have to be converted to euros. There are many price differentials for the same products sold in different European countries, which makes it difficult for the consumer to compare prices and recognise a good or bad deal. The growth in e-commerce and the introduction of the euro will make price comparisons easier for the consumer.

Finally, two other issues have to be tackled successfully if e-commerce is to flourish. The first is the security of the Internet, as people purchase goods with credit cards. Encrypted software and sites will have to be continually developed to keep ahead of the fraudsters. Secondly, it is likely that there will be changes in legislation relating to business-to-business transactions, as at the moment an on-screen document is not legally binding as it does not incorporate a signature.

## Conclusion

This short chapter on future trends should have provided an overview of the new issues that organisations are facing and offered the reader a starting point for thinking further about them.

## Discussion question

Choose one of the four functions of business: marketing, operations management, finance or human resource management. Research the future trends for the function you have chosen and in your findings comment on how organisations will be affected by and react to the future trends you have identified. Present your findings in a 2000-word essay or 15-minute presentation.

## References

1 Prusak, L (1999) 'Making knowledge visible', Mastering Information Management, *Financial Times*, 8 March.
2 Ibid.
3 *Business Café*, BBC2, 21 February 1999.
4 *Open Minds*, BBC2, 19 June 1999.

# Index